Swagatam Das, Ajith Abraham and Amit Konar

Metaheuristic Clustering

Studies in Computational Intelligence, Volume 178

Editor-in-Chief
Prof. Janusz Kacprzyk
Systems Research Institute
Polish Academy of Sciences
ul. Newelska 6
01-447 Warsaw
Poland
E-mail: kacprzyk@ibspan.waw.pl

Further volumes of this series can be found on our homepage: springer.com

Vol. 156. Dawn E. Holmes and Lakhmi C. Jain (Eds.)
Innovations in Bayesian Networks, 2008
ISBN 978-3-540-85065-6

Vol. 157. Ying-ping Chen and Meng-Hiot Lim (Eds.)
Linkage in Evolutionary Computation, 2008
ISBN 978-3-540-85067-0

Vol. 158. Marina Gavrilova (Ed.)
Generalized Voronoi Diagram: A Geometry-Based Approach to Computational Intelligence, 2009
ISBN 978-3-540-85125-7

Vol. 159. Dimitri Plemenos and Georgios Miaoulis (Eds.)
Artificial Intelligence Techniques for Computer Graphics, 2009
ISBN 978-3-540-85127-1

Vol. 160. P. Rajasekaran and Vasantha Kalyani David
Pattern Recognition using Neural and Functional Networks, 2009
ISBN 978-3-540-85129-5

Vol. 161. Francisco Baptista Pereira and Jorge Tavares (Eds.)
Bio-inspired Algorithms for the Vehicle Routing Problem, 2009
ISBN 978-3-540-85151-6

Vol. 162. Costin Badica, Giuseppe Mangioni, Vincenza Carchiolo and Dumitru Dan Burdescu (Eds.)
Intelligent Distributed Computing, Systems and Applications, 2008
ISBN 978-3-540-85256-8

Vol. 163. Pawel Delimata, Mikhail Ju. Moshkov, Andrzej Skowron and Zbigniew Suraj
Inhibitory Rules in Data Analysis, 2009
ISBN 978-3-540-85637-5

Vol. 164. Nadia Nedjah, Luiza de Macedo Mourelle, Janusz Kacprzyk, Felipe M.G. França and Alberto Ferreira de Souza (Eds.)
Intelligent Text Categorization and Clustering, 2009
ISBN 978-3-540-85643-6

Vol. 165. Djamel A. Zighed, Shusaku Tsumoto, Zbigniew W. Ras and Hakim Hacid (Eds.)
Mining Complex Data, 2009
ISBN 978-3-540-88066-0

Vol. 166. Constantinos Koutsojannis and Spiros Sirmakessis (Eds.)
Tools and Applications with Artificial Intelligence, 2009
ISBN 978-3-540-88068-4

Vol. 167. Ngoc Thanh Nguyen and Lakhmi C. Jain (Eds.)
Intelligent Agents in the Evolution of Web and Applications, 2009
ISBN 978-3-540-88070-7

Vol. 168. Andreas Tolk and Lakhmi C. Jain (Eds.)
Complex Systems in Knowledge-based Environments: Theory, Models and Applications, 2009
ISBN 978-3-540-88074-5

Vol. 169. Nadia Nedjah, Luiza de Macedo Mourelle and Janusz Kacprzyk (Eds.)
Innovative Applications in Data Mining, 2009
ISBN 978-3-540-88044-8

Vol. 170. Lakhmi C. Jain and Ngoc Thanh Nguyen (Eds.)
Knowledge Processing and Decision Making in Agent-Based Systems, 2009
ISBN 978-3-540-88048-6

Vol. 171. Chi-Keong Goh, Yew-Soon Ong and Kay Chen Tan (Eds.)
Multi-Objective Memetic Algorithms, 2009
ISBN 978-3-540-88050-9

Vol. 172. I-Hsien Ting and Hui-Ju Wu (Eds.)
Web Mining Applications in E-Commerce and E-Services, 2009
ISBN 978-3-540-88080-6

Vol. 173. Tobias Grosche
Computational Intelligence in Integrated Airline Scheduling, 2009
ISBN 978-3-540-89886-3

Vol. 174. Ajith Abraham, Rafael Falcón and Rafael Bello (Eds.)
Rough Set Theory: A True Landmark in Data Analysis, 2009
ISBN 978-3-540-89886-3

Vol. 175. Godfrey C. Onwubolu and Donald Davendra (Eds.)
Differential Evolution: A Handbook for Global Permutation-Based Combinatorial Optimization, 2009
ISBN 978-3-540-92150-9

Vol. 176. Beniamino Murgante, Giuseppe Borruso and Alessandra Lapucci (Eds.)
Geocomputation and Urban Planning, 2009
ISBN 978-3-540-89929-7

Vol. 177. Dikai Liu, Lingfeng Wang and Kay Chen Tan (Eds.)
Design and Control of Intelligent Robotic Systems, 2009
ISBN 978-3-540-89932-7

Vol. 178. Swagatam Das, Ajith Abraham and Amit Konar
Metaheuristic Clustering, 2009
ISBN 978-3-540-92172-1

Swagatam Das
Ajith Abraham
Amit Konar

Metaheuristic Clustering

Springer

Swagatam Das
Department of Electronics and
Telecommunication Engineering (ETCE)
Jadavpur University
Raja S. C. Mullick Road
Jadavpur, Calcutta - 700032
India

Amit Konar
Department of Electronics and
Telecommunication Engineering (ETCE)
Jadavpur University
Raja S. C. Mullick Road
Jadavpur, Calcutta - 700032
India

Ajith Abraham
Norwegian Center of Excellence
Center of Excellence for Quantifiable
Quality of Service
Norwegian University of Science
and Technology
O.S. Bragstads plass 2E
NO-7491 Trondheim
Norway

ISBN 978-3-540-92172-1 e-ISBN 978-3-540-93964-1

DOI 10.1007/978-3-540-93964-1

Studies in Computational Intelligence ISSN 1860949X

Library of Congress Control Number: 2008942042

Typeset & Cover Design: Scientific Publishing Services Pvt. Ltd., Chennai, India.

Printed in acid-free paper

9 8 7 6 5 4 3 2 1

springer.com

Foreword

Indisputably in the oceans of data surrounding us, clustering has gained a central position as a conceptual and algorithmic framework that helps the user make sense of data and reveal some underlying structure that is hidden behind overwhelming streams of numbers.

There are thousands of clustering techniques one can encounter in the literature. We will be seeing far more methods arising over the passage of time. Just the recent search using Google Scholar (dated November 18, 2008) has returned about 1,510,000 hits. This number speaks to the dynamics and omnipresence of the clustering paradigm and its numerous applications.

What we have started seeing more vividly are the two fundamental clustering challenges one has to deal with in an effective manner. First, it becomes apparent that clustering is a processes guided by several objectives (objective functions) rather than a single and somewhat isolated goal. This has led us to the concept of multiobjective clustering. Likewise we have started to realize that to make clustering more user-centric, one needs to fully accommodate some prior domain knowledge and this line of pursuit has resulted in a so-called knowledge-based clustering. Second, there is an acute need for optimization tools that are of *global* nature and in this way may help realize a comprehensive search which is of structural as well as of parametric character. The role evolutionary computing has been already acknowledged in this particular context yet there is a large unexplored research territory where we can anticipate a great deal of interesting findings.

The treatise authored by Professors Das, Abraham, and Konar tackles a very fundamental and practically highly relevant research topic: how to make clustering more efficient and very much in rapport with the reality of multifaceted data and diversified needs of the end users. The notion of metaheuristics used in the title of the book is very much reflective of its very content.

The reader is carefully navigated through the efficacies of clustering, evolutionary optimization and a hybridization of the both. The exposure of the material is lucid. Quite complicated concepts are presented in a clear and convincing way which can be attributed to the expertise of Professors Das, Abraham, and Konar. While Evolutionary Computing has been recognized as a viable optimization platform, it has been noted quite early that a number of well-known techniques such as e.g., Genetic Algorithms and Evolutionary Algorithms come with a substantial computational overhead which becomes difficult to accept in case of problems of higher dimensionality. From this standpoint, the alternative of Differential Evolution (DE) pursued by the authors is indeed a very fortunate choice.

In the exposure of the material, the authors have achieved a sound balance between the theory and practice. We witness a wealth of fundamental and far reaching results, especially when it comes to the analysis of the dynamics of Differential Evolution. We can appreciate the applied facets of the monograph where the algorithmic setting established in the book stresses applicability or leads directly to interesting and well-rounded applications in data analysis.

All in all, this is not only a very timely and badly needed volume but also an outstanding, comprehensive and authoritative treatise of the important subject of metaheuristics clustering.

Professor, Canada Research Chair, IEEE Fellow Witold Pedrycz
University of Alberta, Canada
November 2008

Preface

Cluster analysis means the organization of an unlabeled collection of objects (or *patterns*) into separate groups based on their similarity. Each valid group, called a 'cluster', should consist of objects that are similar among themselves and dissimilar to objects of other groups. As human beings, we resort to clustering as one of our most primitive mental activities for organizing the data we receive every day, so that we may draw important conclusions from them. It is well nigh impossible to process every piece of such data as a single entity. Thus, humans tend to categorize entities (i.e. objects, persons, events) into clusters. Each cluster is then characterized by the common attributes (features) of the entities that belong to that cluster.

Human beings possess the natural ability of clustering objects. Given a box full of marbles of four different colors say red, green, blue, and yellow, even a child may separate these marbles into four clusters based on their colors. However, making a computer solve this type of problems is quite difficult and demands the attention of computer scientists and engineers all over the world till date. The major hurdle in this task is that the functioning of the brain is much less understood. The mechanisms, with which it stores huge amounts of information, processes them at lightning speeds and infers meaningful rules, and retrieves information as and when necessary have till now eluded the scientists. A question that naturally comes up is: what is the point in making a computer perform clustering when people can do this so easily? The answer is far from trivial. The most important characteristic of this information age is the abundance of data. Advances in computer technology, in particular the Internet, have led to what some people call "data explosion": the amount of data available to any person has increased so much that it is more than he or she can handle. In reality the amount of data is vast and in addition, each data item (an abstraction of a real-life object) may be characterized by a large number of attributes (or *features*), which are based on certain measurements taken on the real-life objects and may be numerical or non-numerical. Mathematically we may think of a mapping of each data item into a point in the multi-dimensional feature space (each dimension corresponding to one feature) that is beyond our perception when number of features exceed just 3. Thus it is nearly impossible for human beings to partition tens of thousands of data items, each coming with several features (usually much greater than 3), into meaningful clusters within a short interval of time. Nonetheless, the task is of paramount importance for organizing and summarizing huge piles of data and discovering useful knowledge from them. So, can we devise some means to generalize to arbitrary dimensions of what humans perceive in two or three dimensions, as densely connected "patches" or "clouds" within data space? The entire research on cluster analysis may be considered as an effort to find satisfactory answers to this fundamental question.

The task of computerized data clustering has been approached from diverse domains of knowledge like graph theory, statistics (multivariate analysis), artificial neural networks, fuzzy set theory, and so on. One of the most popular approaches in this direction has been the formulation of clustering as an optimization problem, where the best partitioning of a given dataset is achieved by minimizing/maximizing one (single-objective clustering) or more (multi-objective clustering) objective functions. The objective functions are usually formed capturing certain statistical-mathematical relationship among the individual data items and the candidate set of representatives of each cluster (also known as cluster-centroids). The clusters are either hard, that is each sample point is unequivocally assigned to a cluster and is considered to bear no similarity to members of other clusters, or fuzzy, in which case a membership function expresses the degree of belongingness of a data item to each cluster.

Most of the classical optimization-based clustering algorithms (including the celebrated hard c-means and fuzzy c-means algorithms) rely on local search techniques (like iterative function optimization, Lagrange's multiplier, Picard's iterations etc.) for optimizing the clustering criterion functions. The local search methods, however, suffer from two great disadvantages. Firstly they are prone to getting trapped in some local optima of the multi-dimensional and usually multimodal landscape of the objective function. Secondly performances of these methods are usually very sensitive to the initial values of the search variables.

Although many respected texts of pattern recognition describe clustering as an unsupervised learning method, most of the traditional clustering algorithms require a prior specification of the number of clusters in the data for guiding the partitioning process, thus making it not completely unsupervised. On the other hand, in many practical situations, it is impossible to provide even an estimation of the number of naturally occurring clusters in a previously unhandled dataset. For example, while attempting to classify a large database of handwritten characters in an unknown language; it is not possible to determine the correct number of distinct letters beforehand. Again, while clustering a set of documents arising from the query to a search engine, the number of classes can change for each set of documents that result from an interaction with the search engine. Data mining tools that predict future trends and behaviors for allowing businesses to make proactive and knowledge-driven decisions, demand fast and fully automatic clustering of very large datasets with minimal or no user intervention. Thus it is evident that the complexity of the data analysis tasks in recent times has posed severe challenges before the classical clustering techniques.

Starting from early 1960s, a keen observation of the underlying relation between optimization and biological evolution has led to the development of an important paradigm of computational intelligence – the evolutionary computing

(EC) - for performing very complex search and optimization. Evolutionary computing harnesses the power of natural selection to turn computers into automatic optimisation and design tools. This volume investigates the application of a recently developed evolutionary computing algorithm, well-known as the Differential Evolution (DE), to develop robust, fast and fully automatic clustering techniques that can circumvent the problems with several classical clustering schemes, as illustrated earlier.

Since its advent in 1995, DE has drawn the attention of the practitioners in optimization all over the globe due to its high degrees of robustness, convergence speed, and accuracy in real parameter optimization problems. A very simple algorithm to code with so few (typically 3 in classical DE) adjustable control-parameters, DE has been shown to outperform several veteran members of the EC family like the Genetic Algorithms (GA), Evolutionary Strategies (ES), and Memetic Algorithms (MA) over both benchmark and real-world problems. Unlike GAs, however, the application of DE to clustering problems has not been much investigated.

In this Volume, we illustrate the performance of DE, when applied to both single and multi-objective clustering problems, where the number of clusters is not known beforehand and must be determined on the run. We first undertake a statistical analysis of the search operators and the convergence behaviour of DE near an isolated equilibrium point in the search space. Taking a cue from the analysis mentioned earlier, we propose a few parameter automation strategies that improve the performance of classical DE without imposing any serious additional computational burden. Next we develop a new DE-based crisp clustering algorithm, which can not only correctly partition the data in appropriate clusters but also find the optimal number of clusters automatically. The proposed algorithm incorporates a new real-coded scheme for search variable representation that makes room for several possible choices of the number of clusters in the dataset. An extensive comparison with several other state-of-the-art clustering algorithms over many synthetic and real-life datasets reflects the statistically superior performance of the proposed scheme in terms of final accuracy, speed and robustness. We also applied the proposed clustering method to an interesting problem of automatic image pixel clustering and land cover study from satellite images. The proposed clustering technique is next extended to the fuzzy clustering in kernel induced feature space, for tackling more complex clusters, which are linearly non-separable and overlapping in nature. A new DE-variant with balanced exploration and exploitation abilities has been proposed for optimizing the clustering objectives in higher dimensional kernel space. The new DE variant is shown to perform better than the classical DE and many other recently developed algorithms for kernel-based clustering in a statistically significant fashion. Finally the Volume compares four most recently proposed multi-objective (MO) variants of the DE with two other state-of-the-art MO clustering methods over ten datasets of widely varying ranges of complexity. A novel framework for multi-objective automatic clustering is proposed for the multi-objective DE variants, one or more of which is always seen to find statistically better result than their other state-of-the-art contestants. An interesting application of the multi-objective DE based clustering to gene expression data of yeast is also investigated in this context.

The most important characteristics of the algorithms proposed in the Volume are:

1) They can optimally cluster a previously unhandled dataset (with numerical features) into correct number of clusters through one shot of optimization. As opposed to the classical local search based optimization techniques, they are able to locate the global optima of the multi-modal landscape of clustering objective function quickly.

2) Their computational speeds are faster than those of the clustering techniques based on other evolutionary and swarm intelligence algorithms.

3) They are fairly robust against different initial conditions and can produce nearly similar results (with small standard deviations) over repeated runs,

4) Owing to the characteristics of DE, they have very few control parameters and can yield good final accuracy over a large variety of clustering problems with minimal or no hand tuning.

The Volume is organized in 7 Chapters. The first Chapter presents a detailed review of the evolutionary clustering algorithms. The Chapter begins with a formal overview of the clustering problem, similarity and dissimilarity measures between patterns and the various methods of clustering. It then addresses a few classical clustering algorithms, pertinent to the present work. Next the Chapter discusses the relevance of evolutionary computing techniques in pattern clustering and outlines the most promising evolutionary clustering methods. The Chapter ends with a discussion on automatic clustering problem, which remains largely unsolved by most of the traditional clustering algorithms.

Chapter 2 presents a conceptual outline of the DE algorithm in sufficient details. It then reviews six prominent variants of DE, including DE with trigonometric mutation, DE with arithmetic recombination, DE/rand/1/either-or, self-adaptive DE, opposition-based DE, binary DE, DE with adaptive local search and finally a new family of DE-variants based on neighborhood-based mutation. An interesting algorithm resulting from the synergy of DE with an important swarm intelligence algorithm, well known as Particle Swarm Optimization (PSO) is also addressed in the Chapter.

Chapter 3 investigates the dynamics of a canonical DE algorithm with DE/rand/1 type mutation and binomial crossover. The Chapter develops a simple mathematical model of the underlying evolutionary dynamics of a one-dimensional DE-population. The model relates the search mechanism of DE to that of the classical gradient descent search. The stability and convergence-behavior of the proposed dynamics is then analyzed with the help of Lyapunov's stability theorems. The mathematical model, developed in this Chapter, provides important insights into the search mechanism of DE in a near neighborhood of an isolated optimum. The Chapter also presents empirical simulation results over simple objective functions to validate the theoretical analyses.

Chapter 4 describes a DE-based algorithm for the automatic crisp clustering of large unlabeled datasets. In contrast to most of the existing clustering techniques, the algorithm, proposed by the chapter, requires no prior knowledge of the data to be classified. Rather, it determines the optimal number of clusters in the data 'on the run'. Superiority of the new method has been demonstrated by comparing it with two recently developed partitional clustering techniques and one popular hierarchical clustering algorithm. The partitional algorithms are based on two powerful optimization algorithms well-known as the Genetic Algorithm (GA) and the Particle Swarm Optimization (PSO). The Chapter also reports an interesting practical application of the proposed method to automatic segmentation of gray-scale images in their intensity space.

Chapter 5 extends the work reported in Chapter 4 to the fuzzy clustering of complex and linearly non-separable datasets in kernel-induced feature space. The proposed method is based on a modified version of the classical DE algorithm, which uses a novel neighborhood-based mutation strategy. It also employs a kernel-induced similarity measure instead of the conventional sum-of-squares distance. Use of the kernel function makes it possible to cluster data that is linearly non-separable in the original input space into homogeneous groups in a transformed high-dimensional feature space. The vector representation scheme remains identical to that described in Chapter 4. The performance of the proposed method has been extensively compared with a few state of the art clustering techniques over a test-suite of several artificial and real life datasets. Based on experiment results, the Chapter also provides some empirical guidelines for selecting the suitable parameters of the modified DE algorithm

Chapter 6 considers the task of fuzzy clustering in a multi-objective optimization (MO) framework. It compares the performances of four recently developed multi-objective variants of DE over the fuzzy clustering problem, where two conflicting fuzzy validity indices are simultaneously optimized. The resultant Pareto optimal set of solutions from each algorithm consists of a number of non-dominated solutions, from which the user can choose the most promising ones according to the problem specifications. A real-coded representation of the search variables, proposed in Chapter 4, is used for DE to accommodate variable number of cluster centers. The performances of four DE variants have also been contrasted to that of two most well-known schemes of MO clustering namely the NSGA II (Non Dominated Sorting GA) and MOCK (Multi-Objective Clustering with an unknown number of clusters K). Experimental results over four artificial and four real life datasets (including a gene expression dataset of yeast sporulation) of varying range of complexities indicates that DE holds immense promise as a candidate algorithm for devising MO clustering schemes.

Finally Chapter 7 concludes the Volume with a discussion on the possible extensions of the works undertaken and projects a possible direction of future research.

We are very much grateful to our friends and colleagues for reviewing the different parts of the manuscript and for providing us valuable feedback. The first author would like to thank the swarm of his undergraduate and graduate students including Mr. Sambarta Dasgupta, Mr. Arijit Biswas, and Ms. Sudeshna Sil for helping him in performing several simulation experiments reported in the book.

Authors would like to thank Dr. Thomas Ditzinger, Springer Engineering Inhouse Editor, Studies in Computational Intelligence Series, Professor Janusz Kacprzyk (Editor-in-Chief, Springer Studies in Computational Intelligence Series) and Ms. Heather King (Editorial Assistant, Springer Verlag, Heidelberg) for the editorial assistance and excellent cooperative collaboration to produce this important scientific work. We hope that the reader will share our excitement to present this volume on 'Metaheuristic Clustering' and will find it useful.

Swagatam Das, Ajith Abraham* and Amit Konar

Department of Electronics and Telecommunication Engineering, Jadavpur University, Kolkata 700032, India
**Center of Excellence for Quantifiable Quality of Service (Q2S), Norwegian University of Science and Technology, Trondheim, Norway*

Contents

Chapter 1
Metaheuristic Pattern Clustering – An Overview

Abstract. This chapter provides a comprehensive overview to the data clustering techniques, based on naturally-inspired metaheuristic algorithms. At first the clustering problem, similarity and dissimilarity measures between patterns and the methods of cluster validation are presented in a formal way. A few classical clustering algorithms are also addressed. The chapter then discusses the relevance of population-based approach with a focus on evolutionary computing in pattern clustering and outlines the most promising evolutionary clustering methods. The chapter ends with a discussion on the automatic clustering problem, which remains largely unsolved by most of the traditional clustering algorithms.

1.1 Introduction

Several authentic texts in the field of pattern recognition (e.g. [1 – 3]) define the problem of data clustering informally, e.g., Arabie *et al.* [4] give the following definition of clustering as:

"Those methods concerned in some ways with the identification of homogeneous groups of objects,"

while Everitt [2] provides many different definitions including the following one for a cluster:

"A cluster is a set of entities that are alike, and entities from different clusters are not alike."

The widespread use of informal definitions such as these reveals one of the prevailing and fundamental problems in cluster analysis: the difficulty of providing one single formal (but sufficiently broad) definition of clustering and of the concept of a cluster. In fact, Everitt points out that "such formal definition is not only difficult but may even be misplaced" [2, page 6].

This is because the concept of a cluster is a generalization (to arbitrary dimensions) of what human perceive, in two or three dimensions, as densely connected "patches" or "clouds" within data space, a human intuition that is inherently difficult to capture by means of any single criteria. In fact, the following example, inspired by biology shows that the process of assigning objects to clusters may lead to very different results, depending on the specific criterion used for clustering.

S. Das, A. Abraham, and A. Konar: Metaheuristic Clustering, SCI 178, pp. 1–62.
springerlink.com

Example 1.1

Consider the following animals: sheep, dog, cat (mammals), sparrow, seagull (birds), viper, lizard (reptiles), goldfish, red mullet, blue shark (fish), and frog (amphibians). In order to organize these animals into clusters, we need to define a clustering criterion. Thus, if we employ the way these animals bear their progeny as a clustering criterion, the sheep, dog, cat, and the shark will be assigned to the same cluster, while all the rest will form a second cluster (see Figure 1.1 (a)). If the clustering criterion is the existence of lungs, the goldfish, red mullet and the shark are assigned to the same cluster, while all the other animals are assigned to a second cluster (Figure 1.1 (b)). On the other hand, if the clustering criterion is the environment where the animals live, the sheep, dog, cat sparrow, seagull, viper, and the lizard will form one cluster (animals living outside water), the goldfish, red mullet, and the blue shark will form a second one (animals living only in water), and the third cluster will contain only the frog, since it may live in the water or out of it (Figure 1.1 (c)).It is worth pointing out that if the existence of a vertebral column be the clustering criterion, all the animals listed above will fall in the same cluster. Finally we may use composite clustering criteria as well. For example, if the clustering criterion is the way in which these animals bear their progeny and the existence of lungs then we end up with four clusters as shown in Figure 1.1 (d).

Thus we find that due to the lack of any universal clustering criteria, the clustering task is highly subjective for most of the cases. Consider, for example, Figure 1.2. How many "sensible" ways of clustering can we obtain for these points? The most logical answer seems to be two. The first clustering contains two clusters (surrounded by dashed boundary) while the second clustering contains four clusters (surrounded by solid circles). Which clustering is "correct"? It seems that there is no definite answer. Both clustering results are valid. The best thing to do is give the results to an expert and let the expert decide about the most sensible one. Thus unless the number of clusters are specified beforehand, the final answer to the clustering problem is not unique, but problem-specific and involves the knowledge of human expert in most cases.

From a machine learning perspective, clusters correspond to the hidden patterns in data, the search for clusters is a kind of unsupervised learning, and the resulting system represents a *data concept*. It is important to understand the difference between clustering (unsupervised classification) and supervised classification (discriminant analysis) at this point. In supervised classification, we are provided with a collection of *labeled* (pre-classified) patterns; the problem is to label a newly encountered, yet unlabeled, pattern. Typically, the given labeled (*training*) patterns are used to learn the descriptions of classes, which in turn are used to label a new pattern. In the case of clustering, the problem is to group a given collection of unlabeled patterns into meaningful clusters. In a sense, labels are associated with clusters also, but these category labels are *data driven*; that is, they are obtained solely from the data [5].

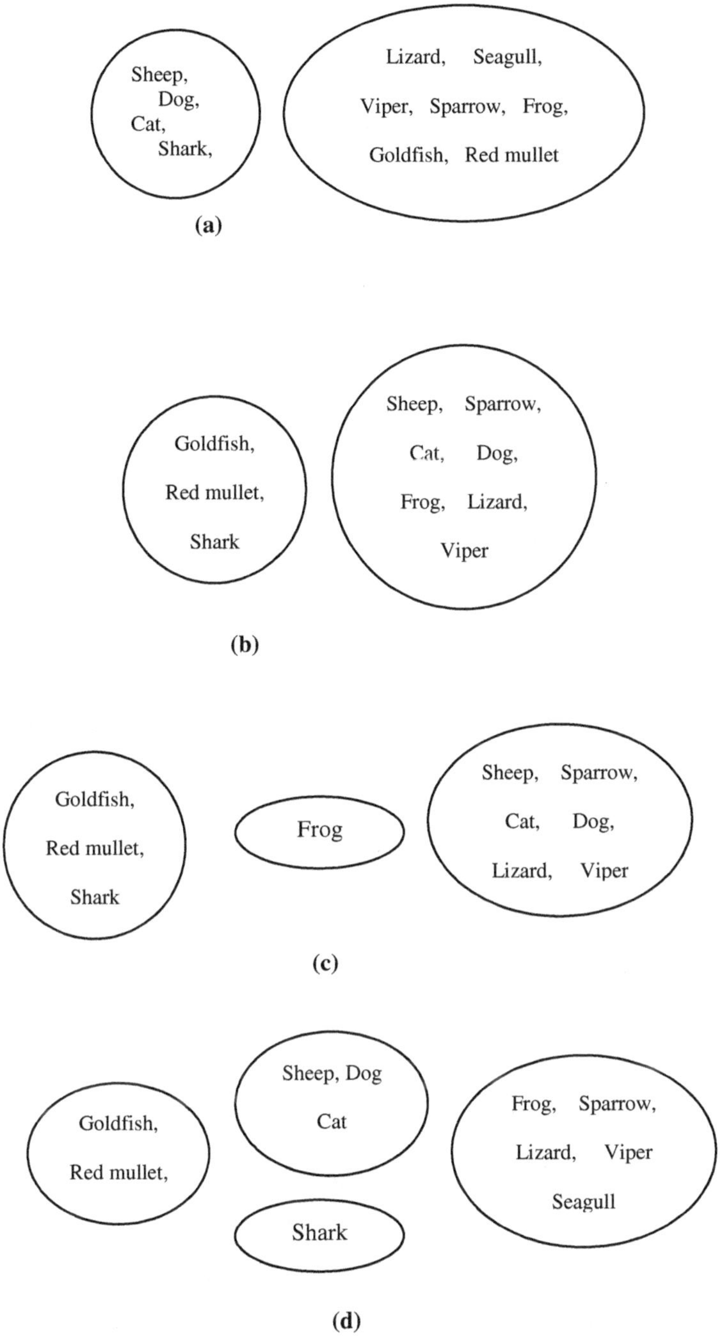

Fig. 1.1 Resulting clusters if the clustering criterion is (a) the way the animals bear their progeny (b) the existence of lungs (c) the environment where the animals live, and (d) the way these animals bear their progeny and the existence of lungs

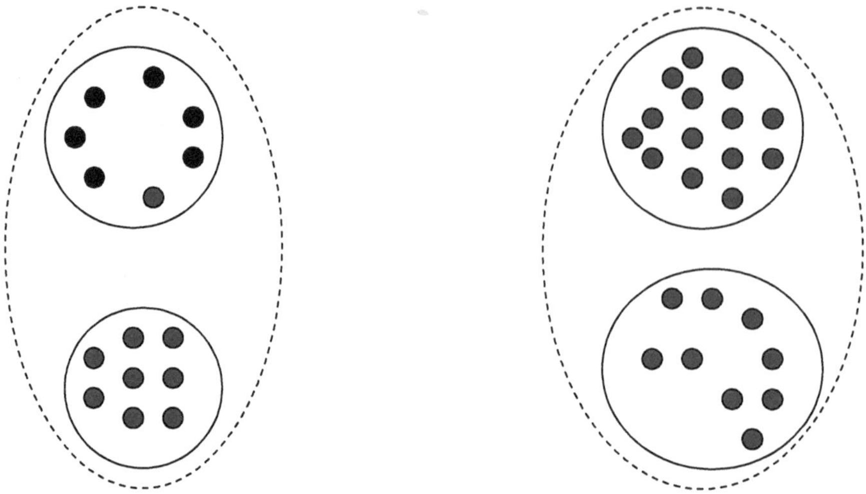

Fig. 1.2 Dashed ellipses show a coarse clustering of the data resulting in two clusters whereas solid circles indicate a finer partitioning of the data resulting into four clusters

A clustering algorithm is expected to discover the natural grouping (pertinent to human notion of clustering) that exists in a set of patterns or data points. Each pattern can be identified as a point in the hyperspace, called *feature space*, spanned by the features associated with it. The input to the clustering algorithm is a set of such points in the multi-dimensional feature space. Figure 1.3 (a) shows such a set of unlabelled points in a two-dimensional feature space. An ideal crisp clustering algorithm should present as its output, the label for each pattern, i.e. the cluster to which each point must belong. Figure 1.3 (b) depicts the situation after clustering of the dataset shown in Figure 1.3 (a) by an ideal clustering algorithm.

The problem of data clustering has been approached from diverse fields of knowledge like statistics (multivariate analysis) [6], graph theory [7], expectation maximization algorithms [8], artificial neural networks [9 - 11], evolutionary computing [12, 13], any colony clustering [118] and so on. Despite being an age old problem, clustering has been attracting the researchers from diverse fields of knowledge. People from the emerging areas of statistics, graph theory, and soft computing are proposing new algorithms, on a regular basis, to meet the increasing complexity of vast real-world datasets. Thus, it seems well nigh impossible to review the huge and multifaceted literature on clustering in the scope of this Volume. We here, instead, shall remain confined mainly to the field of population-based pattern clustering with a focus on evolutionary algorithms, where we attempt to develop some novel clustering methods.

Evolutionary Computing is the collective name for a range of problem-solving techniques based on principles of biological evolution, like natural selection and genetic inheritance. Evolutionary computation techniques use iterative progress, such as growth or development in a population of the candidate solutions of the

Fig. 1.3 An example of clustering with an ideal algorithm

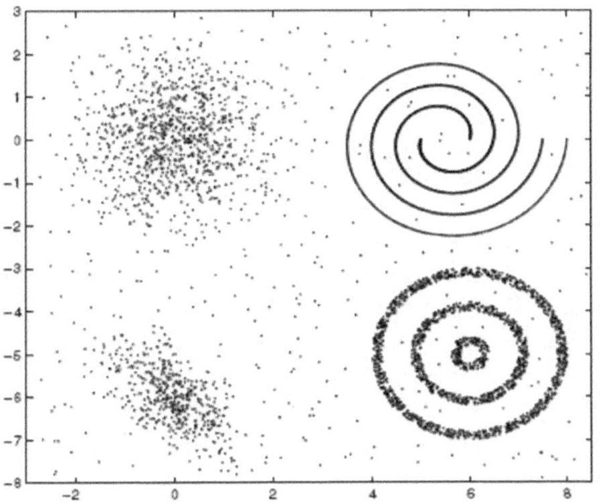

(a) Unlabelled data set with background noise

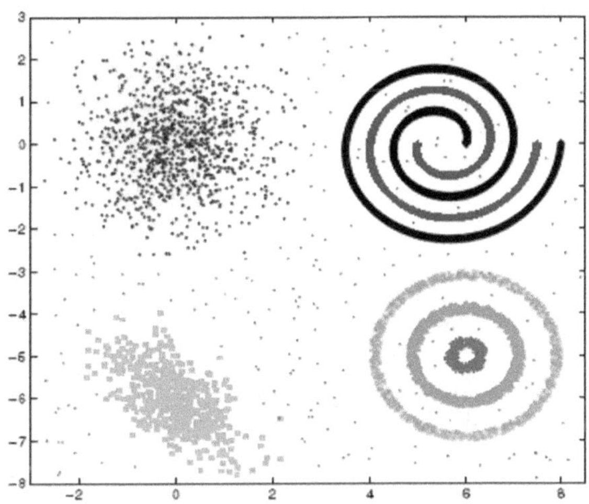

(b) Expected result after clustering

given problem. This population is then selected in a guided random search, using parallel processing to achieve the desired end. Since the past few decades, algorithms from this field have attracted the researchers from the field of pattern recognition and clustering. Clustering techniques based on evolutionary computing algorithms have reportedly outperformed many classical methods of partitioning a dataset. In this Chapter, we explore the applicability of these bio-inspired approaches to the development of self-organizing, evolving, adaptive, and autonomous clustering techniques, which will meet the requirements of next-generation data mining systems, such as diversity, scalability, robustness, and resilience.

1.2 The Clustering Problem

Let us now try to give some definitions of "clustering", which, although may not be universal, give us an idea of what clustering is, in a systematic manner. We present the clustering concept mathematically following works like [2, 4, and 5] and explain the necessary terms and concepts associated with the clustering problems addressed in this Volume.

1.2.1 Basic Definitions

The following terms are used throughout the thesis:

Definition 1.1: A ***Pattern*** (or ***feature vector***) is a physical or abstract structure of objects, which are to be grouped properly by the clustring algorithm [5].

Definition 1.2: A ***Feature*** (or ***attribute***) is an individual component of a pattern. It represents one of the traits based on which the patterns are to be classified [4] .

Definition 1.3: A ***Cluster*** is a well defined collection of similar patterns and patterns from two different clusters must be dissimilar [5].

Definition 1.4: A ***Hard*** (or ***crisp***) clustering algorithm assigns each pattern to one and only one cluster [5].

Definition 1.5: A ***Fuzzy*** clustering algorithm assigns each pattern to each cluster with a certain degree of membership [5].

Definition 1.6: A ***Distance Measure*** is a metric based on which the dissimilarity of the patterns are evaluated [5].

Now we may formalize the definition of the clustering problem in the following way.

Let $\mathbf{P} = \{P_1, P_2, \ldots\ldots, P_n\}$ be a set of n patterns each having d features. These patterns can also be represented by a profile data matrix $\mathbf{Z}_{n \times d}$ having n d-dimensional row vectors. The i-th row vector $\vec{Z}_i$ characterises the i-th object from the set $\mathbf{P}$ and each element $z_{i,j}$ in $\vec{Z}_i$ corresponds to the j-th real value feature (j = $1, 2, \ldots\ldots, d$) of the i-th pattern ($i = 1,2,\ldots.., n$). Given such an $\mathbf{Z}_{n \times d}$ a partitional clustering algorithm tries to find out a partition $\mathbf{C} = \{\mathbf{C}_1, \mathbf{C}_2, \ldots\ldots, \mathbf{C}_k\}$ such that similarity of the patterns in the same cluster $\mathbf{C}_i$ is maximum and patterns from different clusters differ as far as possible. The partitions should maintain the following properties:

1) Each cluster should have at least one pattern assigned. i.e. $C_i \neq \Phi$ $\forall i \in \{1,2,...,k\}$
2) Two different clusters should have no pattern in common. i.e. $C_i \cap C_j = \Phi$ $\forall i \neq j$ and $i, j \in \{1,2,...,k\}$
3) Each pattern should definitely be attached to a cluster i.e. $\bigcup_{i=1}^{k} C_i = \mathbf{P}$

Since the given dataset can be partitioned in a number of ways maintaining all of the above properties, a fitness function or in other words some measure of the adequacy of the partitioning must be defined. Then the problem turns out to be one of finding a partition $\mathbf{C}^*$ of optimal or near optimal adequacy as compared to all other feasible solutions $\mathcal{C} = \{\mathbf{C}^1, \mathbf{C}^2, \ldots\ldots, \mathbf{C}^{N(n,k)}\}$ where $N(n,k) = \frac{1}{k!}\sum_{i=1}^{k}(-1)^i \binom{k}{i}(k-i)^i$ is the number of feasible partitions. This is same as

$$\underset{C}{\text{optimize}}\ f(Z_{n\times d}, C) \tag{1.1}$$

where C is a single partition from the set $\mathcal{C}$ and f is a fitness function that quantifies the goodness of a partition on the basis of the similarity or dissimilarity measures of the patterns. We shall come back to the function f in a much more detailed fashion shortly. Brucker [14] illustrated that the clustering problem is NP-hard when the number of clusters exceeds 3.

In the case of fuzzy clustering, a pattern may belong to all the classes with a certain fuzzy membership grade for each class. So, in this case we need to evolve an appropriate partition matrix $\mathbf{U} = [u_{ij}]_{k\times n}$, where $u_{ij} \in [0, 1]$, such that u_{ij} denotes the grade of membership of the i-th element to the j-th cluster. In fuzzy partitioning of the data, the following conditions hold:

$$\left.\begin{aligned}
&0 < \sum_{j=1}^{n} u_{ij} < n \qquad \text{for } i = 1, 2, \ldots\ldots, k.\\
&\sum_{i=1}^{k} u_{ij} = 1 \qquad \text{for } j = 1, 2, \ldots\ldots, n.\\
&\sum_{i=1}^{k}\sum_{j=1}^{n} u_{ij} = n
\end{aligned}\right\} \tag{1.2}$$

1.2.2 Proximity Measures

Two kinds of proximity measures (between two patterns) are commonly found in the literature on clustering. The first type, known as *dissimilarity measure(DM)*, takes larger values when the two patterns differ from each other to a large extent. On the other hand, the second type, called *similarity measure(SM)*, tends to become large when the two patterns have sufficiently close feature values.

To measure the extent of dissimilarity among the patterns one of the most popular dissimilarity measure is the Euclidean distance. Euclidean distance between two data vectors $\vec{Z}_u$ and $\vec{Z}_v$ may be computed as,

$$d(\vec{Z}_u, \vec{Z}_v) = \sqrt{\sum_{i=1}^{d} (z_{u,i} - z_{v,i})^2} = \left\|\vec{Z}_u - \vec{Z}_v\right\| \quad (1.3)$$

Euclidean distance is a special case (for $\alpha = 2$) of the more generalized Minkowski metric [9] defined as,

$$d^{\alpha}(\vec{Z}_u, \vec{Z}_v) = (\sum_{i=1}^{k} (z_{u,i} - z_{v,i})^{\alpha})^{1/\alpha} = \left\|\vec{Z}_u - \vec{Z}_v\right\|^{1/\alpha} . \quad (1.4)$$

When $\alpha = 1$, the above measure is referred as the Manhattan distance [2]. However, in high-dimensional feature space, Minkowski metric becomes inefficient as the ditance between the patterns increase with increasing dimensionality and the concepts of near and far become weaker [15]. Further in Minkowski metric the large scaled features dominate over the others. Another popular dissimilarity measure, found in the literature is Mahalanobis distance, which may be defined as:

$$d_M(\vec{Z}_u, \vec{Z}_v) = (\vec{Z}_u - \vec{Z}_v)^T \Sigma^{-1} (\vec{Z}_u - \vec{Z}_v), \quad (1.5)$$

where Σ is the covariance matrix of the patterns. If the covariance matrix is the identity matrix, the Mahalanobis distance reduces to the Euclidean distance. If the covariance matrix is diagonal, then the resulting distance measure is called the normalized Euclidean distance.

Among the most common SMs for real-valued vectors used in practice, one is the cosine distance (basically the dot product of two data vectors) defined as,

$$\vec{Z}_i^T . \vec{Z}_j = \frac{\sum_{i=1}^{d} Z_{u,i} . Z_{v,i}}{\|Z_u\| \|Z_v\|} \quad (1.6)$$

Cosine distance measures the angular difference of the two data vectors and not the difference of their magnitudes. Pearson's correlation coefficient [16] is another popular SM, which may be expressed as:

$$r_{Pearson}(\vec{Z}_u, \vec{Z}_v) = \frac{\vec{Z}_{ul}^T . \vec{Z}_{vl}}{\left\|\vec{Z}_{ul}\right\| . \left\|\vec{Z}_{vl}\right\|}, \tag{1.7}$$

where $\vec{Z}_{ul} = [Z_{u,1} - \overline{Z}_u, ..., Z_{u,d} - \vec{Z}_u]^T$ and $\vec{Z}_{vl} = [Z_{v,1} - \overline{Z}_v, ..., Z_{v,d} - \vec{Z}_v]^T$, with $\overline{Z}_u = \frac{1}{d}\sum_{i=1}^{d} Z_{u,i}$ and $\overline{Z}_v = \frac{1}{d}\sum_{i=1}^{d} Z_{v,i}$. Usually $\vec{Z}_{ul}$ and $\vec{Z}_{vl}$ are called difference vectors. Evidently, this SM takes values between +1 and -1.

1.2.3 Clustering Validity Indices

In most of the cases, a class of statistical-mathematical functions, based on the notion of similarity or dissimilarity between the data points, is employed for judging the soundness of the clustering solutions provided by an algorithm. The functions are collectively known as the *cluster validity indices* (*CVI*). Generally, a cluster validity index serves two purposes. First, it can be used to determine the number of clusters, and secondly, it finds out the corresponding best partition. One traditional approach for determining the optimum number of classes is to run the algorithm repeatedly with different number of classes as input and then to select the partitioning of the data resulting in the best validity measure [17]. Ideally, a validity index should take care of the following aspects of the partitioning:

1) ***Cohesion:*** Patterns in one cluster should be as similar to each other as possi ble. The fitness variance of the patterns in a cluster is an indication of the cluster's cohesion or compactness.
2) ***Separation:*** Clusters should be well separated. Distance among the cluster centers, (may be their Euclidean distance) gives an indication of cluster separation.

For crisp clustering, some of the well-known indices available in the literature are the Dunn's index (DI) [18], Calinski-Harabasz index [19], Davis-Bouldin (DB) index [20], PBM index [21], and the CS measure [22]. Similarly for fuzzy clustering, the most prominent indices are Partition Coefficient [23], Partition Entropy [24], Xie-Beni index [25] etc. All these indices behave somewhat like an objective function, minimization/maximization of which is expected to lead to an optimal partitioning of the dataset under test. Because of their optimizing character, the cluster validity indices are best used in association with any optimization algorithm for finding out the appropriate clusters in a dataset. In the following Sections, we discuss some validity indices, pertinent to clustering problems.

1.2.3.1 The Davis-Bouldin (DB) Index

This measure is a function of the ratio of the sum of within-cluster scatter to between-cluster separation, and it uses both the clusters and their sample means. First, we define the *within i-th cluster scatter* and the *between i-th and j-th cluster distance* respectively as,

$$S_{i,q} = \left[\frac{1}{N_i} \sum_{\vec{X} \in C_i} \left\| \vec{Z} - \vec{m}_i \right\|_2^r \right]^{1/r} \tag{1.8}$$

$$d_{ij,t} = \left\{ \sum_{p=1}^{d} \left| m_{i,p} - m_{j,p} \right|^t \right\}^{1/t} = \left\| \vec{m}_i - \vec{m}_j \right\|_t \tag{1.9}$$

where $\vec{m}_i$ is the i-th cluster center, r and t are integers and q, t can be selected independently. N_i is the number of elements in the i-th cluster C_i. Next $R_{i,rt}$ is defined as,

$$R_{i,rt} = \max_{j \in K, j \neq i} \left\{ \frac{S_{i,r} + S_{j,r}}{d_{ij,t}} \right\} \tag{1.10}$$

Finally, we define the DB measure as,

$$DB(k) = \frac{1}{k} \sum_{i=1}^{k} R_{i,rt} \tag{1.11}$$

The smallest *DB (k)* indicates a valid optimal partition.

1.2.3.2 The Dunn and Dunn Like Indices

The Dunn's index tries to maximize inter cluster distances (i.e. increase separation) while minimize intra-cluster distances (i.e. increase compactness). This index is defined as:

$$D = \min_{i=1,2,\ldots,k} \left\{ \min_{j=i+1,\ldots,k} \left(\frac{d(C_i, C_j)}{\max\limits_{p=1,2,\ldots,k} diam(C_p)} \right) \right\}, \tag{1.12}$$

where $d(C_i, C_j) = \min\limits_{\vec{Z}_r \in C_i, \vec{Z}_s \in C_j} \{ d(\vec{Z}_r, \vec{Z}_s) \}$ and $diam(C_p) = \max\limits_{\vec{Z}_u, \vec{Z}_v \in C_p} \{ d(\vec{Z}_u, \vec{Z}_v) \}$.

If a data set contains well-separated clusters, the distances among the clusters are usually large and the diameters of the clusters are expected to be small [17]. Therefore larger value means better cluster configuration. The main disadvantages of the Dunn index are the following: the calculation of the index is time consuming and this index is very sensitive to noise (as the maximum cluster diameter can be large in a noisy environment). Several Dunn-like indices have been proposed [26]. These indices use different definition for cluster distance and cluster diameter.

1.2.3.3 S_Dbw Validity Index

This validity index, proposed by Halkidi and Vazirgiannis [17], is based on cluster compactness and separation but it also takes into consideration the density of the clusters. Formally the S_Dbw index measures the intra-cluster variance and the inter-cluster variance. The intra cluster variance measures the average scattering of clusters. Inter – cluster density is defined as follows:

$$Dens\_bw = \frac{1}{k(k-1)} \sum_{i=1}^{k} \left(\sum_{\substack{j=1 \\ i \neq j}}^{k} \frac{density(b_{ij})}{\max\{density(\vec{m}_i), density(\vec{m}_j)\}} \right), \tag{1.13}$$

where b_{ij} is the middle point of the line segment that is defined by the $\vec{m}_i$ and $\vec{m}_j$ cluster-centers. The density function around a point is defined as follows: it counts the number of points in a hyper-sphere whose radius is equal to the average standard deviation of clusters. The average standard deviation of clusters is defined as:

$$stddev = \frac{1}{k} \sqrt{\sum_{i=1}^{k} \|\sigma(\vec{m}_i)\|}, \tag{1.14}$$

where the cluster variance may be given by,

$$\sigma_{\vec{m}_i}^{p} = \frac{1}{|C_i|} \sum_{j=1}^{|C_i|} (Z_{j,p} - m_{i,p})^2$$

$$\text{and} \quad \sigma(\vec{m}_i) = \begin{bmatrix} \sigma_{\vec{m}_i}^{1} \\ . \\ . \\ \sigma_{\vec{m}_i}^{d} \end{bmatrix}, \tag{1.15}$$

$|C_i|$ indicating the number of data points in cluster C_i. The definition of S_Dbw indicates that both criteria of "good" clustering are properly combined and it enables reliable evaluation of clustering results. Lower index value indicates better clustering schema.

1.2.3.4 Partition Coefficient

Bezdek [23] designed the partition coefficient (*PC*) to measure the amount of "overlap" between clusters in fuzzy clustering. The partition coefficient (*PC*) is defined as follows.

$$PC(k) = \frac{1}{n} \sum_{i=1}^{k} \sum_{j=1}^{n} (u_{ij})^2, \tag{1.16}$$

where u_{ij} $(i = 1,2,\ldots,k;\ j = 1,2,\ldots,n)$ is the membership of the *j*-th data point in *i*-th cluster. Disadvantages of the partition coefficient are its monotonic decreasing with *k* and the lack of direct connection to some property of the data themselves.

1.2.3.5 Classification Entropy

Classification entropy is defined as:

$$CE(k) = -\frac{1}{n}\sum_{i=1}^{k}\sum_{j=1}^{n} u_{ij}\log(u_{ij}) \tag{1.17}$$

Bezdek proved the relation $0 \le 1 - PC(k) \le CE(k)$ for all probabilistic cluster partitions *k*. It is basically a measure for the fuzziness of the cluster partition only, which is similar to the partition coefficient.

1.2.3.6 Xie-Beni Index

This index, due to Xie and Beni [25], is meant for judging the quality of fuzzy clustering. It may be expressed as:

$$XB = \frac{\sum_{i=1}^{k}\sum_{j=1}^{n} u_{ij}^2 \left\|\vec{Z}_j - \vec{m}_i\right\|^2}{n \times \min_{i \ne j}\left\|\vec{m}_i - \vec{m}_j\right\|^2} \tag{1.18}$$

Note that the numerator of the Xie-Beni index is basically the objective function minimized by the FCM algorithm. The denominator of *XB* is basically the separation between the *i*-th and *j*-th clusters. The more separated the clusters are, the larger is the denominator and hence the smaller is the value of *XB*. Thus, the optimal number of clusters can be obtained by minimizing the *XB* value.

1.2.3.7 The PS Measure

This measure rests upon the concept of a *point symmetry distance, which* was shown to be very effective for clustering by Chou *et al.* [27, 28]. The point symmetry distance is defined as follows the "point symmetry distance" between a pattern $\vec{Z}_j$ and the reference vector (i.e. cluster centroid) $\vec{m}$ is defined as:

$$d_s(\vec{Z}_j, \vec{m}) = \min_{\substack{i=1,2,\dots,n \\ and\, i \ne j}} \left\{\frac{\left\|(\vec{Z}_j - \vec{m}) + (\vec{Z}_i - \vec{m})\right\|}{\left(\left\|\vec{Z}_j - \vec{m}\right\| + \left\|\vec{Z}_i - \vec{m}\right\|\right)}\right\} \tag{1.19}$$

Chou *et al.* [29] defined a clustering validity index based on the point symmetry measure as

$$PS(k) = \frac{1}{k}\sum_{i=1}^{k}\left[\frac{1}{|C_i|}\sum_{j \in C_i}\frac{d_s(\vec{Z}_j, \vec{m}_i).d_e(\vec{Z}_j, \vec{m}_i)}{\min_{\substack{p,q=1,2,\dots,k \\ p \ne q}}\{d_e(\vec{m}_p, \vec{m}_q)\}}\right] \tag{1.20}$$

where d_e is the Euclidean distance between two points in the d-dimensional feature space, $|C_i|$ is the number of data points belonging to cluster C_i. While computing the point symmetry distance d_s according to equation (1.17) we keep in mind that the vectors $\vec{X}_j$ and $\vec{X}_k$ belong to the same class C_i. Finally, the smallest *PS(k)* indicates a valid optimal partition with the optimal cluster number *k*. Su and Chou [30] illustrated the usefulness of the PS measure over many difficult non-image datasets having clusters of widely different geometrical shapes.

1.2.3.8 The PBMF Index

Recently Pakhira *et al.* [31] have proposed a new clustering validity index, the hard version of which may be given as,

$$PBM(k) = (\frac{1}{k} \times \frac{E_1}{E_k} \times D_k)^2 \tag{1.21}$$

where *k* is , as usual, the number of clusters and

$$E_k = \sum_{i=1}^{k} E_i \tag{1.22}$$

and

$$E_i = \sum_{j=1}^{n} u_{ij} \left\| \vec{Z}_j - \vec{m}_i \right\|^2 \tag{1.23}$$

and

$$D_k = \max_{i,j=1}^{k} \left\| \vec{m}_i - \vec{m}_j \right\| \tag{1.24}$$

$\vec{m}_i$ being the *i*-th cluster center. u_{ij} is the *(i, j)*-th entry of the partition matrix **U** = $[u_{ij}]_{k \times n}$. In [32], Pakhira *et al.* have also proposed a fuzzy version of this validity measure:

$$PBMF(k) = (\frac{1}{k} \times \frac{E_1}{J_m} \times D_k), \tag{1.25}$$

where

$$J_m = \sum_{j=1}^{n} \sum_{i=1}^{k} (u_{ij})^m \left\| \vec{Z}_j - \vec{m}_i \right\| \tag{1.26}$$

In equation (1.23) the term E_1 usually appears as a constant for a particular dataset. The maximum value of the PBMF index is supposed to give the optimal number of clusters in a dataset.

1.2.3.9 The CS Measure

Recently, Chou *et al.* [32] proposed the CS measure for evaluating the validity of a clustering scheme. Before applying the CS measure, centroid of a cluster is computed by averaging the data vectors belonging to that cluster using the formula,

$$\vec{m}_i = \frac{1}{N_i} \sum_{\vec{Z}_j \in C_i} \vec{Z}_j \tag{1.27}$$

A distance metric between any two data points $\vec{Z}_p$ and $\vec{Z}_r$ is denoted by $d(\vec{Z}_p, \vec{Z}_r)$. Then the CS measure can be defined as,

$$CS(k) = \frac{\frac{1}{k}\sum_{i=1}^{k}[\frac{1}{|C_i|}\sum_{\vec{Z}_p \in C_i} \max_{\vec{Z}_y \in C_i}\{d(\vec{Z}_p, \vec{Z}_y)\}]}{\frac{1}{k}\sum_{i=1}^{k}[\min_{j \in k, j \neq i}\{d(\vec{m}_i, \vec{m}_j)\}]}$$

$$= \frac{\sum_{i=1}^{k}[\frac{1}{|C_i|}\sum_{\vec{Z}_p \in C_i} \max_{\vec{Z}_y \in C_i}\{d(\vec{Z}_p, \vec{Z}_y)\}]}{\sum_{i=1}^{k}[\min_{j \in k, j \neq i}\{d(\vec{m}_i, \vec{m}_j)\}]} \tag{1.28}$$

As can easily be perceived, this measure is a function of the ratio of the sum of within-cluster scatter to between-cluster separation and has the same basic rationale as the DI and the DB measures. According to Chou *et al.* CS measure is more efficient in tackling clusters of different densities and/or sizes than the other popular validity measures, the price being paid in terms of high computational load with increasing k and n.

1.3 The Classical Clustering Algorithms

Most of the existing data clustering algorithms can be either *hierarchical* or *partitional* [5,16]. Within each of the types, there exists a wealth of subtypes and different algorithms for finding the clusters. Below we outline the most important traditional clustering algorithms briefly.

1.3.1 Hierarchical Clustering Algorithms

In hierarchical clustering, the data are not partitioned into a particular cluster in a single step. Instead, a series of partitions takes place, which may run from a single cluster containing all objects to k clusters each containing a single object. Hierarchical Clustering is subdivided into agglomerative methods, which proceed by series of fusions of the n objects into groups, and divisive methods, which separate n objects successively into finer groupings. Agglomerative techniques are more

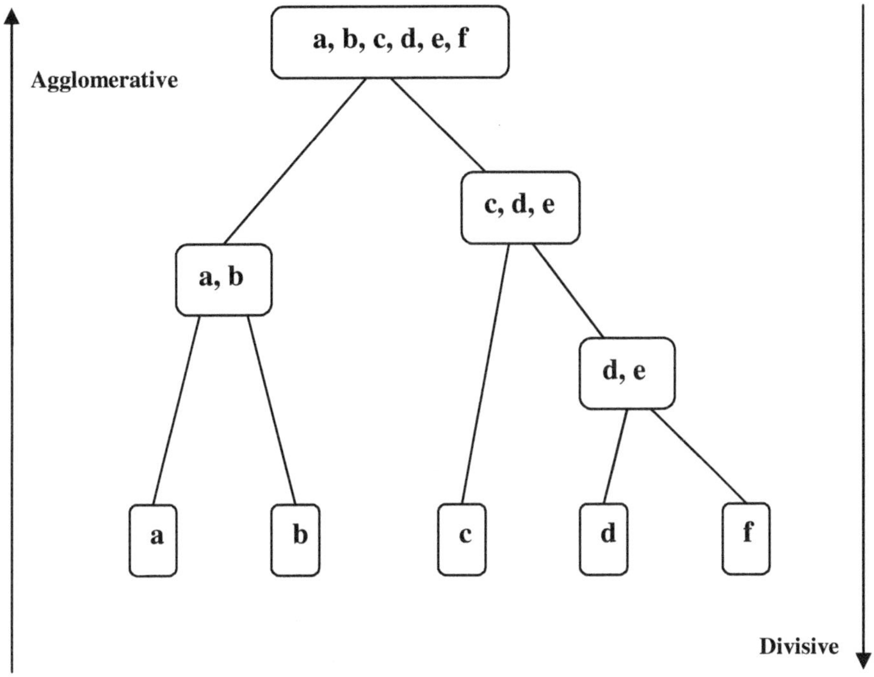

Fig. 1.4 An example dendrogram in hierarchical clustering

commonly used. Hierarchical clustering may be represented by a two dimensional diagram known as dendrogram, which illustrates the fusions or divisions made at each successive stage of analysis. An example of such a dendrogram is shown in Figure 1.4.

At each particular stage the agglomerative algorithm joins together the two clusters which are closest together (most similar). (At the first stage, of course, this amounts to joining together the two objects that are closest together, since at the initial stage each cluster has one object.) Differences between methods arise because of the different ways of defining distance (or similarity) between clusters. This can be done in mainly three different methods: Single-Link, Complete-Link, and the Average-Link methods.

In Single-Link clustering (also called the *connectedness* or *minimum* method), we consider the distance between one cluster and another cluster to be equal to the shortest distance from any member of one cluster to any member of the other cluster. If the data consist of similarities, we consider the similarity between one cluster and another cluster to be equal to the greatest similarity from any member of one cluster to any member of the other cluster. In Complete-Link clustering (also called the *diameter* or *maximum* method), we consider the distance between one cluster and another cluster to be equal to the greatest distance from any member of one cluster to any member of the other cluster. In Average-Linkage clustering, we consider the distance between one cluster and another cluster to be equal to the average distance from any member of one cluster to any member of the other cluster. The Single-Link method usually creates elongated clusters, and the Complete-Link method usually results in more compact clusters. The Average-link acts in a midway basis yielding clusters somewhere between the two previous methods. Theodoridis and Koutroumba [16], presented a nice overview of the probabilistic interpretation of these classical agglomerative clustering methods.

Among the most popular hierarchical clustering algorithms, *BIRCH* [33] uses a hierarchical data structure called CF-tree for partitioning the incoming data points in an incremental and dynamic way. CF-tree is a height balanced tree, which stores the clustering features and it is based on two parameters: *branching factor B* and *threshold T*, which referred to the diameter of a cluster (the diameter (or radius) of each cluster must be less than *T*). *BIRCH* can typically find a good clustering with a single scan of the data and improve the quality further with a few additional scans. It is also the first clustering algorithm to handle noise effectively [33]. However, it does not always correspond to a natural cluster, since each node in CF-tree can hold a limited number of entries due to its size. Moreover, it is order-sensitive as it may generate different clusters for different orders of the same input data.

CURE [34] represents each cluster by a certain number of points that are generated by selecting well-scattered points and then shrinking them toward the cluster centroid by a specified fraction. It uses a combination of random sampling and partition clustering to handle large databases. *ROCK* [35] is a robust clustering algorithm for Boolean and categorical data. It introduces two new concepts, that are a point's neighbors and links, and it is based on them in order to measure the similarity/proximity between a pair of data points.

In general, the hierarchical algorithms have the following two basic advantages [5].

1) The number of classes need not be specified a prior.
2) They are independent of the initial conditions.

However, the hierarchical methods suffer from the following drawbacks as well:

1) They are computationally expensive (runtime complexity is $O(n^2 \log n)$ and space complexity is $O(n^2)$), a feature which makes them unsuitable for very large datasets.
2) They are static, i.e. data-points assigned to a cluster can not move to another cluster.
3) They may fail to separate overlapping clusters due to lack of information about the global shape or size of the clusters

1.3.2 Partitional Clustering Algorithms

Partitional clustering algorithms attempt to decompose the dataset directly into a set of disjoint clusters. They try to optimize certain criterion function that may emphasize the local structure of the data, as by assigning clusters to peaks in the probability density function, or the global structure. Typically, the global criteria involve minimizing some measure of dissimilarity in the samples within each cluster, while maximizing the dissimilarity of different clusters. The advantages of the hierarchical algorithms are the disadvantages of the partitional algorithms and vice versa. Partitional algorithms are generally iterative in nature and converge to some local optima. The general steps of an iterative clustering algorithm as illustrated by Hammerly and Elkan [36] may be given as:

Iterative Partitional Clustering

Step 1: Randomly initialize k cluster centroids.
Step 2: **Repeat**

(a) **for** each pattern $\vec{Z}_i$ in the dataset **do**

Compute its membership $u(\vec{m}_j \mid \vec{Z}_i)$ to each centroid and its weight $w(\vec{Z}_i)$

end

(b) Recalculate the k cluster centroids using

$$\vec{m}_j = \frac{\sum_{\forall \vec{Z}_i} u(\vec{m}_j \mid \vec{Z}_i).w(\vec{Z}_i).\vec{Z}_i}{\sum_{\forall \vec{Z}_i} u(\vec{m}_j \mid \vec{Z}_i).w(\vec{Z}_i)} \tag{1.29}$$

until a stopping condition is satisfied.

In the above algorithm, $u(\vec{m}_j \mid \vec{Z}_i)$ is the membership function, which quantifies the membership of pattern $\vec{Z}_i$ to j-th cluster C_j. The membership function must satisfy the following constraints:

1) $u(\vec{m}_j \mid \vec{Z}_i) \geq 1, \quad j = 1,2,...,k$ and $i = 1,2,....,n$.

2) $\sum_{j=1}^{k} u(\vec{m}_j \mid \vec{Z}_i) = 1, \quad j = 1,2,...,k$.

Crisp clustering algorithms use a *hard* membership function (i.e. $u(\vec{m}_j \mid \vec{Z}_i) \in \{0,1\}$) whereas fuzzy clustering algorithms use a *soft* membership function (i.e. $u(\vec{m}_j \mid \vec{Z}_i) \in [0,1\}]$) [67]. The weight function $w(\vec{Z}_i)$, proposed by Zhang [37], in (1.7) indicates how much influence pattern $\vec{Z}_i$ has in re-computing the cluster centroids [45].

An iterative clustering algorithm may come up with various stopping criteria such as:

1) Stop when the change in centroid values over subsequent iterations is smaller than a use specified value.
2) Stop when the quantization error is small enough, or
3) Stop when a maximum number of iterations has been exceeded.

In the following Section, we shall briefly review some of the most popular partitional clustering algorithms of current interest.

1.3.2.1 The k-Means Algorithm

The most widely used iterative k-means algorithm for partitionl clustering dates back to mid-sixties [38]. k-means algorithm aims at minimizing the ICS (Intra Cluster Spread), which for k cluster centers can be defined as,

$$ICS(C_1, C_2, ..., C_k) = \sum_{j=1}^{k} \sum_{\vec{Z}_i \in C_i} d^2(\vec{Z}_i, \vec{m}_j), \tag{1.30}$$

where $\vec{m}_j$ represents the centroid of the j-th cluster C_j and $\vec{Z}_i$ is a data point belonging to cluster C_j. The k-means (or Hard c-means) algorithm starts with k cluster-centroids (these centroids are initially selected randomly or derived from some priori information). Each pattern in the data set is then assigned to the closest cluster-centre. The centroids are finally updated by using the mean of the associated patterns. The process is repeated until some stopping criterion is met. The membership and weight functions for k-means are defined as:

$$u(\vec{m}_j \mid \vec{Z}_i) = \begin{cases} 1 \text{ if } d^2(\vec{Z}_i, \vec{m}_j) = \arg\min_j \{d^2(\vec{Z}_i, \vec{m}_j)\} \\ 0 \text{ otherwise.} \end{cases} \tag{1.31}$$

and

$$w(\vec{Z}_i) = 1 \tag{1.32}$$

Thus k-means has a hard membership function. Furthermore, k-means has a constant weight function, i.e. all patterns belonging to a cluster have equal influence in computing the centroid of the cluster. The k-means has two main advantages [5] viz,

1) It is very easy to implement
2) The time complexity is only $O(n)$ (n being the number of data points), which makes it suitable for large data sets.

However it suffers from the following disadvantages [5]-

1) The user has to specify the number of classes in advance.
2) The performance of the algorithm is data-dependent.
3) The algorithm uses a greedy approach and is heavily dependent on the initial conditions. This often leads k-means to converge to sub-optimal solutions.

At this point, we would like to mention another k-prototype algorithm called *k-mode* [40]. It is based on k-means algorithm, but it aims aim at clustering categorical data.

1.3.2.2 The k-Medoids Algorithm

In the k-medoids algorithm [16], on the other hand, each cluster is represented by one of the representative objects in the cluster located near the center. Partitioning Around Medoids (PAM) [41] starts from an initial set of medoids, and iteratively replaces one of the medoids by one of the non-medoids if it improves the total distance of the resulting clustering. Although PAM works effectively for small data, it does not scale well for large datasets [5].

The objective of PAM is to determine a representative object (*medoid*) for each cluster, that is, to find the most centrally located objects within the clusters. The algorithm begins by selecting an object as medoid for each of k clusters. Then, each of the non-selected objects is grouped with the medoid to which it is the most similar. PAM swaps medoids with other non-selected objects until all objects qualify as medoid. It is clear that PAM is an expensive algorithm as regards finding the medoids, as it compares an object with entire dataset [41].

CLARA (*Clustering Large Applications*), is an implementation of PAM in a subset of the dataset. It draws multiple samples of the dataset, applies PAM on samples, and then outputs the best clustering out of these samples [41]. *CLARANS* (*Clustering Large Applications based on Randomized Search*) [41], combines the sampling techniques with PAM. The clustering process can be presented as searching a graph where every node is a potential solution, that is, a set of k medoids. The clustering obtained after replacing a medoid is called the *neighbor* of the current clustering. CLARANS selects a node and compares it to a user-defined number of their neighbors searching for a local minimum. If a better neighbor is found (i.e., having lower-square error), CLARANS moves to the neighbor's node and the process start again; otherwise the current clustering is a local optimum. If the local optimum is found, CLARANS starts with a new randomly selected node in search for a new local optimum.

1.3.2.3 The Fuzzy c-Means Algorithm

Bezdek proposed a fuzzy version of the k-means algorithm, now well-known as the fuzzy c-means (FCM) (or sometimes called the fuzzy k-means) algorithm[42]. It seems to be the most popular algorithm in the field of fuzzy clustering. The FCM is based on a fuzzy extension of the least-square error criterion. In the classical FCM algorithm, a *within cluster sum* function J_q is minimized to evolve the proper cluster centers:

$$\vec{m}_i = \frac{\sum_{j=1}^{n} (u_{ij})^q . \vec{Z}_j}{\sum_{j=1}^{n} (u_{ij})^q} \tag{1.33}$$

where $\vec{m}_j$ is the j-th cluster center, $\vec{Z}_i$ is the i-th d-dimensional data vector, and u_{ij} is the membership of the i-th data point to j-th cluster. Here q ($q > 1$) is any real number that influences the membership grade. Increasing the value of q makes the algorithm even fuzzier. Given k classes, we can determine their cluster centers $\vec{m}_j$ for $i = 1$ to k by means of the following expression:

$$\vec{m}_i = \frac{\sum_{j=1}^{n} (u_{ij})^q \vec{X}_j}{\sum_{j=1}^{n} (u_{ij})^q} \tag{1.34}$$

Now differentiating the performance criterion with respect to $\vec{m}_i$ (treating u_{ij} as constants) and with respect to u_{ij} (treating $\vec{m}_i$ as constants) and setting them to zero the following relation can be obtained:

$$u_{ij} = \left[\sum_{p=1}^{k} \left(\frac{\|\vec{Z}_i - \vec{m}_j\|^2}{\|\vec{Z}_i - \vec{m}_p\|^2} \right)^{1/(q-1)} \right]^{-1} \tag{1.35}$$

In general FCM performs better than the k-means algorithm especially when the clusters are overlapping in nature [15, 16]. However, like k-means here we also need to specify the number of possible classes a priori. In addition, FCM may converge to some local optima of its fitness landscape [5]. Several modifications of the classical FCM algorithm can be found in [43-45].

1.3.2.4 The Expectation-Maximization Algorithm

Another popular approach to partitional clustering comes through the Expectation-Maximization (EM) algorithm [46, 47]. The EM algorithm has its roots in statistics and is used for finding maximum likelihood estimates of parameters in probabilistic models, where the model depends on unobserved latent variables. EM partitions a dataset into clusters by determining a mixture of Gaussian probability density functions (PDFs) fitting the data. Each Gaussian PDF has a mean and a covariance matrix. The objective function that the EM optimizes as defined by Hamerly and Elkan [48] is given by:

$$J_{EM} = -\sum_{i=1}^{n} \log\left(\sum_{j=1}^{k} p(\vec{Z}_i \mid \vec{m}_j).p(\vec{m}_j) \right), \tag{1.36}$$

where $p(\vec{Z}_i \mid \vec{m}_j)$ is the probability of $\vec{Z}_i$ given that it is generated by a Gaussian distribution function with centroid $\vec{m}_j$, and $p(\vec{m}_j)$ is the prior probability of centroid $\vec{m}_i$. The membership and weight functions for EM are defined as [48]:

$$u(\vec{m}_j \mid \vec{Z}_i) = \frac{p(\vec{Z}_i \mid \vec{m}_j)}{p(\vec{Z}_i)} \tag{1.37}$$

$$w(\vec{Z}_i) = 1 \tag{1.38}$$

Hence, EM has a soft membership function and a constant weight function. The algorithm starts with an initial estimate of the parameters. Then, an expectation step is performed where the known data values are used to compute expected values of the unknown data. This is followed by a maximization step where the known and expected values of the data are used to generate new estimate of the parameters. The expectation and maximization steps are repeated until convergence.

Empirical results [49, 36] and analysis illustrate that the k-means algorithm performs comparable to EM [66]. Furthermore, Alldrin *et al.* [50] reports that EM fails on high dimensional data due to numerical precision problems. They also observe that the Gaussians often collapse into delta functions. In addition EM is sensitive to the initial conditions [36, 51] and needs the number of clusters to be specified beforehand. Finally EM assumes the underlying probability density of each cluster is Gaussian, which may not be true in practice [52].

1.3.2.5 The k-Harmonic Means Algorithm

Recently Zhang *et al.* [53, 54] proposed a novel partitional algorithm called the k-harmonic means (KHM) algorithm, with promising results. In KHM, the harmonic mean of the distance of each cluster center or centroid from every pattern is calculated. The cluster centroids are then updated accordingly. The KHM optimizes the following objective function:

$$J_{KHM} = \sum_{i=1}^{n} \frac{k}{\sum_{p=1}^{k} \frac{1}{\left\| \vec{Z}_i - \vec{m}_p \right\|^{\alpha}}}, \tag{1.39}$$

where α is a user-defined parameter and $\alpha \geq 2$. The membership and weight functions for KHM are given by:

$$u(\vec{m}_j \mid \vec{Z}_i) = \frac{\left\| \vec{Z}_i - \vec{m}_j \right\|^{-\alpha-2}}{\sum_{p=1}^{k} \left\| \vec{Z}_i - \vec{m}_p \right\|^{-\alpha-2}} \tag{1.40}$$

$$w(\vec{Z}_i) = \frac{\sum_{p=1}^{k} \left\| \vec{Z}_i - \vec{m}_p \right\|^{-\alpha-2}}{\left(\sum_{p=1}^{k} \left\| \vec{Z}_i - \vec{m}_p \right\|^{-\alpha} \right)^2} \tag{1.41}$$

Thus, KHM has a soft membership and a varying weight function. KHM assigns higher weights to the patterns, which are far away from all the centroids to help the centroids cover the entire data. Unlike the k-means, KHM is less sensitive to the initial conditions and does not have the problem of collapsing Gaussians like EM [36].

1.3.3 Density-Based Clustering Algorithms

Density based algorithms typically regard clusters as dense regions of objects in the data space that are separated by regions of low density. The main idea of density-based approach is to find regions of high density and low density, with high-density regions being separated from low-density regions. These approaches can make it easy to discover arbitrary clusters. A common way is to divide the high-dimensional space into density-based grid units. Units containing relatively high densities are the cluster centers and the boundaries between clusters fall in the regions of low-density units [55].

Density based spatial clustering of applications with noise, DBSCAN [56], rely on a density-based notion of clusters, which is designed to discover clusters of arbitrary shape and also have ability to handle noise. The main task of this algorithm is class identification, i.e. the grouping of the objects into meaningful subclasses. In [57] another density-based clustering algorithm, DENCLUE, is proposed. This algorithm introduces a new approach to cluster large multimedia databases. The basic idea of this approach is to model the overall point density analytically as the sum of influence functions of the data points. The influence function can be seen as a function, which describes the impact of a data point within its neighborhood. Then clusters can be identified by determining density attractors. Density attractors are local maximum of the overall density function. In addition, clusters of arbitrary shape can be easily described by a simple equation based on overall density function. The main advantages of DENCLUE are that it has good clustering properties in data sets with large amounts of noise and it allows a compact mathematically description of arbitrary shaped clusters in high-dimensional data sets. However, DENCLUE clustering is based on two parameters and as in most other approaches the quality of the resulting clustering depends on the choice of them. These parameters are [57]: i) parameter N which determines the influence of a data point in its neighborhood and ii) The other parameter describes whether a density-attractor is significant, allowing a reduction of the number of density-attractors and helping to improve the performance.

One of the density-based clustering algorithms, other than the well-known DBSCAN, is the mean-shift algorithm [58]. The mean shift algorithm is a powerful technique for image segmentation. The algorithm recursively moves to the kernel smoothed centroid for every data point. The quadratic computational complexity of the algorithm is a significant barrier to the scalability of this algorithm to practical applications.

1.3.4 Grid-Based Clustering Algorithms

Recently a number of clustering algorithms have been presented for spatial data, known as grid-based algorithms. These algorithms quantize the space into a finite number of cells and then do all operations on the quantized space. *STING* (Statistical Information Grid-based method) [59] is representative of this category. It divides the spatial area into rectangular cells using a hierarchical structure. *STING* goes through the data set and computes the statistical parameters (such as mean, variance, minimum, maximum and type of distribution) of each numerical feature of the objects within cells. Then it generates a hierarchical structure of the grid cells so as to represent the clustering information at different levels. Based on this structure *STING* enables the usage of clustering information to search for queries or the efficient assignment of a new object to the clusters. *WaveCluster* [60] is the latest grid-based algorithm proposed in literature. It is based on signal processing techniques (wavelet transformation) to convert the spatial data into frequency domain. More specifically, it first summarizes the data by imposing a multidimensional grid structure onto the data space [61]. Each grid cell summarizes the information of a group of points that map into the cell. Then it uses a wavelet transformation to transform the original feature space. In wavelet transform, convolution with an appropriate function results in a transformed space where the natural clusters in the data become distinguishable. Thus, we can identify the clusters by finding the dense regions in the transformed domain. A-priori knowledge about the exact number of clusters is not required in *WaveCluster*.

1.3.5 A Comparative View of the Traditional Clustering Algorithms

This Section offers an overview of the main characteristics of the clustering algorithms presented in Sections 1.3.1 to 1.3.4, in a comparative fashion. Tables 1.1 to 1.4 summarize the main concepts and the characteristics of the most representative clustering algorithms. More specifically the comparison is based on the following features of the algorithms: i) the type of the data that an algorithm supports (numerical, categorical, or spatial), ii) the shape of clusters, iii) ability to handle noise and outliers, iv) the clustering criterion and, v) the runtime complexity. Moreover, we present the input parameters of the algorithms while we study the influence of these parameters to the clustering results. Finally, we describe the type of algorithms results, i.e., the information that an algorithm gives so as to represent the discovered clusters in a data set.

As Table 1.1 depicts, partitional algorithms are applicable mainly to numerical data sets. However, there are some variants of k-means such as k-mode, which handle categorical data. k-mode is based on k-means method to discover clusters while it adopts new concepts in order to handle categorical data. Thus, the cluster centers are replaced with "modes", a new dissimilarity measure used to deal with categorical objects. Another characteristic of partitional algorithms is that they are unable to handle noise and outliers and they are not suitable to discover clusters

Table 1.1 The main characteristics of the partitional clustering algorithms

Category		Hierarchical				
Name	Type of Data	Runtime complexity order	Favored Cluster Shape	Outliers, noise	Input Parameters	Final Results
BIRCH	Numerical	$O(n)$	Non-convex	Yes	Radius of clusters, branching factor.	CF = (number of points in the cluster *N*, linear sum of points in the cluster *LS*, the square sum of *N* data *SS*) points
CURE	Numerical	$O(n^2 \log n)$	Arbitrary	Yes	Number of clusters, number of cluster representatives	Assignment of data values to clusters
ROCK	categorical	$O(n^2 + n.m_m.m_a + n^2 \log n)$, where m_m is the maximum number of neighbors for a data point and m_a is the average number of neighbors for the same data point.	Arbitrary	Yes	Number of clusters	Assignment of data values to clusters

Table 1.2 The main characteristics of the hierarchical clustering algorithms

Category		Partitional				
Name	Type of Data	Runtime complexity order	Favored Cluster Shape	Outliers, noise	Input Parameters	Final Results
k-means	Numerical	$O(n)$	Non-convex	No	Number of clusters	Cluster Centers
k-modes	Categorical	$O(n)$	Non-convex	No	Number of clusters	Modes of clusters
PAM	Numerical	$O(k.(n-k)^2)$	Non-convex	No	Number of clusters	Medoids of clusters
CLARA	Numerical	$O(k.(40+k)^2 + k.(n-k))$	Non-convex	No	Number of clusters	Medoids of clusters
CLARANS	Numerical	$O(k.n^2)$	Non-convex	No	Number of clusters, Maximum number of neighbors examined.	Medoids of clusters
Fuzzy c-means	Numerical	$O(n^2)$	Non-convex	No	Number of clusters	Cluster Centroids, membership values or beliefs

Table 1.3 The main characteristics of the density-based clustering algorithms

Category		Density-based				
Name	Type of Data	Runtime complexity order	Favored Cluster Shape	Outliers, noise	Input Parameters	Final Results
DBSCAN	Numerical	$O(n.\log n)$	Arbitrary	Yes	Cluster radius, minimum number of objects	Assignment of data values to clusters
DENCLUE	Numerical	$O(n.\log n)$	Arbitrary	Yes	Cluster radius, minimum number of objects	Assignment of data values to clusters

Table 1.4 The main characteristics of the grid-based clustering algorithms

Category		Grid-based				
Name	Type of Data	Runtime complexity order	Favored Cluster Shape	Outliers, noise	Input Parameters	Final Results
Wave-cluster	Spatial	$O(n)$	Arbitrary	Yes	Wavelets, the number of grid cells for each dimensions, the number of application of wavelet transform	Clustered objects
STING	Spatial	$O(g)$, where g is the number of grid cells at the lowest level.	Arbitrary	Yes	Number of objects in a cell	Clustered objects

with non-convex shapes. Moreover, they are based on certain assumption to partition a data set. Thus, they need to specify the number of clusters in advance except for CLARANS, which needs as input the maximum number of neighbors of a node as well as the number of local minima that will be found in order to define a partitioning of a dataset. The result of clustering process is the set of representative points of the discovered clusters. These points may be the centers or the medoids (most centrally located object within a cluster) of the clusters depending on the algorithm.

A summarized view of the characteristics of *hierarchical clustering* methods is presented in Table 1.3. The algorithms of this category create a hierarchical decomposition of the database represented as dendrogram. They are more efficient in handling noise and outliers than partitional algorithms. However, they break down due to their non-linear time complexity (typically, complexity $O(n^2)$, where n is the number of points in the dataset) and huge *input/output* cost when the number of input data points is large.

From Table 1.3, we note that, DBSCAN requires the user to specify the radius of the neighborhood of a point *Eps*, and the minimum number of points in the neighborhood, *MinPts*. Then, it is obvious that DBSCAN is very sensitive to the parameters *Eps* and *MinPts*, which are difficult to determine. Similarly, DENCLUE requires careful selection of its input parameters' value (i.e., σ and ξ), since such parameters may influence the quality of clustering results. However, the major advantage of DENCLUE in comparison with other clustering algorithms are [61]: i) it has a solid mathematical foundation and generalized other clustering methods, such as partitional, hierarchical, ii) it has good clustering properties for data sets with large amount of noise, iii) it allows a compact mathematical description of arbitrary shaped clusters in high-dimensional data sets, iv) it uses grid cells and only keeps information about the cells that actually contain points. It manages these cells in a tree-based access structure and thus it is significant faster than some influential algorithms such as DBSCAN. In general terms the complexity of density based algorithms is $O(n \log n)$. They do not perform any sort of sampling, and thus they could incur substantial I/O costs. Finally, density-based algorithms may fail to use random sampling to reduce the input size, unless sample's size is large. This is because there may be substantial difference between the density in the sample's cluster and the clusters in the whole data set.

From Table 1.4, we see that, since STING goes through the database once to compute the statistical parameters of the cells, it is generally an efficient method for generating clusters and its time complexity is $O(n)$However, STING uses a multi resolution approach to perform cluster analysis and thus the quality of its clustering results depends on the granularity of the lowest level of grid. Moreover, STING does not consider the spatial relationship between the children and their neighboring cells to construct the parent cell. The result is that all cluster boundaries are either horizontal or vertical and thus the quality of clusters is questionable [60]. On the other hand, WaveCluster efficiently achieves to detect arbitrary shape clusters at different scales exploiting well-known signal processing techniques. It does not require the specification of input parameters (e.g. the number of clusters or a neighborhood radius), though a-priori estimation of the expected number of clusters helps in selecting the correct resolution of clusters.

1.4 Population Based Optimization Techniques

1.4.1 Optimization Algorithms

Scientists and engineers from all disciplines often have to deal with the classical problem of search and optimization. Optimization means the action of finding the best-suited solution of some problem within the given constraints and flexibilities. While optimizing a system performance we aim at finding out such a set of values of the system parameters for which the overall performance of the system will be simply the best under some given conditions. Usually the parameters governing the system performance are represented in a vector like:

$$\vec{X} = [x_1, x_2, x_3, ..., x_D]^T$$

To measure how far the 'best' performance we have achieved, an objective function (or fitness function) is designed for the system. This function takes the parameter vector as its argument and returns a value from which we can judge the quality of the parameters (or solutions of the problem) in making the system response optimal. Obviously the best parameter vector is one for which the objective function value is either minimized or maximized. The task of optimization is basically a search for such the parameter vector $\vec{X}^*$ which minimizes (or maximizes) an objective function $f(\vec{X})$ i.e. $f(\vec{X}^*) < f(\vec{X})$ for all $\vec{X} \in \Omega$ where Ω is a possibly large finite set serving as the domain of the search. For now we restrict ourselves to *Single Objective Optimization* problems where the objective function is a scalar function of multidimensional parameter vectors. The above description may be put in a formal language in the following manner [62].

Definition 1.7 **(Optimization Problem).** Let S be a search Space and $\Omega \subseteq S$ be the feasible part and f be the objective function (also known as the fitness function). Then the problem of optimization amounts to finding out an $\vec{X} \in \Omega$, such that:

$$f(\vec{X}) \leq f(\vec{Y}) \qquad \text{for minimization problems, and}$$

$$f(\vec{X}) \geq f(\vec{Y}) \qquad \text{for maximization problems}$$

for every $\vec{Y} \in \Omega$. In that case $\vec{X}$ is the global optimum. The objective function *f* is either numerical $(f : S \to \Re)$ or ordinal $(f : S \times S \to \Re)$. If *f* is numerical then the above definition describes a numerical optimization problem.

Now we define a local optimum.

***Definition 1.2* (Local Optimum).** First a distance measure between the two solutions is defined $dist : S \times S \to \Re$. Then for all $\vec{X} \in S$ a neighborhood of $\vec{X}$ may be defined as

$$N(\vec{X}) = \{\vec{Y} \in S \mid dist(\vec{X}, \vec{Y}) \leq \varepsilon$$

Finally a solution $\vec{X} \in \Omega$ is said to be a local minimum if $f(\vec{X}) \leq f(\vec{Y})$ for all $\vec{Y} \in N(\vec{X})$ and a local maximum if $f(\vec{X}) \geq f(\vec{Y})$ for all $\vec{Y} \in N(\vec{X})$. A local optimum is a local minimum or maximum. For now we restrict ourselves to Single Objective Optimization problems where the objective function or fitness function is a scalar function of multidimensional parameter vectors.

The literature on optimization is very huge and multifaceted. Instead of making any attempt to review such a vast realm, we only try to represent schematically the different classes of optimization methods only to identify the place of the evolutionary computing techniques amidst the wide range of optimization algorithms.

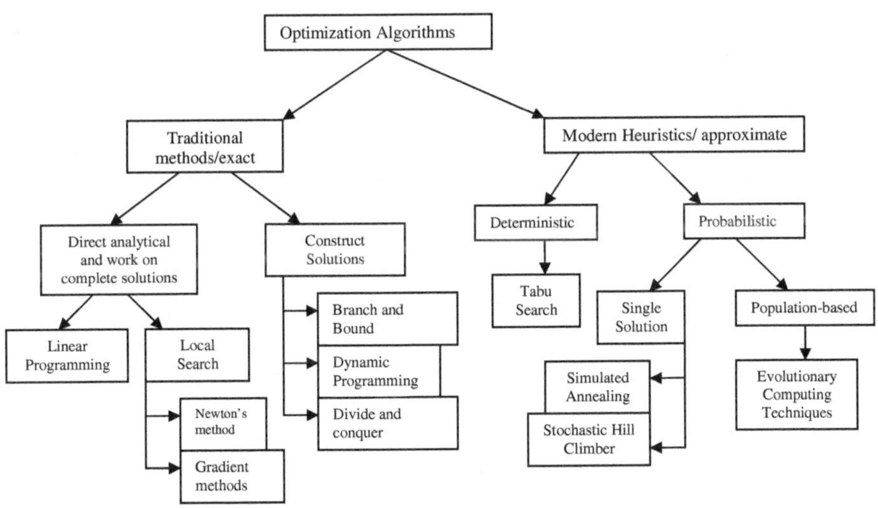

Fig. 1.4 Taxonomy of different optimization methods

In Figure 1.4 we divide the different optimization methods into different classes and placed these in a tree depending on whether they are exact or approximate, deterministic or probabilistic, process complete solutions or construct solutions during the search, and finally whether they maintain a single candidate solution or a have population of solutions. The left branch of Figure 1.4 contains the traditional methods. They all give exact solutions except for local search techniques such as Newton's method and gradient methods, which like modern heuristics, provide approximate solutions. The direct analytical methods can be stopped at any time and return a solution because they process complete solutions, whereas branch and bound, dynamic programming, and divide and conquer all construct solutions during the search, and therefore cannot be stopped to return a complete solution before the whole search is done.

1.4.2 The Evolutionary Computing (EC) Family

To tackle complex search problems (as well as many other complex computational tasks), computer-scientists have been looking into the nature for years – both as model and as metaphor - for inspiration. Optimization is at the heart of many natural processes like Darwinian Evolution itself. Through millions of years, every species had to optimize their physical structures to adapt to the environments they were in. This process of adaptation, this morphological optimization is so perfect that nowadays, the similarity between a shark, a dolphin or a submarine is striking. A keen observation of the underlying relation between optimization and biological evolution has led to the development of a new paradigm of computational intelligence – the evolutionary computing techniques [63 - 65] - for performing very complex search and optimization.

Evolutionary computation uses iterative progress, such as growth or development in a population. This population is then selected in a guided random search using parallel processing to achieve the desired end. Such processes are often inspired by biological mechanisms of evolution. The paradigm of evolutionary computing techniques dates back to early 1950s, when the idea to use Darwinian principles for automated problem solving originated. It was not until the sixties that three distinct interpretations of this idea started to be developed in three different places. Evolutionary Programming was introduced by Lawrence J. Fogel in the USA [66], while John Henry Holland called his method a Genetic Algorithm [67]. In Germany Ingo Rechenberg and Hans-Paul Schwefel introduced the Evolution Strategies [68, 69]. These areas developed separately for about 15 years. From the early nineties on they are unified as different representatives ("dialects") of one technology, called evolutionary computing. Also in the early nineties, a fourth stream following the general ideas had emerged – Genetic Programming [70].

Now-a-days evolutionary computation techniques mostly involve meta-heuristic optimization algorithms such as:

1) Evolutionary algorithms (comprising of genetic algorithms, evolutionary programming, evolution strategy, genetic programming, learning classifier systems, and differential evolution)
2) Swarm intelligence (comprising of ant colony optimization and particle swarm optimization) [71]
   And involved in a lesser extent also in:
3) Self-organization such as self-organizing maps, growing neural gas [72]
4) Artificial life (digital organism) [73]
5) Cultural algorithms [74]
6) Harmony search algorithm [75]
7) Artificial immune systems [76]
8) Learnable evolution model [77].

1.4.3 The Evolutionary Algorithms

Evolutionary Algorithms (EA) are inspired by Darwinian evolution. Natural selection and adaptation in Darwinian evolution are the key sources of inspiration driving the EA candidate solutions towards the optimum by 'survival of the fittest'. To be more precise an EA consists of a population of individuals each having a fitness-value and a genome representing one candidate solution to the given problem. General to all EAs is also the notion of generations as selection removes poor individuals allowing good individuals to survive. EAs also modify the individuals to refine the population of candidate solutions. Below we illustrate the general principle of an EA with a simple pseudo-code. Here *P(t)* denotes a population of chromosomes (trial solutions of the problem at hand) at time t. The procedure initializes a population *P(t)* randomly at iteration $t = 0$. The function: Evaluate *P(t)* determines the fitness of the chromosomes by employing a specially constructed fitness measuring function. The *while*-loop includes three main steps. First it increases the iteration index by 1. Next it selects a population *P(t)* from *P(t-1)* based on the results of fitness evaluation. The function: Alter *P(t)* evolves *P(t)* by some complex non-linear operations. The while loop then re-evaluates *P(t)* for the next iteration, and continues evolution until the terminating condition is reached.

```
Procedure Evolutionary-Computation
Begin
  t←0;
  Initialize P(t);
  Evaluate P(t);
  While (terminating condition not reached) do
   Begin
    t←t+1;
    Select P(t) from P(t-1);
    Alter P(t);
    Evaluate P(t);
   End While;
End.
```

EAs have been described by several researchers [78]. In this Section we briefly outline the most prominent members of the EA family of algorithms.

1.4.3.1 Evolutionary Strategies (ESs)

Evolution strategy (ES) [79, 80] is an optimization technique based on ideas of adaptation and evolution. Evolution strategies use natural problem-dependent representations, and primarily mutation and selection as search operators. As common with evolutionary algorithms, the operators are applied in a loop. An iteration of the loop is called a generation. The sequence of generations is continued until a termination criterion is met.

As far as real-valued search spaces are concerned, mutation is normally performed by adding a normally distributed random value to each vector component. The step size or mutation strength (i.e. the standard deviation of the normal distribution) is often governed by self-adaptation (see evolution window). Individual step sizes for each coordinate or correlations between coordinates are either governed by self-adaptation [81] or by covariance matrix adaptation (CMA-ES) [82].

The (environmental) selection in evolution strategies is deterministic and only based on the fitness rankings, not on the actual fitness values. The simplest ES operates on a population of size two: the current point (parent) and the result of its mutation. Only if the mutant has a higher fitness than the parent, it becomes the parent of the next generation. Otherwise the mutant is disregarded. This is a (1+1)-ES. More generally, λ mutants can be generated and compete with the parent, called $(1 + \lambda)$-ES. In a $(1, \lambda)$-ES the best mutant becomes the parent of the next generation while the current parent is always disregarded.

Contemporary derivatives of evolution strategy often use a population of μ parents and also recombination as an additional operator (called $(\mu/\rho+, \lambda)$-ES). This is believed to make them less prone to get stuck in local optima.

1.4.3.2 Evolutionary Programming (EP)

Evolutionary Programming (EP) [66, 83] use a tree representation to develop automata recognizing strings in formal languages only altering the population of individual trees by mutation. Later the operators of selection and recombination were incorporated, specialized to work on the tree encoding (see Figure 1.5). A more general encoding scheme was also developed and EP is suitable for problems like numerical or real-world optimization.

Currently evolutionary programming is a wide evolutionary computing dialect with no fixed structure or (representation), in contrast with some of the other dialects. It is becoming harder to distinguish from evolutionary strategies. Some of its original variants are quite similar to the later genetic programming, except that the program structure is fixed and its numerical parameters are allowed to evolve. Its main variation operator is mutation; members of the population are viewed as part of a specific species rather than members of the same species therefore each parent generates an offspring, using a $(\mu + \mu)$ survivor selection.

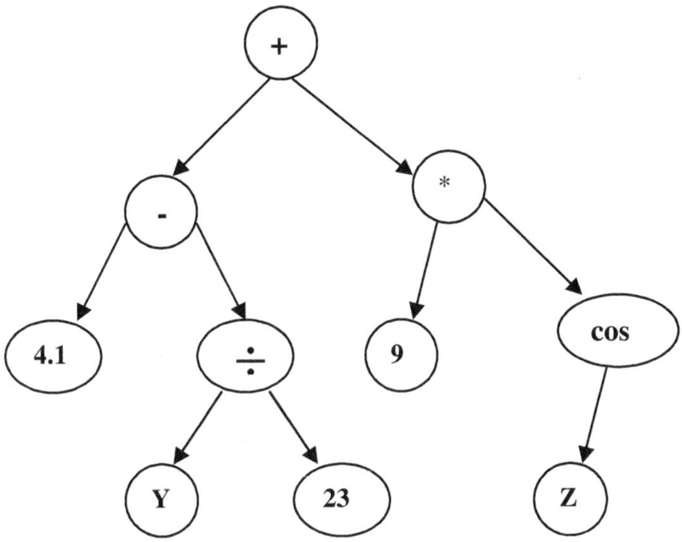

$$\left(4.1-\left(\frac{Y}{23}\right)\right)+\left(9*\cos(Z)\right)$$

Fig. 1.5 A program tree for representing some mathematical function

1.4.3.3 Genetic Algorithms (GAs)

John Holland [67] proposed an attractive class of computational models, called Genetic Algorithms (GAs) that mimic the biological evolution process for solving problems in a wide domain. The mechanisms under GA have been analyzed and explained later by Goldberg [84], De Jong [85], Davis [86], Muehlenbein [87], Fogel [65], Vose [88] and many others. GA has three major applications, namely, intelligent search, optimization and machine learning.

A GA operates through a simple cycle of stages [89]:

i) Creation of a "population" of strings,
ii) Evaluation of each string,
iii) Selection of best strings and
iv) Genetic manipulation to create new population of strings.

The flowchart of a genetic algorithm is presented below in Figure 1.6.

Selection plays a central role in GAs determining how individuals compete for gene survival. Selection weeds out the bad solutions and keeps the good ones. This can be done by *fitness proportional selection* that assigns a selection probability in proportion to the fitness of the given individual [65]. This however tends

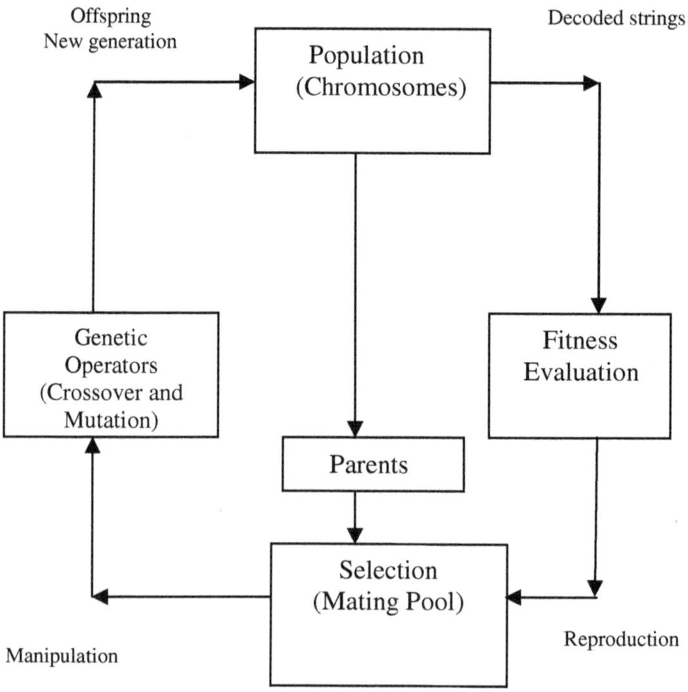

Fig. 1.6 Flowchart of a genetic algorithm

to be sub-optimal as the effective selection strength can be changed by adding an offset. More commonly used is *tournament selection*, where a number of randomly picked individuals are compared to each other [84]. The individual with the best fitness is then selected to be a part of the next generation. The tournament size determines how many individuals are to be compared per selected individual. Selection in GAs is usually done on the whole original population and usually repeated for all individuals in the population. This typical kind of selection allows for an individual to be selected several times and thus results in a loss of diversity. Because of the randomness in selection most techniques can not guarantee survival of the current best solution. Neither can most traditional recombination or mutation operators. *Elitism* provides this guarantee by explicitly selecting the best individual or group of individuals. Both the typical selection and the additional elitism technique can lead to duplicates, however of good individuals. Population diversity decreases with duplicates, but the search can benefit when it comes to the recombination of individuals with good genes.

Recombination of individuals is done to investigate the performance of new individuals that resemble exiting ones. This is done on the genotype level of the individuals and leads to the construction of new intermediate solutions. The notion of generations arises as parent individuals recombine their genes to create offspring. Usually the parents are removed to make room for the offspring carrying some of their genes. Recombination is often done by *crossover*, creating two offspring individuals from two parent individuals by giving each offspring a part of each parental gene. For binary encoded GAs one can use *1-point crossover*, which creates one of two offspring individuals by copying the gene of the first parent up to a random point and then copying the gene of the second parent. Vice versa is done for the second offspring individual. This technique has been generalized creating the *n-point crossover*, with 1-point crossover being most frequently used. For real encoded GAs one often uses *arithmetic crossover* that per genome entry assigns a weighted arithmetic mean of each parent to each offspring. The crossover probability is usually a variable GA parameter.

Mutation happens infrequently both in nature and in GAs to the genome of an individual altering it slightly. Mutation can maintain population diversity and contribute to the exploration of the search space. Exploration by mutation is often slow compared to recombination, but in problems where recombination is disruptive mutation can be an important way to explore the landscape. Even if most of the search is being performed by recombination, mutation can be vital to provide the diversity which recombination needs. The probability of mutation is usually a variable GA parameter.

1.4.3.4 Genetic Programming (GPs)

Genetic programming (GP) is an evolutionary algorithm based methodology inspired by biological evolution to find computer programs that perform a user-defined task. It is a specialization of genetic algorithms where each individual is a computer program. The first results on the GP methodology were reported by Smith [90] and Cramer [91]. Koza has however, remained the main proponent of GP and has pioneered the application of genetic programming in various complex optimization and search problems [70].

GP evolves computer programs, traditionally represented in memory as tree structures (similar to the one shown in Figure 1.5). Trees can be easily evaluated in a recursive manner. Every tree node has an operator function and every terminal node has an operand, making mathematical expressions easy to evolve and evaluate. Thus traditionally GP favors the use of programming languages that naturally embody tree structures (for example, Lisp; other functional programming languages are also suitable). Non-tree representations have been suggested and successfully implemented, such as the simpler linear genetic programming which suits the more traditional imperative languages (see, for example, [92]). The commercial GP software Discipulus uses AIM, automatic induction of binary machine code to achieve better performance [93].

1.4.4 Swarm Intelligence Algorithms

The behavior of a single ant, bee, termite and wasp often is too simple, but their collective and social behavior is of paramount significance. A look at National Geographic TV Channel reveals that advanced mammals including lions also enjoy social lives, perhaps for their self-existence at old age and in particular when they are wounded. The collective and social behavior of living creatures motivated researchers to undertake the study of today what is known as *Swarm Intelligence*. Historically, the phrase Swarm Intelligence (SI) was coined by Beni and Wang in late 1980s [94] in the context of cellular robotics. A group of researchers in different parts of the world started working almost at the same time to study the versatile behavior of different living creatures and especially the social insects. The efforts to mimic such behaviors through computer simulation finally resulted into the fascinating field of SI. SI systems are typically made up of a population of simple agents (an entity capable of performing/executing certain operations) interacting locally with one another and with their environment. Although there is normally no centralized control structure dictating how individual agents should behave, local interactions between such agents often lead to the emergence of global behavior. Below we outline two SI algorithms, which find important applications in data clustering in these days.

1.4.4.1 The Particle Swarm Optimization (PSO)

The concept of Particle Swarms, although initially introduced for simulating human social behaviors, has become very popular these days as an efficient search and optimization technique. The Particle Swarm Optimization (PSO) [95, 96], as it is called now, does not require any gradient information of the function to be optimized, uses only primitive mathematical operators and is conceptually very simple. In PSO, a population of conceptual 'particles' is initialized with random positions $\vec{X}_i$ and velocities $\vec{V}_i$, and a function, *f*, is evaluated, using the particle's positional coordinates as input values. In an *D*-dimensional search space, $\vec{X}_i = (x_{i1}, x_{i2}, ..., x_{iD})^T$ and $\vec{V}_i = (v_{i1}, v_{i2}, ..., v_{iD})^T$. In literature, the basic equations for updating the *d*-th dimension of the velocity and position of the *i*-th particle for PSO are presented most popularly in the following way:

$$\begin{aligned} v_{i,d}(t) = \omega.v_{i,d}(t-1) + \varphi_1.rand1_{i,d}(0,1).(p^l_{i,d} - x_{i,d}(t-1)) + \\ \varphi_2.rand2_{i,d}(0,1).(p^g_d - x_{i,d}(t-1)) \end{aligned} \tag{1.42}$$

$$x_{i,d}(t) = x_{i,d}(t-1) + v_{i,d}(t) \tag{1.43}$$

Please note that in (1.42) and (1.43), φ_1 and φ_2 are two positive numbers known as the *acceleration coefficients*. The positive constant ω is known as *inertia factor*. $rand1_{i,d}(0,1)$ and $rand2_{i,d}(0,1)$ are the two uniformly distributed random numbers in the range of [0, 1]. While applying PSO, we define a maximum velocity $\vec{V}_{\max} = [v_{\max,1}, v_{\max,2}, ..., v_{\max,D}]^T$ of the particles in order to control their convergence behavior near optima. If $|v_{i,d}|$ exceeds a positive constant value $v_{\max,d}$ specified by the user, then the velocity of that dimension is assigned to $\text{sgn}(v_{i,d}).v_{\max,d}$ where *sgn* stands for the signum function and is defined as:

$$\begin{aligned} \operatorname{sgn}(x) &= 1, && \text{if } x > 0 \\ &= 0, && \text{if } x = 0 \\ &= -1, && \text{if } x < 0 \end{aligned} \tag{1.44}$$

While updating the velocity of a particle using (1.42), different dimensions will have different values for $rand1$ and $rand2$. Some researchers, however, prefer to use the same values of these random coefficients for all dimensions of a given particle. They use the following formula to update the velocities of the particles:

$$\begin{aligned} v_{i,d}(t) = \omega.v_{i,d}(t-1) + \varphi_1.rand1_i(0,1).(p_{i,d}^l(t) - x_{i,d}(t-1)) + \\ \varphi_2.rand2_i(0,1).(p_d^g(t) - x_{i,d}(t-1)) \end{aligned} \tag{1.45}$$

Comparing the two variants in (1.42) and (1.45), the former can have a larger search space due to independent updating of each dimension, while the second is dimension-dependent and has a smaller search space due to the same random numbers being used for all dimensions.

1.4.4.2 The Ant Colony Optimization (ACO)

The ant colony optimization algorithm (ACO) [99 – 99], introduced by Dorigo in 1992 [97], is a probabilistic technique for solving computational problems, which can be reduced to finding good paths through graphs. They are inspired by the behaviour of ants in finding paths from the colony to food.

In the real world, ants (initially) wander randomly, and upon finding food return to their colony while laying down pheromone trails. If other ants find such a path, they are likely not to keep traveling at random, but to instead follow the trail; returning and reinforcing it if they eventually find food (see Ant communication). Over time, however, the pheromone trail starts to evaporate, thus reducing its attractive strength. The more time it takes for an ant to travel down the path and back again, the more time the pheromones have to evaporate. A short path, by comparison, gets marched over faster, and thus the pheromone density remains high as it is laid on the path as fast as it can evaporate. Pheromone evaporation has also the advantage of avoiding the convergence to a locally optimal solution. If there were no evaporation at all, the paths chosen by the first ants would tend to be excessively attractive to the following ones. In that case, the exploration of the solution space would be constrained.

Thus, when one ant finds a good (i.e. short) path from the colony to a food source, other ants are more likely to follow that path, and positive feedback eventually leads all the ants following a single path. The idea of the ant colony algorithm is to mimic this behavior with "simulated ants" walking around the graph representing the problem to solve.

Ant colony optimization algorithms have been used to produce near-optimal solutions to the Travelling Salesman Problem (TSP) [100]. They have an advantage over simulated annealing and genetic algorithm approaches when the graph may change dynamically; the ant colony algorithm can be run continuously and adapt to changes in real time. This is of interest in network routing and urban transportation systems.

1.4.5 Evolutionary Computing (EC) Techniques in Pattern Clustering

The evolutionary computing methods are, in spirit, randomized search and optimization techniques guided by the principles of evolution, natural genetics, and self organization. They are efficient adaptive and robust search processes, producing near optimal solutions and have a large amount of implicit parallelism. Since the problem of data clustering can be easily formulated as a complex optimization problem (see equation (1.1)), the application of EC algorithms for performing certain clustering tasks, which need optimization of computation requirements, and robust, fast, and close approximate solution, appears to be appropriate and natural [101]. Moreover, the errors generated in experiments related to the measurements of data, can be handled with the robust characteristics of these algorithms. To some extent such errors may be regarded as contributing to genetic diversity, a desirable property for the EC techniques [45].

The Evolutionary Algorithms (EAs) are executed iteratively on a set of coded solutions, called population, with three basic operators: selection/ reproduction, crossover and mutation. They use only the payoff (objective function) information and probabilistic transition rules for moving to the next iteration. They are different from most of the normal optimization and search procedures in four ways:

a) EAs work with the coding of the parameter set, not with the parameters themselves
b) EAs work simultaneously with multiple points, and not a single point
c) EAs search via sampling (a blind search) using only the payoff information
d) EAs search using stochastic operators, not deterministic rules

The main advantages using EAs in solving clustering problems are:

a) Several tasks in pattern clustering involve optimization of different criteria (like mean intra-cluster variance, inter-cluster separation, quantization error etc.); thereby making the application of EAs more natural and appropriate.
b) Problems of large-scale data clustering seldom need the exact optimum solution; rather they require robust, fast and close approximate solutions, which most of the EC techniques are known to provide efficiently.
c) EAs can process, in parallel, populations billions times larger than is usual for conventional computation. The usual expectation is that larger populations can sustain larger ranges of genetic variation and thus can generate high-fitness individuals in fewer generations.
d) Measurements of real-world data inherently involve errors. These are more tolerable in executing evolutionary algorithms than in executing deterministic algorithms. (To some extent, errors may be regarded as contributing to genetic diversity, a desirable property.)

1.5 Clustering Methods Based on Evolutionary Algorithms

This Section outlines the most prominent clustering algorithms, based on evolutionary algorithms like GA, ES, EP, and GP in a comprehensive manner.

1.5.1 The GA-Based Partitional Clustering Algorithms - Earlier Approaches

The key idea of the GA-based clustering algorithms, is to create a population of candidate solutions to an optimization problem, which is iteratively refined by alteration and selection of good solutions for the next iteration. Candidate solutions are selected according to a fitness function, which evaluates their quality in respect to the optimization problem. In case of GAs, the alteration consists of mutation to explore solutions in the local neighborhood of existing solutions and crossover to recombine information between different candidate solutions. An important advantage of these algorithms is their ability to cope with local optima by maintaining, recombining and comparing several candidate solutions simultaneously. In contrast, local search heuristics, such as the simulated annealing algorithm [102], only refine a single candidate solution and are notoriously weak in coping with local optima. Deterministic local search, which is used in algorithms like the k-means, always converges to the nearest local optimum from the starting position of the search.

The first application of GAs to clustering was introduced by Raghavan and Birchand [103] and it belongs to the first approach of using a direct encoding of the object-cluster association. The idea in this approach is to use a genetic encoding that allocates directly n objects to k clusters, such that each candidate solution consists of n genes with integer values in the interval [1, k]. For example, for $n=5$ and $k=3$ the encoding "11322" allocates the first and the second object to cluster 1 and third object to cluster 3 and the fourth and fifth object to cluster 2 and therefore the following clusters ({1,2}, {3}, {4,5}) are identified. Based on this problem representation, the GA tries to find the optimal partition according to a fitness function which measures the partition goodness. It has been shown that such an algorithm outperforms *k-means* in the analysis of simulated and real datasets (e.g.see [104]). However, the representation scheme has a major drawback because of its redundancy, for instance, “11322" and "22311" represent the same grouping solution ({1,2}, {3},{4,4}). Falkenauer *et al.* [105] tackled this problem in an elegant way. In addition to the mentioned encoding of n genes representing each object-cluster association, they represent the group labels as additional genes in the encoding and apply *ad hoc* evolutionary operators on them.

The second kind of GA approach to partitional clustering is to encode cluster separating boundaries. Bandyopadhyay *et al.*[106 - 108] used GAs to determine hyperplanes as decision boundaries, which divide the attribute feature space to separate the clusters. For this they encode the location and orientation of a set of hyperplanes with a gene representation of flexible length. Apart from minimizing the number of misclassified objects, their approach tries to minimize thenumber of hyperplanes required. Another interesting and more flexible approach by Bandyopadhyay and Maulik [109] is to determine the boundaries between clusters by connected linear segments instead of rigid planes.

The third way to use GAs in partitional clustering is to encode a representative variable (typically a centroid or medoid) and optionally a set of parameters to describe the extend and shape of the variance for each cluster. Srikanth *et al.* [110]

proposed an approach, which encodes the centre, extend, and orientation of an ellipsoid for each cluster. Moreover, many authors proposed cluster centroids, baricentres, or medoids as representation points to allocate each object to a specific cluster (e.g.[111 - 114]). The idea is to determine a representation point for each cluster and to allocate each object to the cluster with the nearest representation point, where 'nearest' refers to a distance measure, such as Euclidean distance. The fitness of a candidate solution is then computed as the adequacy of the identified partition according to a statistical criterion, such as the Marriott or variance ratio criterion. Many studies have shown that this approach is more robust in converging towards the optimal partition than classic partitional algorithms (e.g.[111 – 114]).

Finally, some authors introduced hybrid clustering algorithms, which combine classic clustering techniques with GAs. For example, Krishna and Murthy [115] introduced a GA with the direct encoding of object-cluster associations as in [103], but applied k-means to determine the quality of the GA candidate solutions. For this, each GA candidate solution is used as a starting point for a k-means run. The quality of the solution found by the k-means run is then used as the fitness of the GA candidate solution.

1.5.2 Clustering Algorithms Based on ES, EP, and GP

Lee and Antonsson [116] used an Evolution Strategy (ES) to automatically cluster a dataset without any prior knowledge of the number of clusters. The proposed ES implemented variable length individuals to search simultaneously for the cluster centroids and number of clusters. In Lee and Antonsson's method, each individual encoded a set of centroids. The length of each individual is randomly chosen from a user-specified range of cluster numbers. The centroids of each individual are then initialized randomly. Mutation is then applied to the individuals by adding/subtracting a Gaussian random variable of zero mean and unit variance. Two-point crossover is also applied as a *length changing operator*. A (10+ 60) ES selection is used where 10 is the number of parents and 60 is the number of offspring generated in each generation. The best 10 individuals from the set of parents and offspring are used for the next generation. Lee and Antonsson used a modified version of the mean-squared-error (MSE) as their fitness function, which took the following form:

$$J_{ES} = \sqrt{k+1} \sum_{j=1}^{k} \sum_{\vec{Z}_i \in C_j} d^2(\vec{Z}_i, \vec{V}_j) \tag{1.46}$$

Please note that the conventional MSE has been multiplied in (1.50) by a constant corresponding to the square root of number of cluster. This modification is done with a view to penalizing the large number of clusters.

An approach to classify a dataset dynamically using Evolutionary Programming (EP) can be found in Sarkar *et al.* [117] where two fitness functions are optimized simultaneously: one gives the optimal number of clusters, whereas the other leads to a proper identification of each cluster's centroid. This algorithm determines the optimum number of clusters and optimal cluster centers in such a way that locally optimal solutions are avoided. Another advantage is that the clustering here is independent of the initial choice of the cluster centers. However, the proposed method is applicable for clustering task where clusters are crisp (non-overlapping) and spherical in nature.

1.6 Clustering Using Swarm Intelligence Algorithms

This section outlines the present state-of-the-art clustering algorithms based on SI algorithms, especially the ACO and PSO algorithms. A comprehensive review of the clustering techniques based on SI techniques can be found in [101].

1.6.1 The Ant Colony Based Clustering Algorithms

Ant colonies provide a means to formulate some powerful nature-inspired heuristics for solving the clustering problems. Among other social movements, researchers have simulated the way, ants work collaboratively in the task of grouping dead bodies so, as to keep the nest clean [71]. In Nature many species actually organize a cemetery. If corpses, or more precisely, sufficiently large parts of corpses are randomly distributed in space at the beginning of the experiment, the workers form cemetery clusters within a few hours, following a behavior similar to aggregation. If the experimental arena is not sufficiently large, or if it contains spatial heterogeneities, the clusters will be formed along the edges of the arena or, more generally, following the heterogeneities. The basic mechanism underlying this type of aggregation phenomenon is an attraction between dead items mediated by the ant workers: small clusters of items grow by attracting workers to deposit more items. It is this positive and auto-catalytic feedback that leads to the formation of larger and larger clusters. In this case, it is therefore the distribution of the clusters in the environment that plays the role of stigmergic variable. It can be observed that, with time the ants tend to cluster all dead bodies in a specific region of the environment, thus forming piles of corpses. Larval sorting and corpse cleaning by ant was first modeled by Deneubourg *et al.* for accomplishing certain tasks in robotics [118]. This inspired the Ant-based clustering algorithm [119]. Lumer and Faieta modified the algorithm using a dissimilarity-based evaluation of the local density, in order to make it suitable for data clustering [120]. This introduced standard Ant Clustering Algorithm (ACA). It has subsequently been used for numerical data analysis [121], data-mining [121], graph-partitioning [122 - 124] and text-mining [125 - 127]. Many authors [125, 128] proposed a number of modifications to improve the convergence rate and to get optimal number of clusters. Monmarche *et al.* hybridized the Ant-based clustering algorithm with k-means algorithm [129] and compared it to traditional k-means on various data sets, using the classification error for evaluation purposes. However, the results obtained with this method are not applicable to ordinary ant-based clustering since it differs significantly from the latter.

Like a standard ACO, ant-based clustering is a distributed process that employs positive feedback. Ants are modeled by simple agents that randomly move in their environment. The environment is considered to be a low dimensional space, more generally a two-dimensional plane with square grid. Initially, each data object that represents a multi-dimensional pattern is randomly distributed over the 2-D space. Data items that are scattered within this environment can be picked up, transported and dropped by the agents in a probabilistic way. The picking and dropping operation are influenced by the similarity and density of the data items within the ant's local neighborhood. Generally, the size of the neighborhood is 3×3. Probability of picking up data items is more when the object are either isolated or surrounded by dissimilar items. They trend to drop them in the vicinity of similar ones. In this way, a clustering of the elements on the grid is obtained.

Kanade and Hall [130] presented a hybridization of the ant systems with the classical FCM algorithm to determine the number of clusters in a given dataset automatically. In their fuzzy ant algorithm, at first the ant based clustering is used to create raw clusters and then these clusters are refined using the FCM algorithm. Initially the ants move the individual data objects to form heaps. The centroids of these heaps are taken as the initial cluster centers and the FCM algorithm is used to refine these clusters. In the second stage the objects obtained from the FCM algorithm are hardened according to the maximum membership criteria to form new heaps. These new heaps are then sometimes moved and merged by the ants. The final clusters formed are refined by using the FCM algorithm.

A number of modifications have been introduced to the basic ant based clustering scheme that improve the quality of the clustering, the speed of convergence and, in particular, the spatial separation between clusters on the grid, which is essential for the scheme of cluster retrieval. A detailed description of the variants and results on the qualitative performance gains afforded by these extensions are provided in [131].

1.6.2 The PSO-Based Clustering Algorithms

Research efforts have made it possible to view data clustering as an optimization problem. This view offers us a chance to apply PSO algorithm for evolving a set of candidate cluster centroids and thus determining a near optimal partitioning of the dataset at hand. PSO-based clustering algorithm was first introduced by Omran *et al.* in [132]. The results of Omran *et al.* [132, 133] depicted that PSO based method outperformed k-means, FCM and a few other state-of-the-art clustering algorithms. Omran *et al.* used a quantization error based fitness measure for judging the performance of a clustering algorithm. The quantization error is defined as:

$$J_e = \frac{\sum_{i=1}^{k} \sum_{\forall \vec{X}_j \in C_i} d(\vec{Z}_j, \vec{V}_i) / N_i}{k} \tag{1.47}$$

where C_i is the i-th cluster center and N_i is the number of data points belonging to the i-th cluster. Each particle in the PSO algorithm represents a possible set of k cluster centroids as:

$$\vec{X}_i(t) = \tag{1.48}$$

$\vec{V}_{i,1}$	$\vec{V}_{i,2}$	....	$\vec{V}_{i,k}$

where $\vec{V}_{i,p}$ refers to the p-th cluster centroid vector of the i-th particle. The quality of each particle is measured by the following fitness function:

$$f(\vec{X}_i, M_i) = w_1 \bar{d}_{\max}(M_i, \vec{Z}_i) + w_2(R_{\max} - d_{\min}(\vec{Z}_i)) + w_3 J_e \tag{1.49}$$

In the above expression, R_{max} is the maximum feature value in the dataset and $\mathbf{M}_i$ is the matrix representing the assignment of the patterns to the clusters of the i-th particle. Each element $m_{i,\,k,\,p}$ indicates whether the pattern $\vec{Z}_p$ belongs to cluster C_k of i-th particle. The user-defined constants w_1, w_2, and w_3 are used to weigh the contributions from different sub-objectives. In addition,

$$\bar{d}_{\max} = \max_{k \in 1,2,\ldots,K} \{ \sum_{\forall \vec{X}_p \in C_{i,K}} d(\vec{Z}_p, \vec{V}_{i,k}) / n_{i,k} \}$$

and,

$$d_{\min}(\vec{Z}_i) = \min_{\forall p,q,p \neq q} \{ d(\vec{V}_{i,p}, \vec{V}_{i,q}) \}$$

is the minimum Euclidean distance between any pair of clusters. In the above, $n_{i,k}$ is the number of patterns that belong to cluster $C_{i,k}$ of particle i. he fitness function is a multi-objective optimization problem, which minimizes the intra-cluster distance, maximizes inter-cluster separation, and reduces the quantization error. Van der Merwe and Engelbrecht hybridized this approach with the k-means algorithm for clustering general datasets [134]. A single particle of the swarm is initialized with the result of the k-means algorithm. The rest of the swarm is initialized randomly. In 2003, Xiao *et al.* used a new approach based on the synergism of the PSO and the Self Organizing Maps (SOM) [135] for clustering gene expression data. They got promising results by applying the hybrid SOM-PSO algorithm over the gene expression data of Yeast and Rat Hepatocytes..

Cui *et al.* [136] proposed a PSO based hybrid algorithm for classifying the text documents. They applied the PSO, K-means and a hybrid PSO clustering algorithm on four different text document datasets. The results illustrate that the hybrid PSO algorithm can generate more compact clustering results over a short span of time than the k-means algorithm.

1.7 Automatic Clustering: Evolutionary Vs. Classical Approaches

Tremendous research effort has gone in the past few years to evolve the clusters in complex datasets through evolutionary computing techniques. However, little work has been taken up to determine the optimal number of clusters at the same time. Most of the existing clustering techniques, based on evolutionary algorithms, accept the number of clusters k as an input instead of determining the same on the run. Nevertheless, in many practical situations, the appropriate number of groups in a previously unhandled dataset may be unknown or impossible to determine even approximately. For example while attempting to classify a large database of handwritten characters in an unknown language; it is not possible to determine the correct number of distinct letters beforehand. Similarly in multi-agent robotic co-operation, a robot has to identify its fellow robots, obstacles, target, etc. from the images of its surroundings grabbed by a digital camera working as the 'robot eye'. Prior to segmentation and localization of different objects, the images taken by a robot need fast and automatic clustering so that similar objects can be marked identically in the image. Again, while clustering a set of documents arising from the query to a search engine, the number of classes k change for each set of documents that result from an interaction with the search engine. Also if the dataset is described by high-dimensional feature vectors (which is very often the case for real-world problems), it may be practically impossible to visualize the data for tracking its number of clusters.

Finding an optimal number of clusters in a large dataset is usually a challenging task. The problem has been investigated by several researchers [137, 16] but the outcome is still unsatisfactory [138]. ISODATA, proposed by Ball and Hall [139], is an enhancement of the k-means algorithm, with addition of the possibility of merging classes and splitting elongated classes. An alternative approach to ISODATA is SYNERACT [140], which combines k-means with hierarchical descending approaches. According to Huang [140], SYNERACT is faster than and almost as accurate as ISODATA. Furthermore, it does not require the number of clusters and initial location of centroids to be specified in advance. Another improvement to the k-means algorithm is proposed by Rosenberger and Chehdi [138], which automatically finds the number of clusters in an image set by using intermediate results. Furthermore, Pelleg and Moore [141] proposed a k-means based algorithm, called X-means that uses model selection. X-means searches over a range of values of K and selects the value with the best Bayesian Information Criterion (BIC) [142] score. Recently, Hamerly and Elkan [15] proposed another wrapper around k-means called G-means. G-means starts with a small value for k, and with each iteration splits up the clusters whose data do not fit a Gaussian distribution. Between each round of splitting, k-means is applied to the entire data set in order to refine the current solution. According to [15], G-means works better than X-means; however, it works only for data having spherical and/or elliptical clusters. G-means is not designed to work for arbitrary-shaped clusters. A program called *snob* [142, 143] uses various methods to assign objects to classes in an intelligent manner. After each assignment, the Wallace Information Measure [144, 145] is calculated and based on this calculation the assignment is accepted or rejected. Hence, snob can split/merge and move points between classes. Bischof *et al.* [146] proposed another algorithm based on k-means which uses a minimum description length (MDL) framework. The algorithm starts with a large value for k and proceeds to remove centroids when this results in a reduction of the description length. k-means is used between the steps that reduce k. Gath and Geva proposed an unsupervised clustering algorithm based on the combination of FCM and fuzzy maximum likelihood estimation [147]. Lorette *et al.* [148] proposed an algorithm based on fuzzy clustering to dynamically determine the number of clusters in a data set. This approach, however, requires a user-defined parameter, which has profound effect on the number of clusters generated (i.e. not fully unsupervised). Frigui and Krishnapuram proposed a robust competitive clustering algorithm, based on the process of competitive agglomeration. Initialization does, however, have a significant effect on the result of this algorithm [149].

Kohonen's Self Organizing Maps (SOM) [150 - 151] can be used to automatically find the number of clusters in a data set. SOM combines competitive learning (in which different nodes in the Kohonen network compete to be the winners when an input pattern is presented) with a topological structuring of nodes, such that adjacent nodes tend to have similar weight vectors (this is done via lateral feedback) [152]. SOM suffers from being dependent on the order in which the data points are presented. To overcome this problem, the choice of data points can be randomized during each epoch [153].

Automatic clustering has also been attempted from the field of evolutionary computing. In this section, we discuss in details of three recently developed evolutionary clustering algorithms, which can partition a given data set without any prior knowledge of the number of clusters present. The first two algorithms are based on GA and the third one uses the PSO.

1.7.2 Genetic Clustering with Unknown Number of Clusters K (GCUK) Algorithm

In 2002, Bandyaopadhyay and Maulik proposed an automatic clustering algorithm [109] based on GA and named it GCUK (Genetic Clustering with Unknown Number of clusters *K*). In this scheme, the chromosomes are made up of real numbers (representing the coordinates of the cluster centers in the feature space) and the *don't care* symbols #. The number of cluster centers k is not fixed and rather determined by the algorithm on the run. Only the upper and lower limits of k viz ($K_{\max}$, $K_{\min}$) are given as inputs. The fitness of the chromosomes is computed through the Davis-Bouldin index described in section 2.3.1. The smallest *DB(k)* yields a valid optimal partition. Hence the fitness function used in [12] is *1/DB(k).* Single point crossover is applied on the chromosomes with fixed crossover probability μ_c. However during crossing over, each cluster center is interpreted as an indivisible gene. Each valid allele (which is not a wild card #) in a chromosome is mutated with constant mutation probability μ_m. A number δ in the range [0, 1] is generated with uniform distribution. If the value at a gene position is v, after mutation it becomes

$$\begin{aligned} &v \times (1 \pm 2\delta) && \text{if } v \neq 0 \\ &\pm 2\delta && \text{if } v = 0 \end{aligned} \tag{1.50}$$

+ or – symbol appear with equal probability. Proportional selection scheme as described in [67] is also applied here. The process of fitness computation, selection, mutation and crossover are repeated up to a maximum number of iterations. The algorithm implements elite in each generation by preserving the best chromosome yielded up to that generation in a location outside the current population. Hence on termination that location contains the centers of the final clusters.

1.7.3 The FVGA Algorithm

The FVGA [155] is a fuzzy extension of the previously described GCUK algorithm. It has many features common to the GCUK algorithm. It, however, uses the Xie-Beni clustering validity index, described in section 2.3.4, as the basis of its fitness function.

A finite size, P, of the initial population is generated randomly. The strings contain different numbers of cluster centers. If we consider d-dimensional data, and if a string represents k clusters, then there are $c \times d$ random values of real numbers in the string. In the initial population, the cluster centers are selected from within the data range. We have considered a minimum string length to be $K_{min} = 2$, which represents a two cluster solution. The choice of K_{max} depends on the data sets used. Following the authors, we selected $K_{max} \leq \sqrt{n}$, where, n is number of patterns present in the data set. Moreover, in the initial population, the positions of the centers within the strings are selected randomly.

Given a chromosome, the centers encoded in it are first extracted. Let the chromosome encode k centers, and let these be denoted as $\vec{m}_1, \vec{m}_2, ..., \vec{m}_k$. The membership values u_{ij} $(i = 1,2,...,k;\ j = 1,2,...,n)$ are computed as follows:

$$u_{ij} = \frac{1}{\sum_{p=1}^{k} \left(\frac{d(\vec{m}_i, \vec{Z}_j)}{d(\vec{m}_p, \vec{Z}_j)} \right)^{\frac{2}{q-1}}}, \tag{1.51}$$

where q is the fuzzifier term, also encountered in FCM. Note that while computing u_{ij} from equation 1.50), if $d(\vec{m}_p, \vec{Z}_j)$ is zero for some p, then u_{ij} is set to zero for all $i = 1,2,...,k$, $i \neq j$ while u_{pj} is set equal to one. The corresponding XB index is computed as in (1.17). The fitness function for a chromosome is then defined as $1/XB$. Note that maximization of the fitness function will ensure minimization of the XB index.

Here like the GCUK, although the strings are variable in length, they are usually converted to a fixed length one by use of don't care ('#') symbols, i.e., the physical length of all the strings are $K_{max} \times d$. Out of these $K_{max} \times d$ locations only

$k \times d$ are used by a string representing k clusters, other locations remain don't care. Therefore, the conventional single point *crossover* can be applied as usual. *Mutation* is performed in the following the same rule as that given in (1.49).

1.7.4 The Dynamic Clustering with Particle Swarm Optimization Algorithm

The Dynamic Clustering with Particle Swarm Optimization (DCPSO) algorithm [156] was first proposed by Omran *et al.* in 2005. The algorithm works as follows: A pool of cluster centroids, $\boldsymbol{M}$, is randomly chosen from the set of data points $\mathbf{Z}$. The swarm of particles, $\boldsymbol{S}$, is then randomly initialized. Binary PSO [34] is then applied to find the "best" set of cluster centroids, $\boldsymbol{M}_\tau$, from $\boldsymbol{M}$. k-means is applied to $\boldsymbol{M}_\tau$ in order to refine the chosen centroids. $\boldsymbol{M}$ is then set to $\boldsymbol{M}_\tau$ plus $\boldsymbol{M}_r$ which a randomly chosen set of centroids from $\mathbf{Z}$ (to add diversity to $\boldsymbol{M}$). The algorithm is then repeated using the new $\boldsymbol{M}$. When the termination criteria are met, $\boldsymbol{M}_\tau$ will be the resulting "optimum" set of cluster centroids and n_τ will be the "optimum" number of clusters in $\mathbf{Z}$. The termination criteria can be a user-defined maximum number of iterations or a lack of progress in improving the best solution found so far for a user-specified consecutive number of iterations, *TC*.

Like GCUK and FVGA, DCPSO can also go with any clustering validity index serving as the basis for its fitness function. Omran *et al.* [156] used the validity index proposed by Turi [51], in order to form the fitness function. Turi's index is given by:

$$V = (c \times N(2,1) + 1) \times \frac{\text{intra}}{\text{inter}} , \tag{1.52}$$

where c is a user specified parameter and $N(2,1)$ is a Gaussian distribution with mean 2 and standard deviation of 1. The *intra* term is the average of all the distances between each data point and its cluster centroid which is defined as:

$$\text{int}\,ra = \frac{1}{n}\sum_{p=1}^{k}\sum_{\vec{Z}_i \in C_p} d^2(\vec{Z}_i, \vec{m}_p) \tag{1.53}$$

This term is used to measure the compactness of the clusters. The *inter* term is the minimum distance between the cluster centroids, which is defined as:

$$\text{inter} = \min\{d^2(\vec{m}_p, \vec{m}_q)\}\forall p = 1,2,\ldots,k-1, \text{and } q = p+1,\ldots,k \tag{1.54}$$

1.8 Clustering with Evolutionary Multi-objective Optimization

1.8.1 Multi-objective Optimization Problem (MOP)

In many practical or real life problems, there are many (possibly conflicting) objectives that need to be optimized simultaneously. Under such circumstances there no longer exists a single optimal solution but rather a whole set of possible solutions of equivalent quality. Consider, for example, the design of an automobile. Possible objectives could be: minimize cost, maximize speed, minimize fuel consumption and maximize luxury. These goals are clearly conflicting and, therefore, there is no single optimum to be found. The field of Multi-objective Optimization (MO) deals with such simultaneous optimization of multiple, possibly competing, objective functions. The MO problems tend to be characterized by a family of alternatives, which must be considered equivalent in the absence of information concerning the relevance of each objective relative to the others.

The family of solutions of an MO problem is composed of the parameter vectors, which cannot be improved in any objective without causing a degradation in at least one of the other objectives. This forms the central idea of *Pareto-optimality*. The concepts of *dominance* and *Pareto-optimality* may be presented more formally as follows [157, 158]:

Definition 1.7: Consider without loss of generality the following multi-objective optimization problem with m decision variables x (parameters) and n objectives y:

$$\text{Maximize } \vec{Y} = f(\vec{X}) = (f_1(x_1,....,x_m),....,f_n(x_1,....,x_m)), \quad (1.55)$$

where $\vec{X} = [x_1,.....,x_m]^T \in P$ and $\vec{Y} = [y_1,....,y_m]^T \in O$ and where $\vec{X}$ is called decision (parameter) vector, P is the parameter space, $\vec{Y}$ is the objective vector, and O is the objective space. A decision vector $\vec{A} \in P$ is said to dominate another decision vector $\vec{B} \in P$ (also written as $\vec{A} \succ \vec{B}$) if and only if:

$$\begin{aligned} &\forall i \in \{1,....,n\}: && f_i(\vec{A}) \geq f_i(\vec{B}) \\ \wedge\ &\exists j \in \{1,.....,n\}: && f_j(\vec{A}) > f_j(\vec{B}) \end{aligned} \quad (1.56)$$

Based on this convention, we can define non-dominated, *Pareto-optimal* solutions as follows:

Definition 1.8: Let $\vec{A} \in P$ be an arbitrary decision vector.

(a) The decision vector $\vec{A}$ *is said to be* non-dominated regarding the set $P' \subseteq P$ if and only if there is no vector in P' which can dominate $\vec{A}$. Formally,

$$\nexists\ \vec{A}' \in P' : P' \succ P \quad (1.57)$$

(b) The decision (parameter) vector $\vec{A}$ is called Pareto-optimal if and only if $\vec{A}$ is non-dominated regarding the whole parameter space P.

1.8.2 Evolutionary Multi-objective Optimization (EMO)

Evolutionary Algorithms (EAs) are well suited to multi-objective optimization problems as they are fundamentally based on biological processes which are inherently multi-objective. During 1993-2005, plethora of different evolutionary algorithms was suggested to solve multi-objective optimization problems and an interested reader may consult [158 - 160]. We here briefly describe two well-known EMO algorithms namely, PAES [161] and NSGA-II [162].

Knowles and Corne [161] suggested a simple MOEA using a single parent, single child evolutionary algorithm, similar to (1+1) ES. Instead of using real parameters, binary strings and bit-wise mutation are used in this algorithm to create the offspring. After creating the child and evaluating its objectives, it is compared with respect to the parent. If the child dominates the parent, then the child is accepted as the next parent and the iteration continues. On the other hand if parent dominates the child, the child is discarded and a new mutated solution (a new solution) is generated from the parent. However if the parent and the child are non-dominating to each other, then the choice between the child and the parent is resolved by comparing them with an archive of best solutions found so far. The child is compared with all members of the archive to check if it dominates any member of the archive. If yes, the child is accepted as the new parent and all the dominated solutions are eliminated from the archive. If the child does not dominate any member of the archive, both the parent and the child are checked for their *nearness* with the solutions of the archive. If the child resides in a less crowded region in the parameter space, it is accepted as a parent and a copy is added to the archive. Generally this crowding concept is implemented by dividing the whole solution space into d^n subspaces where d is the depth parameter and n is the number of objective functions. The subspaces are updated dynamically.

The other popular algorithm for multi-objective optimization is Non-dominated Sorting Genetic Algorithm (NSGA-II) proposed by Deb *et al.* [162]. In NSGA II, initially a random parent population G_0 of size N is created. Then the population is sorted based on the non-domination relation. Each solution of the population is assigned a fitness that is equal to its non-domination level. A child population H_0 is created from the parent population G_0 by using binary tournament selection, recombination, and mutation operators. Generally according to this algorithm, initially a combined population $R_t = G_t + H_t$ is formed of size R_t, which is $2N$. Now all the solutions of R_t are sorted based on their non-domination status. If the total number of solutions belonging to the best non-dominated set F_1 is smaller than N, F_1 is completely included into G_{t+1}. The remaining members of the population G_{t+1} are chosen from subsequent non-dominated fronts in the order of their ranking. To choose exactly N solutions, the solutions of the last included front are sorted using the crowded comparison operator and the best among them (i.e., those with larger values of the crowding distance) are selected to fill in the available slots in G_{t+1}. The new population G_{t+1} is now used for selection, crossover, and mutation to create a new population H_{t+1} of size N and the process continues. The crowding distance operator is also used in the parent selection phase in order to break a tie in the binary tournament selection. This operator is instrumental in maintaining diversity in the Pareto front.

1.8.3 Clustering Using EMO Algorithms (EMOAs)

Despite their wide applications for feature selection, EMOAs have not been applied on a large scale for the actual clustering task by other researchers. Several ensemble methods [163 – 167] have, however, been proposed to combine results of clustering with different and possibly conflicting clustering objectives. In order to construct clustering ensembles, different clustering results are retrieved by repeatedly running the same algorithm (using different initializations, parametrizations, bootstrapping, or a varying number of clusters) or several complementary methods (e.g. agglomerative algorithms based on diverse linkage criteria such as single link and average link). The resulting partitions are then united to an ensemble clustering using graph-based approaches [166], expectation maximization, or co-association methods [164].

Law *et al.* [168] used a multi-objective approach towards clustering synthetic and real life datasets of varying levels of complexity. The proposed multi-objective clustering method used a two-step process. It includes detection of clusters by a set of candidate objective functions as well as their integration into the target partition. A key ingredient of the approach is a cluster goodness function that evaluates the utility of multiple clusters using re-sampling techniques. Multi-objective data clustering is obtained as a solution to a discrete optimization problem in the space of clusters.

Bandyopadhyay *et al.* [169] proposed a non-automatic multi-objective scheme for clustering the pixels of remote sensing satellite images into several fuzzy partitions. They employed the NSGA II algorithm to optimize a number of fuzzy cluster validity indices simultaneously. The resultant set of near-Pareto-optimal solutions contained a number of non-dominated solutions, which the user could judge relatively and pick up the most promising one according to the problem requirements. Real-coded encoding of the cluster centers was used for this purpose.

Handl and Knowles proposed a multi-objective clustering scheme known as Voronoi Initialized Evolutionary Nearest-Neighbor Algorithm (VIENNA [170]), which is based on PESA II [171] and simultaneously optimizes two objectives. It employs a straightforward encoding of a clustering, with a gene for each data item and its allele value specifying the cluster to which the data item should belong. VIENNA needed an advanced initialization scheme based on Voronoi cells and directed mutation to make up for deficiencies in its encoding. In addition, it is non-automatic and does not provide any means to select good solutions from the final Pareto front.

Handl and Knowles [172, 173] proposed an improved EMO-based clustering algorithm, which they named *Multi-Objective Clustering with Automatic k Determination* (MOCK). They fine-tuned one of the objectives used in VIENNA and found a better encoding that does not fix the number of clusters and because of good locality and heritability, allows a much more effective exploration of the search space via suitable operators. They also developed a method for selection of best solutions from the Pareto front based on a null model, thus also determining the number of clusters, automatically. MOCK was also extended in [174] for improving its scalability to large, high-dimensional datasets and data with large number of clusters. Handle and Knowles also introduced MOCK-around-medoids, which allows for the clustering of similarity data [175] (as opposed to vectorial data, i.e. points in a metric space).

1.9 Innovation and Research: Main Contributions of This Volume

Evolutionary clustering algorithms usually try to yield an optimal partitioning of the given data set by maximizing some fitness function. The fitness function is built around certain statistical-mathematical functions, collectively called the clustering validity indices. These functions depend on the candidate cluster centers, the proximity-measure employed, and the number of clusters. They are multimodal in nature (containing many local optima) and usually take too many independent parameters (for even a two-dimensional dataset with two clusters) to visualize. In order to give a visual feel of the complex fitness landscapes that frequently appear in clustering problems, consider a very simple one-dimensional dataset shown in Figure 1.7. In each of the intervals [−1, 1] and [4, 6] we place 11 equidistant data points and thus we have two data clusters centered on 0 and 5.

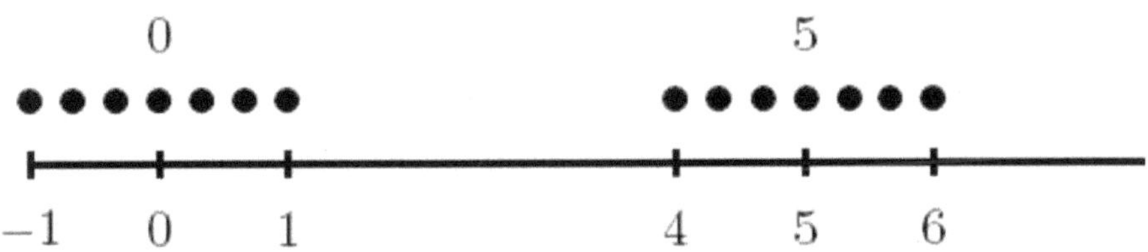

Fig. 1.7 Example of an extremely simple one-dimensional dataset

Now a fitness function based on the reciprocal of the ICS given in equation (1.29) for the k-means (Hard c-means algorithm) is plotted against the two cluster centers in Figure 1.8.

Note that the fitness function is symmetric with respect to the main diagonal. When we exchange the two cluster prototypes, the value of the corresponding function-value will be the same. In all diagrams we can see strong maxima at approximately (0, 5) and (5, 0). This is exactly what we expect: The prototypes are correctly positioned into the centers 0 and 5 (or vice versa) of the data clusters. Now consider the situation when a few noise points have been added to the same dataset of Figure 1.7, in the interval [9.0, 11.0]. In this case the dataset and the corresponding fitness function have been shown in Figures 1.9 (a) and 1.9 (b) respectively.

A close inspection of Figure 1.9 (b) reveals that, in addition to these desired (global) minima, there are now other undesired local minima, namely at approximately (0, 10), (5, 10), (10, 0), (10, 5). For these local minima one prototype covers one of the data clusters, while the other prototype is mislead to the noise cluster.

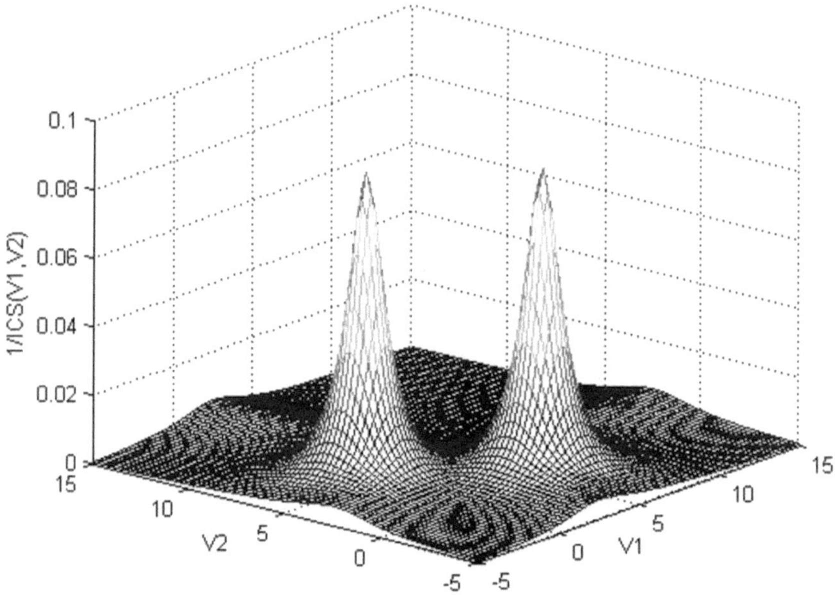

Fig. 1.8 Fitness function (Reciprocal of ICS) plot of the hard c-means algorithm for dataset of Figure 1.7

From the examples presented above, it is intuitively evident that for higher dimensional datasets, corrupted with measurement noise and outliers, and having clusters of various shape and density, the fitness landscapes are very complex and riddled with multiple local optima. Since the final accuracy of the clustering results depend on how efficiently the global optima of such fitness functions are discovered, the evolutionary algorithms employed for this purpose, should be equipped with very efficient search operators. The population members or agents, scattered over the fitness landscape, must explore the search region vividly within a small time and then quickly settle to the global optima by exchanging information among them. Convergence to a good solution is an important issue in evolutionary clustering problems. Some well-known clustering algorithms employ recurrent expressions, (which are updates over iterations until convergence, indicated by no further change in systems states) is observed. Convergence of such clustering algorithms depends greatly on the initial choice of the search variables (i.e. prototypes) and the dynamics followed.

Differential Evolution (DE) [176, 177] has recently emerged as a simple and efficient scheme for global optimization over continuous spaces. DE, by this time, has proved itself as a very fast and robust algorithm in comparison to the other evolutionary population-based algorithms like Genetic Algorithm (GA) and Particle Swarm Optimization (PSO). A statistical analysis undertaken by Zaharie [34], shows that the population variance in DE increases at a faster rate in comparison to the Evolutionary strategy (ES) and this is indicative of the greater explorative power of DE, a feature beneficial for a coarse search over a big search space.

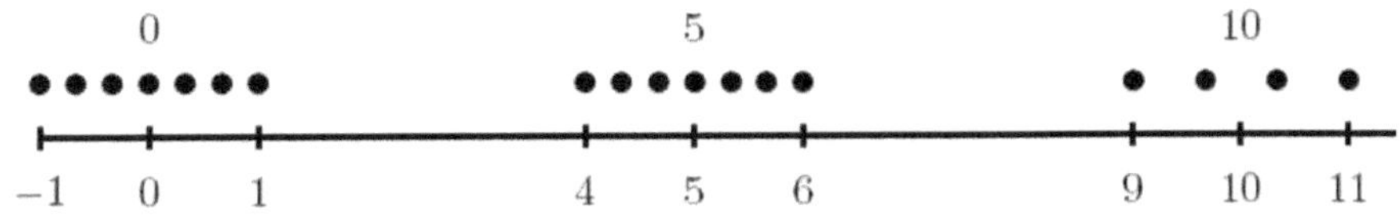

(**a**) Data set of Figure 1.7, but now with some noise points

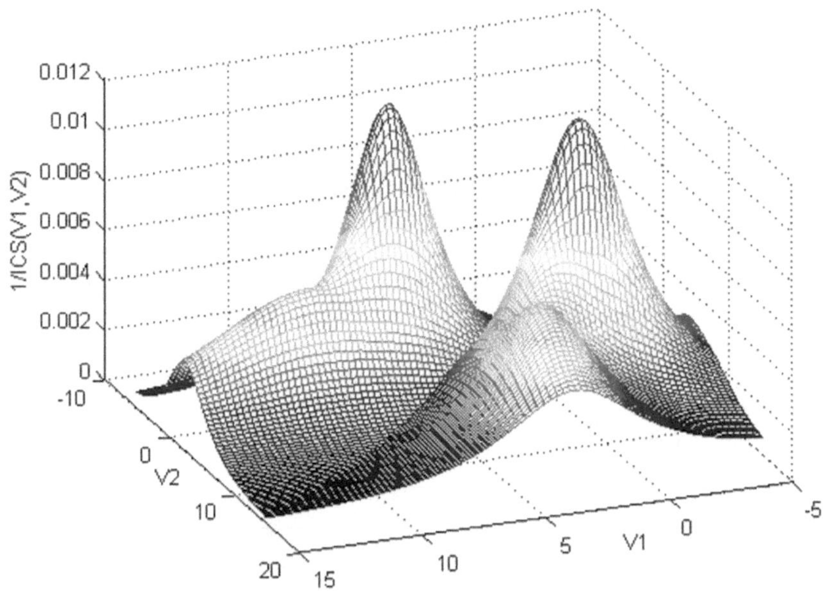

(**b**) Fitness function (Reciprocal of ICS) plot of the hard c-means algorithm for the above dataset

Fig. 1.9 Change in Fitness landscape due to presence of noise in dataset.

Naturally, if DE can be used for pattern clustering, it can justify its significance from the point of view of both final accuracy and convergence speed. Although a plethora of evolutionary clustering algorithms involving GAs and some using PSO have been reported in the literature, very little research has been undertaken to apply DE in clustering problems.

In this Volume, we illustrate the performance of the DE algorithm, when applied to the data clustering problem. We first attempt to gain an insight into the search mechanism and convergence behavior of DE, through simple mathematical and statistical analyses. More precisely, it creates a mathematical model of the population-dynamics found in DE. The model reveals that, the parameter vectors change their positions over the search space, following a rule close in spirit to the classical gradient descent search [178], albeit DE uses only some kind of estimate of the gradient (may be the difference vector) and not any analytical expression of the function-gradient itself. It has been shown that the behavior of the population members in DE is inherently stable near an optimum of the search space, from the concept of Lyapunov stability theorems [179]. Unlike the particles in PSO, the population members in DE show no oscillatory behavior around an optimum, The stability analysis also indicates how the settling time of the population members (time required to reach a given distance from the point of equilibrium) depends on the Crossover Rate *CR*, which is one of the main control parameters in DE. Taking a cue from the analysis mentioned earlier, some parameter automation strategies have been proposed for DE, in order to improve its performance without imposing any additional burden in terms of number of Function Evaluations (FEs).

Further we propose a novel algorithm for automatically partitioning a large unlabelled dataset into an optimal number of clusters, by using the improved DE algorithm. It also proposes a new representation scheme for the search variables in order to determine the optimal number of clusters. The algorithm has been shown to perform very well over the real world datasets taken from the UCI public data repository [180]. However, the performance of the algorithm deteriorates for some artificial datasets, which are linearly non separable and in some cases have overlapping clusters. The performance of the proposed scheme tends to be marginally better than the k-means algorithm over these cases.

In order to overcome such limitations, the DE-based clustering scheme has been used in conjunction with a kernel-induced similarity measure instead of the conventional sum-of-squares distance. Use of the kernel function makes it possible to cluster data that is linearly non-separable in the original input space into homogeneous groups in a transformed high-dimensional feature space [181][182][183]. A kernelized version of the Xie-Beni index has been derived to form the basis of the fitness function for DE. Through empirical experiments, an upper limit to the number of dimensions of the dataset has been derived, below which the DE-based kernel clustering method achieves significantly good performance as compared to its other state-of-the-art clustering algorithms.

Finally, we illustrate an interesting application of the DE-based clustering schemes to the image segmentation problem. Segmentation forms a fundamental step towards several complex computer-vision and image analysis applications including digital mammography, remote sensing and land cover study. Segmentation of non-trivial images is one of the most difficult tasks in image processing. Image segmentation can be treated as a clustering problem where the features describing each pixel correspond to a pattern, and each image region (i.e., segment) corresponds to a cluster [11]. The proposed DE based clustering algorithms have been applied to synthetic, natural and multi-spectral satellite images. The segmentation results have been evaluated on a quantitative basis by comparing the segmented images with corresponding ground truths (i.e. manually segmented image) using certain statistical scoring functions. After performing an extensive comparative study over various datasets, our research indicates that the DE family of algorithms should receive primary attention in the field of evolutionary pattern clustering, instead of the genetic algorithm or particle swarm optimization algorithm.

1.10 Conclusions

This chapter introduced the problem of pattern clustering in a lucid style. We discussed various proximity measures between patterns and the most prominent clustering validity functions found in literature. It then presented a brief overview of a number of traditional clustering algorithms, dividing them into four basic categories: hierarchical, partitional, density-based, and grid-based. A comparative study of the algorithms have also been undertaken, based on their input/outputs, runtime complexity, nature of data, effect of noise and outliers and the geometric shape of the clusters yielded by them. The family of evolutionary algorithms and swarm intelligence approaches were introduced and the scope of these algorithms in devising dynamic, scalable and resilient clustering algorithms was also examined. Several clustering techniques based on the evolutionary and swarm-intelligence techniques have been presented next, in a comprehensible manner. We also discussed the various issues related to automatic clustering algorithms, where the number of clusters is not known a priori.

References

[1] Duda, R.O., Hart, P.E.: Pattern Classification and Scene Analysis. John Wiley and Sons, Chichester (1973)

[2] Everitt, B.S.: Cluster Analysis, 3rd edn. Halsted Press (1993)

[3] Jain, A.K., Dubes, R.C.: Algorithms for Clustering Data. Prentice-Hall, Englewood Cliffs (1988)

[4] Arabie, P., Hubert, L.J., De Soete, G. (eds.): Clustering and Classification, River Edge. World Scientific Publishing, Singapore (1996)

[5] Jain, A.K., Murty, M.N., Flynn, P.J.: Data clustering: a review. ACM Computing Surveys 31(3), 264–323 (1999)

[6] Forgy, E.W.: Cluster Analysis of Multivariate Data: Efficiency Versus Interpretability of classification. Biometrics 21, 768–769 (1965)

[7] Zahn, C.T.: Graph-theoretical methods for detecting and describing gestalt clusters. IEEE Transactions on Computers C-20, 68–86 (1971)

[8] Mitchell, T.: Machine Learning. McGraw-Hill, Inc., New York (1997)

[9] Mao, J., Jain, A.K.: Artificial neural networks for feature extraction and multivariate data projection. IEEE Trans. Neural Network 6, 296–317 (1995)

[10] Pal, N.R., Bezdek, J.C., Tsao, E.C.-K.: Generalized clustering networks and Kohonen's self-organizing scheme. IEEE Trans. Neural Networks 4, 549–557 (1993)

[11] Kohonen, T.: Self-Organizing Maps. Springer Series in Information Sciences, vol. 30. Springer, Heidelberg (1995)

[12] Falkenauer, E.: Genetic Algorithms and Grouping Problems. John Wiley and Son, Chichester (1998)

[13] Paterlini, S., Minerva, T.: Evolutionary approaches for bluster analysis. In: Bonarini, A., Masulli, F., Pasi, G. (eds.) Soft Computing Applications, pp. 167–178. Springer, Berlin (2003)

[14] Brucker, P.: On the complexity of clustering problems. In: Beckmenn, M., Kunzi, H.P. (eds.) Optimization and Operations Research. Lecture Notes in Economics and Mathematical Systems, vol. 157, pp. 45–54. Springer, Berlin (1978)
[15] Hamerly, G., Elkan, C.: Learning the k in k-means. In: Proceedings of the Seventeenth Annual Conference on Neural Information Processing Systems (NIPS), pp. 281–288 (December 2003)
[16] Theodoridis, S., Koutroumbas, K.: Pattern Recognition, 2nd edn. Elsevier Academic Press, Amsterdam (2003)
[17] Halkidi, M., Vazirgiannis, M.: Clustering validity assessment: Finding the optimal partitioning of a data set. In: Proceedings of the 2001 IEEE International Conference on Data Mining (ICDM 2001), San Jose, California, USA, pp. 187–194 (2001)
[18] Dunn, J.C.: Well Separated Clusters and Optimal Fuzzy Partitions. J. Cybern. 4, 95–104 (1974)
[19] Calinski, R.B., Harabasz, J.: Adendrite Method for Cluster Analysis. Commun. Statistics, 1–27 (1974)
[20] Davies, D.L., Bouldin, D.W.: A cluster separation measure. IEEE Transactions on Pattern Analysis and Machine Intelligence 1, 224–227 (1979)
[21] Pakhira, M.K., Bandyopadhyay, S., Maulik, U.: Validity index for crisp and fuzzy clusters. Pattern Recognition Letters 37, 487–501 (2004)
[22] Chou, C.H., Su, M.C., Lai, E.: A new cluster validity measure and its application to image compression. Pattern Analysis and Applications 7(2), 205–220 (2004)
[23] Bezdek, J.C.: Numerical taxonomy with fuzzy sets. Journal of Math. Biol., 157–171 (1974)
[24] Bezdek, J.C.: Cluster validity with fuzzy sets. Journal of Cybernetics (3), 58–72 (1974)
[25] Xie, X., Beni, G.: Validity measure for fuzzy clustering. IEEE Trans. Pattern Anal. Machine Learning 3, 841–846 (1991)
[26] Pal, N.R., Biswas, J.: Cluster validation using graph theoretic concepts. Pattern Recognition 30(6), 847–857 (1997)
[27] Su, M.-C., Chou, C.-H., Lai, E.: A new cluster validity measure for clusters with different densities. In: IASTED International Conference on Intelligent Systems & Control, Salzburg, Austria, pp. 276–281 (2003)
[28] Su, M.-C., Chou, C.-H.: A competitive learning algorithm using symmetry. IEICE Transactions on Fundamentals of Electronics, Communications and Computer Sciences E82-A(4), 680–687 (1999)
[29] Su, M.-C., Liu, Y.-C.: A new approach to clustering data with arbitrary shapes. Pattern Recognition 38, 1887–1901 (2005)
[30] Su, M.-C., Chou, C.-H.: A modified version of the K-means algorithm with a distance based on cluster symmetry. IEEE Trans. on Pattern Analysis and Machine Intelligence 23(6), 674–680 (2001)
[31] Pakhira, M.K., Bandyopadhyay, S., Maulik, U.: A study of some fuzzy cluster validity indices, genetic clustering and application to pixel classification. Fuzzy Sets and Systems 155, 191–214 (2005)
[32] Chou, C.H., Su, M.C., Lai, E.: A new cluster validity measure and its application to image compression. Pattern Analysis and Applications 7(2), 205–220 (2004)
[33] Zhang, T., Ramakrishnman, R., Linvy, M.: BIRCH: An efficient method for very large databases. In: ACM SIGMOD, Montreal, Canada (1996)

[34] Guha, S., Rastogi, R., Shim, K.: CURE: An efficient clustering algorithm for large databases. In: Proceedings of the ACM SIGMOD Conference (1998)
[35] Guha, S., Rastogi, R., Shim, K.: ROCK: A robust clustering algorithm for categorical attributes. In: Proceedings of the IEEE Conference on Data Engineering (1999)
[36] Hammerly, G., Elkan, C.: Alternatives to the k-means algorithm that find better clusterings. In: Proc. ACM on Information and Knowledge Management, pp. 600–607 (November 2002)
[37] Zhang, T.: Convergence of large margin separable linear classification. In: NIPS 2000, pp. 357–363 (2001)
[38] MacQueen, J.: Some methods for classification and analysis of multivariate observations. In: Proceedings of the Fifth Berkely Symposium on Mathematical Statistics and Probability, pp. 281–297 (1967)
[39] Kaufman, L., Rousseeuw, P.J.: Finding Groups in Data: An Introduction to Cluster Analysis. Wiley, New York (1990)
[40] Huang, Z.: A Fast Clustering Algorithm to Cluster very Large Categorical Data Sets in Data Mining. DMKD (1997)
[41] Ng, R., Han, J.: Efficient and effective clustering methods for spatial data mining. In: Proceeding's of the 20th VLDB Conference, Santiago, Chile (1994)
[42] Bezdek, J.C.: Pattern Recognition with Fuzzy Objective Function Algorithms, New York, Plenum (1981)
[43] Wang, X., Wang, Y., Wang, L.: Improving fuzzy c-means clustering based on feature-weight learning. Pattern Recognition Letters 25, 1123–1132 (2004)
[44] Krishnapuram, R., Keller, J.: The possibilistic c-means algorithm: insights and recommendations. IEEE Trans. on Fuzzy Systems 4, 385–393 (1996)
[45] Bezdek, J.C., Keller, J., Krishnampuram, R., Pal, N.R.: Fuzzy Models and Algorithms for Pattern Recognition and Image Processing. Kluwer Academic Publishers, Dordercht (1999)
[46] Dempster, A., Laird, N., Rubin, D.: Maximum likelihood from incomplete data via the EM algorithm. Journal of the Royal Statistical Society, Series B 39(1), 1–38 (1977)
[47] McLachlan, G., Krishnan, T.: The EM algorithm and extensions. Wiley series in probability and statistics. John Wiley & Sons, Chichester (1997)
[48] Hamerly, G., Elkan, C.: Alternatives to the k-Means algorithm that find better clustering (pdf). In: Proceedings of the Eleventh International Conference on Information and Knowledge Management (CIKM 2002), pp. 600–607 (November 2002)
[49] Veenman, C.J., Reinders, M.J.T., Backer, E.: A maximum variance cluster algorithm. IEEE Transactions on Pattern Analysis and Machine Intelligence 24(9), 1273–1280 (2002)
[50] Alldrin, N., Smith, A., Turnbull, D.: Clustering with EM and K-means, unpublished Manuscript (2003), `http://louis.ucsd.edu/~nalldrin/research/cse253\_wi03.pdf`
[51] Turi, R.H.: Clustering-based colour image segmentation. PhD Thesis. Monash University, Australia (2001)
[52] Ng, A., Jordan, M., Weiss, Y.: On spectral clustering: analysis and an algorithm. In: Proceedings of eural Information Processing Systems (NIPS 2001) (2001)
[53] Zhang, B., Hsu, M., Dayal, U.: K-Harmonic means - A data clustering algorithm. Technical Report HPL-1999-124. Hewlett-Packard Labs (1999)

[54] Zhang, B.: Generalized K-Harmonic means - boosting in unsupervised learning., Technical Report HPL-2000-137. Hewlett-Packard Labs (2000)
[55] Bradley, P.S., Fayyad, U.M.: Refining Initial Points for K-Means Clustering. In: ICML 1998, pp. 91–99 (1998)
[56] Ester, M., Kriegel, H.-P., Sander, J., Xu, X.: A density-based algorithm for discovering clusters in large spatial databases with noise. In: Proceeding of 2nd Int. Conf. on Knowledge Discovery and Data Mining, Portland, pp. 226–23 (1996)
[57] Hinneburg, A., Keim, D.: An efficient approach to clustering in large multimedia databases with noise. In: Proceedings of KDD Conference (1998)
[58] Comaniciu, D., Meer, P.: Mean shift: A robust approach toward feature space analysis. IEEE Transactions on Pattern Analysis and Machine Intelligence 24, 603–619 (2002)
[59] Wang, W., Yang, J., Muntz, R.: STING: A statistical Information grid approach to spatial data mining. In: Proceedings of 23rd VLDB Conference (1997)
[60] Sheikholeslami, C., Chatterjee, S., Zhang, A.: WaveCluster: A-multi resolution clustering approach for very large spatial database. In: Proceedings of 24th VLDB Conference, New York, USA (1998)
[61] Han, J., Kamber, M.: Data Mining: Concepts and Techniques. Morgan Kaufmann Publishers, USA (2001)
[62] Back, T., Fogel, D.B., Michalewicz, Z. (eds.): Handbook of Evolutionary Computation. Oxford University Press, Oxford (1997)
[63] De Jong, K.A.: Evolutionary Computation: A Unified Approach. MIT Press, Cambridge (2006)
[64] Eiben, A.E., Smith, J.E.: Introduction to Evolutionary Computing. Springer, Heidelberg (2003)
[65] Fogel, D.B.: Evolutionary Computation: Toward a New Philosophy of Machine Intelligence. IEEE Press, Piscataway (1995)
[66] Fogel, L.J., Owens, A.J., Walsh, M.J.: Artificial Intelligence through Simulated Evolution. John Wiley, New York (1966)
[67] Holland, J.H.: Adaptation in Natural and Artificial Systems. University of Michigan Press, Ann Arbor (1975)
[68] Rechenberg, I.: Evolutionsstrategie - Optimierung Technischer Systeme nach Prinzipien der Biologischen Evolution (PhD thesis, 1971); Reprinted by Fromman-Holzboog (1973)
[69] Schwefel, H.-P.: Numerische Optimierung von Computer-Modellen (PhD thesis) (1974); Reprinted by Birkhäuser (1977)
[70] Koza, J.R.: Genetic Programming: On the Programming of Computers by Means of Natural Evolution. MIT Press, Massachusetts (1992)
[71] Bonabeau, E., Dorigo, M., Theraulaz, G.: Swarm Intelligence: From Natural to Artificial Systems. Oxford University Press Inc., Oxford (1999)
[72] Martinetz, T.M., Schulten, K.J.: A neural-gas network learns topologies. In: Kohonen, T., Mäkisara, K., Simula, O., Kangas, J. (eds.) Artificial Neural Networks, pp. 397–402. North-Holland, Amsterdam (1991)
[73] Langton, C.G. (ed.): Artificial Life: An Overview. MIT Press, Cambridge (1995)
[74] Kobti, Z., Reynolds, R., Kohler, T.: A multi-agent simulation using cultural algorithms: The effect of culture on the resilience of social systems. In: IEEE Congress on Evolutionary Computation, Canberra, Australia, December 5-12 (2003)

[75] Lee, K.S., Geem, Z.W.: A new meta-heuristic algorithm for continuous engineering optimization: harmony search theory and practice. Computer Methods in Applied Mechanics and Engineering (2005)
[76] Dasgupta, D. (ed.): Artificial Immune Systems and Their Applications. Springer, Berlin (1999)
[77] Wojtusiak, J., Michalski, R.S.: The LEM3 Implementation of learnable evolution model and Its testing on complex function optimization problems. In: Proceedings of Genetic and Evolutionary Computation Conference, GECCO 2006, Seattle, WA, July 8-12 (2006)
[78] Michalewicz, Z., Fogel, D.B.: How to Solve It: Modern Heuristics. Springer, Heidelberg (2000)
[79] Beyer, H.-G., Schwefel, H.-P.: Evolution Strategies: A Comprehensive Introduction. Journal Natural Computing 1(1), 3–52 (2002)
[80] Beyer, H.-G.: The Theory of Evolution Strategies. Springer, Heidelberg (2001)
[81] Kita, H.: A comparison study of self-adaptation in volution etrategies and real-coded genetic algorithms. Evolutionary Computation 9(2), 223–241 (2001)
[82] Hansen, N., Ostermeier, A.: Completely de-randomized self-adaptation in evolution strategies. Evolutionary Computation 9(2), 159–195 (2001)
[83] Fogel, D.B.: Evolving Artificial Intelligence, Ph.D. dissertation, Univ. California, San Diego, CA (1992)
[84] Goldberg, D.E.: Genetic Algorithms in Search, Optimization and Machine Learning. Addison-Wesley, Reading (1989)
[85] DeJong, K.A.: An analysis of behavior of a class of genetic adaptive systems. Doctoral Dissertation, University of Michigan (1975)
[86] Davis, T.E., Principa, J.C.: A Markov chain framework for the simple genetic algorithm. Evolutionary Computation 1(3), 269–288 (1993)
[87] Muehlenbein, H., Chakraborty, U.K.: Gene pool recombination genetic algorithm and the onemax function. Journal of Computing and Information Technology 5(3), 167–182 (1997)
[88] Vose, M.D., Liepins, G.E.: Punctuated equilibrium in genetic search. Complex Systems 5, 31–44 (1991)
[89] Filho, J.L.R., Treleven, P.C.: Genetic Algorithm Programming Environment, pp. 28–43. IEEE Computer Society Press, Los Alamitos (1994)
[90] Smith, S.F.: A Learning System Based on Genetic Adaptive Algorithms, PhD dissertation (University of Pittsburgh) (1980)
[91] Cramer, N.L.: A representation for the adaptive generation of simple sequential programs. In: John, J. (ed.) Proceedings of an International Conference on Genetic Algorithms and the Applications, Grefenstette, Carnegie Mellon University (1985)
[92] Banzhaf, W., Nordin, P., Keller, R.E., Francone, F.D.: Genetic programming: An Introduction: On the Automatic Evolution of Computer Programs and Its Applications. Morgan Kaufmann, San Francisco (1998)
[93] Nordin, J.P.: Evolutionary Program Induction of Binary Machine Code and its Application. Krehl Verlag, Muenster, Germany (1997)
[94] Beni, G., Wang, U.: Swarm intelligence in cellular robotic systems. In: NATO Advanced Workshop on Robots and Biological Systems, Il Ciocco, Tuscany, Italy (1989)
[95] Kennedy, J., Eberhart, R.C.: Particle swarm optimization. In: Proc. of the IEEE Int. Conf. on Neural Networks, Piscataway, NJ, pp. 1942–1948 (1995)

[96] Kennedy, J., Eberhart, R.C.: Swarm Intelligence. Morgan Kaufmann, San Francisco (2001)
[97] Dorigo, M.: Optimization, Learning and Natural Algorithms, PhD thesis, Politecnico di Milano, Italy (1992)
[98] Dorigo, M., Maniezzo, V., Colorni, A.: Ant System: optimization by a colony of cooperating agents. IEEE Transactions on Systems, Man, and Cybernetics–Part B 26(1), 29–41 (1996)
[99] Dorigo, M., Gambardella, L.M.: Ant Colony System: A cooperative learning approach to the traveling salesman problem. IEEE Transactions on Evolutionary Computation 1(1), 53–66 (1997)
[100] Johnson, D.S., Mc Geoch, L.A.: The Traveling Salesman Problem: A Case Study in Local Optimization. In: Aarts, E.H.L., Lenstra, J.K. (eds.) Local Search in Combinatorial Optimization, pp. 215–310. John Wiley and Sons Ltd., Chichester (1997)
[101] Abraham, A., Das, S., Roy, S.: Swarm intelligence algorithms for data clustering. In: Maimon, O., Rokach, L. (eds.) Soft Computing for Knowledge Discovery and Data Mining, pp. 279–313. Springer, Germany (2007)
[102] Kirkpatrick, S., Gelatt, C.D., Vecchi, M.P.: Optimization by Simulated Annealing. Science 220(4598), 671–680 (1983)
[103] Raghavan, V.V., Birchand, K.: A clustering strategy based on a formalism of the reproductive process in a natural system. In: Proceedings of the Second International Conference on Information Storage and Retrieval, pp. 10–22
[104] Murthy, C.A., Chowdury, N.: Search of optimal clusters using genetic algorithm. Pattern Recognition Letters 17, 825–832 (1996)
[105] Falkenauer, E.: Genetic Algorithms and Grouping Problems. John Wiley and Son, Chichester (1998)
[106] Bandyopadhyay, S., Murthy, C.A., Pal, S.K.: Pattern classification with genetic algorithms. Pattern Recognition Letters 16, 801–808 (1995)
[107] Bandyopadhyay, S., Murthy, C.A., Pal, S.K.: Pattern classification using genetic algorithm: determination of H. Pattern Recognition Letters 19, 1171–1181 (1998)
[108] Bandyopadhyay, S., Murthy, C.A., Pal, S.K.: Theoretic performance of genetic pattern classifier. Journal of the Franklin Institute 336, 387–422 (1999)
[109] Bandyopadhyay, S., Maulik, U.: Genetic clustering for automatic evolution of clusters and application to image classification. Pattern Recognition 35, 1197–1208 (2002)
[110] Srikanth, R., George, R., Warsi, N., Prabhu, D., Petri, F.E., Buckles, B.P.: A variable-length genetic algorithm for clustering and classification. Pattern Recognition Letters 16, 789–800 (1995)
[111] Maulik, U., Bandyopadhyay, S.: Genetic algorithm-based clustering technique. Pattern Recognition 33, 1455–1465 (2000)
[112] Chiou, Y.C., Lan, L.W.: Theory and methodology genetic clustering algorithms. European Journal of operational Research 135, 413–427 (2001)
[113] Bandyopadhyay, S., Maulik, U.: An evolutionary technique based on K-means algorithm for optimal clustering in RN. Information Sciences 146, 221–237 (2002)
[114] Paterlini, S., Minerva, T.: Evolutionary approaches for cluster analysis. In: Bonarini, A., Masulli, F., Pasi, G. (eds.) Soft Computing Applications, pp. 167–178. Springer, Berlin (2003)
[115] Krishna, K., Murty, M.N.: Genetic K-means algorithm. IEEE Transaction on Systems, Man and Cybernetics 29 (1999)

[116] Lee, C.-Y., Antonsson, E.K.: Dynamic partitional clustering using evolution strategies. In: Proceedings of the Third Asia Pacific Conference on Simulated Evolution and Learning (2000), http://citeseer.ist.psu.edu/501113.html

[117] Sarkar, M., Yegnanarayana, B., Khemani, D.: A clustering algorithm using an evolutionary programming-based approach. Pattern Recognition Letters 18, 975–986 (1997)

[118] Deneubourg, J.L., Goss, S., Franks, N., Sendova-Franks, A., Detrain, C., Chetien, L.: The dynamics of collective sorting: Robot-like ants and ant-like robots. In: Meyer, J.A., Wilson, S.W. (eds.) Proceedings of the First International Conference on Simulation of Adaptive Behaviour: From Animals to Animats 1, pp. 356–363. MIT Press, Cambridge (1991)

[119] Handl, J., Knowles, J., Dorigo, M.: Ant-based clustering: a comparative study of its relative performance with respect to k-means, average link and 1D-SOM., Technical Report TR/IRIDIA/2003-24, IRIDIA, Universite Libre de Bruxelles, Belgium (2003)

[120] Lumer, E., Faieta, B.: Diversity and adaptation in populations of clustering ants. In: Proceedings Third International Conference on Simulation of Adaptive Behavior: from Animals to Animates 3, pp. 499–508. MIT press, Cambridge (1994)

[121] Lumer, E., Faieta, B.: Exploratory Database Analysis via Self-Organization, unpublished manuscript (1995)

[122] Emergent colonization and graph partitioning. In: Proceedings of the Third International Conference on Simulation of Adaptive Behaviour: From Animals to Animats 3, pp. 494–500. MIT Press, Cambridge (1994)

[123] Kuntz, P., Snyers, D.: New results on an ant-based heuristic for highlighting the organization of large graphs. In: Proceedings of the 1999 Congress on Evolutionary Computation, pp. 1451–1458. IEEE Press, Piscataway (1999)

[124] Kuntz, P., Snyers, D., Layzell, P.: A Stochastic Heuristic for Visualizing Graph Clusters in a bi-dimensional Space Prior to Partitioning. Journal of Heuristics 5(3), 327–351 (1998)

[125] Handl, J., Meyer, B.: Improved ant-based clustering and sorting in a document retrieval interface. In: Guervós, J.J.M., Adamidis, P.A., Beyer, H.-G., Fernández-Villacañas, J.-L., Schwefel, H.-P. (eds.) PPSN 2002. LNCS, vol. 2439, pp. 913–923. Springer, Heidelberg (2002)

[126] Hoe, K., Lai, W., Tai, T.: Homogenous ants for web document similarity modeling and categorization. In: Dorigo, M., Di Caro, G.A., Sampels, M. (eds.) Ant Algorithms 2002. LNCS, vol. 2463, pp. 256–261. Springer, Heidelberg (2002)

[127] Ramos, V., Merelo, J.J.: Self-organized stigmergic document maps: Environments as a mechanism for context learning. In: Proceedings of the First Spanish Conference on Evolutionary and Bio-Inspired Algorithms (AEB 2002), pp. 284–293. Centro Univ. M´erida, M´erida, Spain (2002)

[128] Ramos, V., Muge, F., Pina, P.: Self-organized data and image retrieval as a consequence of inter-dynamic synergistic relationships in artificial ant colonies. Soft Computing Systems: Design, Management and Applications 87, 500–509 (2002)

[129] Monmarche, N., Slimane, M., Venturini, G.: Ant class: discovery of clusters in numeric data by a hybridization of an ant colony with the k means algorithm. Internal Report No. 213, E3i, Laboratoire d'Informatique, Universite de Tours (1999)

[130] Kanade, P.M., Hall, L.O.: Fuzzy ants as a clustering concept. In: Proceedings of the 22nd International Conference of the North American Fuzzy Information Processing Society (NAFIPS 2003), pp. 227–232 (2003)

[131] Tsang, W., Kwong, S.: Ant colony clustering and feature extraction for anomaly intrusion detection. In: Abraham, A., Grosan, C., Ramos, V. (eds.) Swarm Intelligence in Data Mining, pp. 101–121. Springer, Heidelberg (2006)
[132] Omran, M., Salman, A., Engelbrecht, A.P.: Image classification using particle swarm optimization. In: Conference on Simulated Evolution and Learning, vol. 1, pp. 370–374 (2002)
[133] Omran, M., Engelbrecht, A.P., Salman, A.: Particle swarm optimization method for image clustering. International Journal of Pattern Recognition and Artificial Intelligence 19(3), 297–322 (2005)
[134] van der Merwe, D.W., Engelbrecht, A.P.: Data clustering using particle swarm optimization. In: Proceedings of the 2003 IEEE Congress on Evolutionary Computation, Piscataway, NJ, pp. 215–220 (2003)
[135] Xiao, X., Dow, E.R., Eberhart, R.C., Miled, Z.B., Oppelt, R.J.: Gene clustering using self-organizing maps and particle swarm optimization. In: Proc. of the 17th International Symposium on Parallel and Distributed Processing (PDPS 2003). IEEE Computer Society, Washington (2003)
[136] Cui, X., Potok, T.E.: Document clustering analysis based on hybrid PSO + K-means algorithm. Journal of Computer Sciences (Special Issue), 27–33 (2005) ISSN 1549-3636
[137] Halkidi, M., Batistakis, Y., Vazirgiannis, M.: On clustering validation techniques. Journal of Intelligent Information Systems (JIIS) 17(2-3), 107–145 (2001)
[138] Rosenberger, C., Chehdi, K.: Unsupervised clustering method with optimal estimation of the number of clusters: Application to image segmentation. In: Proc. IEEE International Conference on Pattern Recognition (ICPR), vol. 1, Barcelona, pp. 1656–1659 (2000)
[139] Ball, G., Hall, D.: A clustering technique for summarizing multivariate data. Behavioral Science 12, 153–155 (1967)
[140] Huang, K.: A synergistic automatic clustering technique (Syneract) for multispectral image analysis. Photogrammetric Engineering and Remote Sensing 1(1), 33–40 (2002)
[141] Pelleg, D., Moore, A.: X-means: Extending K-means with efficient estimation of the number of clusters. In: Proceedings of the 17th International Conference on Machine Learning, pp. 727–734. Morgan Kaufmann, San Francisco (2000)
[142] Kass, R., Wasserman, L.: A reference Bayesian test for nested hypotheses and its relationship to the Schwarz criterion. Journal of the American Statistical Association 90(431), 928–934 (1995)
[143] Wallace, C.S., Dowe, D.L.: Intrinsic classification by MML – the snob program. In: Proceedings 7th Australian Joint Conference on Artificial Intelligence, UNE, Armidale, NSW, Australia, pp. 37–44 (1994)
[144] Wallace, C.S., Boulton, D.M.: An information measure for classification. The Computer Journal 11, 185–194 (1968)
[145] Oliver, J.J., Hand, D.: Introduction to minimum encoding inference. Technical Report No. 94/205, Department of Computer Science, Monash University, Australia (1994)
[146] Bischof, H., Leonardis, A., Selb, A.: MDL principle for robust vector quantization. Pattern Analysis and Applications 2, 59–72 (1999)
[147] Gath, I., Geva, A.: Unsupervised optimal fuzzy clustering. IEEE Transactions on Pattern Analysis and Machine Intelligence 11(7), 773–781 (1989)

[148] Lorette, A., Descombes, X., Zerubia, J.: Fully unsupervised fuzzy clustering with entropy criterion. In: International Conference on Pattern Recognition (ICPR 2000), vol. 3, pp. 3998–4001 (2000)

[149] Frigui, H., Krishnapuram, R.: A robust competitive clustering algorithm with applications in computer vision. IEEE Transactions on Pattern Analysis and Machine Intelligence 21(5), 450–465 (1999)

[150] Kohonen, T.: Self-Organizing Maps. Springer Series in Information Sciences, vol. 30. Springer, N.Y. (1995)

[151] Vesanto, J., Alhoniemi, E.: Clustering of the self organizing map. IEEE Transactions on Neural Networks 11(3), 586–600 (2000), `http://citeseer.ist.psu.edu/vesanto00clustering.html`

[152] Ultsch, A.: Emergence in Self-organizing feature maps. In: Proceedings Workshop on Self-Organizing Maps (WSOM 2007), Bielefeld, Germany (2007)

[153] Mehrotra, K., Mohan, C., Rakka: Elements of Artificial Neural Networks. MIT Press, Cambridge (1997)

[154] Pandya, A., Macy, R.: Pattern Recognition with Neural Networks in C++. CRC Press, Boca Raton (1996)

[155] Maulik, U., Bandyopadhyay, S.: Fuzzy partitioning using real coded variable length genetic algorithm for pixel classification. IEEE Transactions on Geosciences and Remote Sensing 41(5), 1075–1081 (2003)

[156] Omran, M.G., Engelbrecht, A.P., Salman, A.: Dynamic clustering using particle swarm optimization with application in unsupervised image classification. In: Proceedings of World Academy of Science, Engineering and Technology, vol. 9 (November 2005)

[157] Sawaragi, Y., Nakayama, H., Tanino, T.: Theory of multiobjective optimization. Mathematics in Science and Engineering, vol. 176. Academic Press Inc., Orlando (1985)

[158] Deb, K.: Multi-Objective Optimization using Evolutionary Algorithms. John Wiley & Sons, Chichester (2001)

[159] Coello Coello, C.A., Lamont, G.B., Van Veldhuizen, D.A.: Evolutionary Algorithms for Solving Multi-Objective Problems. Springer, Heidelberg (2007)

[160] Abraham, A., Jain, L.C., Goldberg, R. (eds.): Evolutionary Multiobjective Optimization: Theoretical Advances and Applications. Springer, London (2005)

[161] Knowles, J.D., Corne, D.W.: Approxmating the nondominated front using the pareto archived evolution strategy. Evolutionary Computation 8(2), 149–172 (2000)

[162] Deb, K., Pratap, A., Agarwal, S., Meyarivan, T.: A fast and elitist multiobjective genetic algorithm: NSGA-II. IEEE Transactions on Evolutionary Computation 6(2) (2002)

[163] Minaei-Bidgoli, B., Topchy, A., Punch, W.F.: Ensembles of partitions via data resampling. In: Proc. Int. Conf. Inf. Technol. Coding Comput., pp. 188–192 (2004)

[164] Topchy, A., Jain, A.K., Punch, W.: Clustering ensembles: Models of consensus and weak partitions. IEEE Trans. Pattern Anal. Mach. Intell. 27(12), 1866–1881 (2005)

[165] Topchy, A., Minaei, B., Jain, A.K., Punch, W.: Adaptive clustering ensembles. In: Proc. Int. Conf. Pattern Recognit., pp. 272–275 (2004)

[166] Strehl, A., Ghosh, J.: Cluster ensembles—A knowledge reuse framework for combining multiple partitions. J. Machine Learn. Res. 3, 583–617 (2002)

[167] Topchy, A., Jain, A.K., Punch, W.: A mixture model for clustering ensembles. In: Proc. SIAM Int. Conf. Data Mining, pp. 379–390 (2004)

[168] Law, M., Topchy, A., Jain, A.K.: Multiobjective Data Clustering. In: Proceedings of the IEEE Computer Society Conference on Computer Vision and Pattern Recognition, vol. 2, pp. 424–430 (2004)

[169] Bandyopadhyay, S., Maulik, U., Mukhopadhyay, A.: Multiobjective genetic clustering for pixel classification in remote sensing imagery. IEEE Transactions Geoscience and Remote Sensing (2006)

[170] Handl, J., Knowles, J.: Evolutionary multiobjective clustering. In: Proc. 8th Int. Conf. Parallel Problem Solving from Nature, pp. 1081–1091 (2004)

[171] Corne, D.W., Jerram, N.R., Knowles, J.D., Oates, M.J.: PESA-II: Region-based selection in evolutionary multiobjective optimization. In: Proc. Genetic Evol. Comput. Conf., pp. 283–290 (2001)

[172] Handl, J., Knowles, J.: An evolutionary approach to multiobjective clustering. IEEE Transactions on Evolutionary Computation 11(1), 56–76 (2007)

[173] Handl, J., Knowles, J.: Exploiting the tradeoff—the benefits of multiple objectives in data clustering. In: Proc. 3rd Int. Conf. Evol. Multicriterion Optim., pp. 547–560 (2005)

[174] Handl, J., Knowles, J.: Improvements to the scalability of multiobjective clustering. In: Proc. IEEE Congr. Evol. Comput., vol. 3, pp. 2372–2379 (2005)

[175] Handl, J., Knowles, J.: Multiobjective clustering around medoids. In: Proc. IEEE Congr. Evol. Comput., vol. 1, pp. 632–639 (2005)

[176] Storn, R., Price, K.V.: Differential evolution - a simple and efficient adaptive scheme for global optimization over continuous spaces., Technical Report TR-95-012, ICSI (1995), `http://http.icsi.berkeley.edu/~storn/litera.html`

[177] Storn, R., Price, K.V., Lampinen, J.: Differential Evolution - A Practical Approach to Global Optimization. Springer, Berlin (2005)

[178] Snyman, J.A.: Practical Mathematical Optimization: An Introduction to Basic Optimization Theory and Classical and New Gradient-Based Algorithms. Springer Publishing, Heidelberg (2005)

[179] Hahn, W.: Theory and Application of Liapunov's Direct Method. Prentice-Hall, Englewood Cliffs (1963)

[180] Blake, C., Keough, E., Merz, C.J.: UCI Repository of Machine Learning Database (1998), `http://www.ics.uci.edu/~mlearn/MLrepository.html`

[181] Muller, K.R., Mika, S., Ratsch, G., Tsuda, K., Scholkopf, B.: An introduction to kernel-based learning algorithms. IEEE Trans. Neural Networks 12(2), 181–202 (2001)

[182] Girolami, M.: Mercer kernel-based clustering in feature space. IEEE Trans. Neural Networks 13(3), 780–784 (2002)

[183] Scholkopf, B., Smola, A.J.: Learning with Kernels. The MIT Press, Cambridge (2002)

Chapter 2
Differential Evolution Algorithm: Foundations and Perspectives

Abstract. Differential Evolution (DE) has recently emerged as simple and efficient algorithm for global optimization over continuous spaces. DE shares many features of the classical Genetic Algorithms (GA). But it is much easier to implement than GA and applies a kind of differential mutation operator on parent chromosomes to generate the offspring. Since its inception in 1995, DE has drawn the attention of many researchers all over the world, resulting in a lot of variants of the basic algorithm, with improved performance. This chapter begins with a conceptual outline of classical DE and then presents several significant variants of the algorithm in greater details.

2.1 Introduction

The Differential Evolution (DE), proposed by Storn and Price in [6], [7] may be also seen as a simple real-coded GA. The first written article on DE appeared as a technical report in 1995. Since then, DE has proven itself in competitions like the IEEE's International Contest on Evolutionary Optimization (ICEO) in 1996 and 1997 respectively.

In DE community, the individual trial solutions (which constitute a population) are referred as *parameter vectors* or *genomes*. DE operates through the same computational steps as employed by a standard EA. However, unlike traditional EAs, DE employs difference of the parameter vectors to explore the objective function landscape. In this respect, it owes a lot to its two ancestors namely – the Nelder-Mead algorithm [8, 9], and the Controlled Random Search (CRS) algorithm [10], which also relied heavily on the difference vectors to perturb the current trial solutions. Like other population-based search techniques, DE generates new points (trial solutions) that are perturbations of existing points, but these deviations are neither reflections like those in the CRS and Nelder-Mead methods, nor samples from a predefined probability density function, like those in Evolutionary Strategies (ES) [11, 12]. Instead, DE perturbs current generation vectors with the scaled difference of two randomly selected population vectors. To produce a trial vector in its simplest form DE adds the scaled, random vector difference to a third randomly selected population vector. In the selection stage, the trial vector competes against the population vector of the same index. Once the last trial vector has been tested the survivors of all the pair wise competitions become permanent for the next generation in the evolutionary cycle. In the following Sections, we discuss each of these steps in sufficient details.

S. Das, A. Abraham, and A. Konar: Metaheuristic Clustering, SCI 178, pp. 63–110.
springerlink.com

2.2 Differential Evolution: A First Glance

DE is a simple evolutionary algorithm. It works through a simple cycle of stages, presented in Figure 2.1.

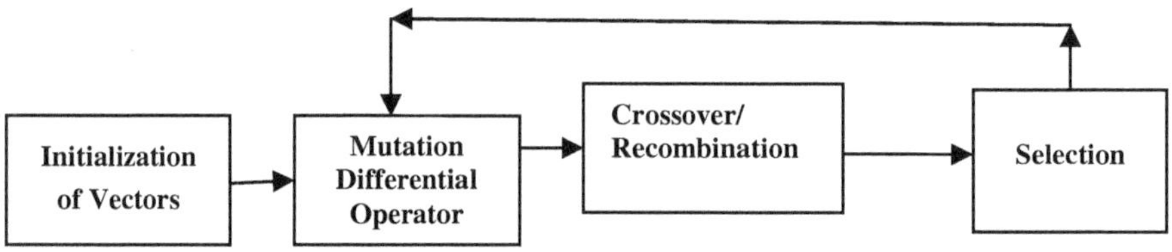

Fig. 2.1 The main stages of differential evolution algorithm

Below we explain each stage separately with sufficient numerical examples.

2.2.1 Initialization of the Parameter Vectors

DE searches for a global optimum point in a D-dimensional continuous hyperspace. It begins with a randomly initiated population of NP D dimensional real-valued parameter vectors. Each vector, also known as *genome/chromosome*, forms a candidate solution to the multi dimensional optimization problem.

We represent subsequent generations in DE by discrete time steps like t = 0, 1, 2 ...t, t+1 etc (in most of the DE literatures, the successive generations are represented by G, G+1, G+2...or g, g+1 etc. [13] but we adopt a slightly different notation in order to remain consistent with the notations used in other chapters of the Volume and also to facilitate the mathematical analysis of DE undertaken here. Since the parameter vectors are likely to be changed over different generations, we adopt the following notation for representing the i-th vector of the population at the current generation (i.e. at time $t = t$) as:

$$\vec{X}_i(t) = [x_{1,1}(t), x_{i,2}(t), \ldots, x_{i,D}(t)]^T, \tag{2.1}$$

with $i = 1, 2, \ldots, NP$.

For each parameter of the problem, there may be a certain range within which value of the parameter should lie for better search results. At the very beginning of a DE run or at $t = 0$, problem parameters or independent variables are initialized somewhere in their feasible numerical range. So, if the j-th parameter of the given problem has its lower and upper bound as $x_{\min,j}$ and $x_{\max,j}$ respectively and $rand_{i,j}(0,1)$ denotes the j-th instantiation of a uniformly distributed random number lying between 0 and 1 for the i-th vector, then we may initialize the j-th component of the i-th population members as,

$$x_{i,j}(0) = x_{\min,j} + rand_{i,j}(0,1).(x_{\max,j} - x_{\min,j}) \tag{2.2}$$

The process is illustrated in Figure 2.3 for 10 parameter vectors in two-dimensional search space. Closed curves in Figure 2.3 denote constant cost contours, which for a given cost function *f*, corresponds to $f(x_1, x_2) = \text{constant}$. In Figure 2.2, we have shown the constant cost contours for the two dimensional sphere functions on the $x_1 - x_2$ parameter plane.

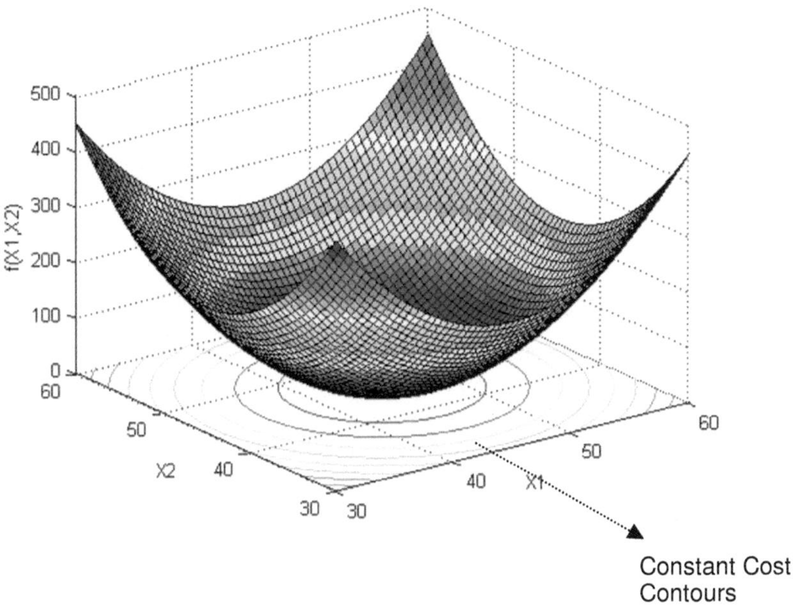

Fig. 2.2 Constant Cost Contours for Sphere Function

Example 2.1

Suppose we are to minimize the five-dimensional sphere function $f(\vec{X}) = \sum_{i=1}^{5} x_i^2$

Suppose we start with a toy population of 5 vectors. If the search range permitted for each variable lie from -10 to +10 then a particular chromosome say the 2nd one may be initialized as follows:

$x_{2,1}(0) = -10 + rand_{2,1}(0, 1).\{10-(-10)\}=-10+0.621\times 20 = 2.42.$

$x_{2,2}(0) = -10 + rand_{2,2}(0, 1).\{10-(-10)\}=-10+0.519\times 20 = 0.38.$

$x_{2,3}(0) = -10 + rand_{2,3}(0, 1).\{10-(-10)\}=-10+0.982\times 20 = 9.64$

$x_{2,4}(0) = -10 + rand_{2,4}(0, 1).\{10-(-10)\}=-10+0.279\times 20 = -4.42$

$x_{2,5}(0) = -10 + rand_{2,5}(0, 1).\{10-(-10)\}=-10+0.025\times 20 = -9.50$

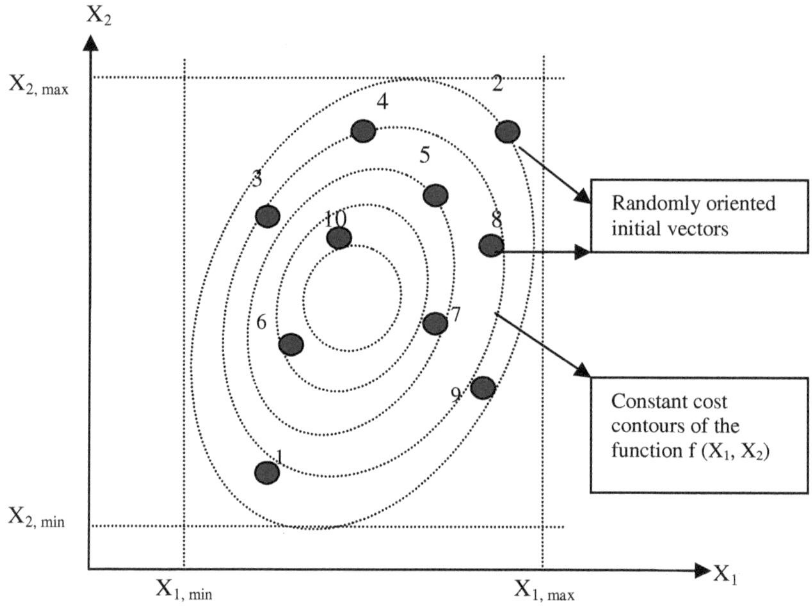

Fig. 2.3 Initializing a DE population of NP = 10, on a two-dimensional parametric space

In this way, all the 5 vectors are initialized and we start with a population, which looks like:

$$\vec{X}_1(0) = [3.82, 4.78, -9.34, 5.36, -3.77]^T$$
$$\vec{X}_2(0) = [2.42, 0.38, 9.64, -4.42, -9.50]^T$$
$$\vec{X}_3(0) = [2.76, 8.41, -0.92, -5.09, -1.32]^T$$
$$\vec{X}_4(0) = [9.12, 7.93, -3.27, -2.08, -9.26]^T$$
$$\vec{X}_5(0) = [5.72, -8.52, -5.04, 3.36, -2.51]^T$$

2.2.2 *Mutation with Differential Operators*

Biologically 'mutation' means a sudden change in the gene characteristics of a chromosome. In the context of the evolutionary computing paradigm, however, mutation is also seen as a change or perturbation with a random element. Most of the real-coded EAs typically simulate the effects mutation with additive increments, which are randomly generated by a predetermined Probability Density Function (PDF) [1, 12]. DE, however, applies a uniform PDF not to generate increments, but to randomly sample vector differences like $\Delta\vec{X}_{r2,r3} = (\vec{X}_{r2} - \vec{X}_{r3})$. In DE, mutation amounts to creating a donor vector $\vec{V}_i(t)$ for changing each population member $\vec{X}_i(t)$ (say), in each generation (or in one iteration of the algorithm). To create $\vec{V}_i(t)$ for each i-th member of the current population (also called the *target* vector), three other distinct parameter vectors, say the vectors $\vec{X}_{r_1^i}$, $\vec{X}_{r_2^i}$, and $\vec{X}_{r_3^i}$ are picked up randomly from the current population. The indices r_1^i, r_2^i and r_3^i are mutually exclusive integers randomly chosen from the range [1, NP], which are also different from the base vector index i. These indices are randomly generated once for each mutant vector. Now the difference of any two of these three vectors is scaled by a scalar number F and the scaled difference is added to the third one whence we obtain the donor vector $\vec{V}_i(t)$. We can express the process as,

$$\vec{V}_i(t) = \vec{X}_{r_1^i}(t) + F.(\vec{X}_{r_2^i}(t) - \vec{X}_{r_3^i}(t)) \tag{2.3}$$

Actually it is the mutation scheme that demarcates among the different kinds of DE schemes. Here presently we discuss one of the most popular schemes for the formation of the donor vector. The process is illustrated in Figure 2.4.

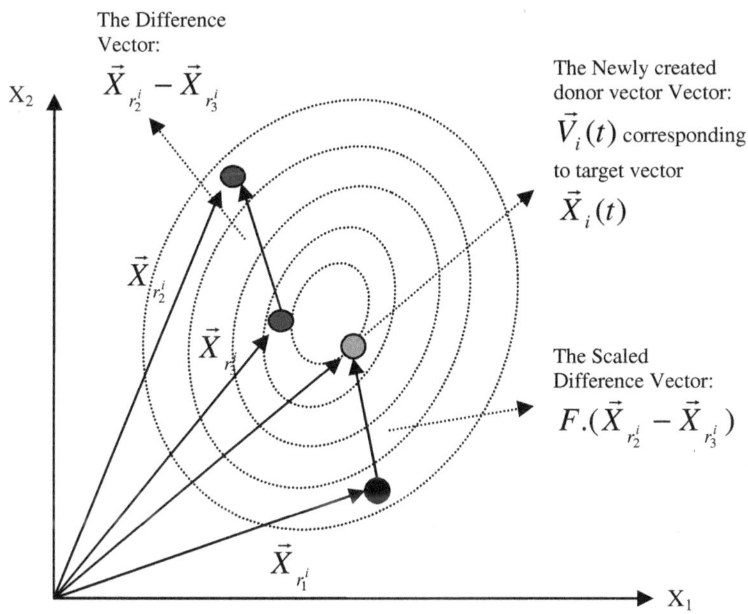

Fig. 2.4 Illustrating a simple DE mutation scheme in two-dimensional parametric space

Example 2.2

Let us get back to the initialized population of chromosomes in Example 2.1. Suppose for the target vector $\vec{X}_1(0)$ in our program randomly selects three other vectors, say $\vec{X}_3(0)$, $\vec{X}_4(0)$ and $\vec{X}_2(0)$. If the scale factor F mentioned in equation (2.3) assume a value 0.8, then, the donor vector $\vec{V}_1(0)$ can be formed as:

$$\begin{bmatrix} v_{1,1}(0) \\ v_{1,2}(0) \\ v_{1,3}(0) \\ v_{1,4}(0) \\ v_{1,5}(0) \end{bmatrix} = \begin{bmatrix} 2.76 \\ 8.41 \\ -0.92 \\ -5.09 \\ -1.32 \end{bmatrix} + 0.8. \left\{ \begin{bmatrix} 9.12 \\ 7.93 \\ -3.27 \\ -2.08 \\ -9.26 \end{bmatrix} - \begin{bmatrix} 2.42 \\ 0.38 \\ 9.64 \\ -4.42 \\ -9.50 \end{bmatrix} \right\} = \begin{bmatrix} 8.12 \\ 14.45 \\ -11.24 \\ -3.22 \\ -1.12 \end{bmatrix}$$

Note that the 2nd and the 3rd elements of the donor vector go out of our previously designated search domain of [-10, 10]. Hence we fix them at 10 and -10 respectively, yielding finally,

$$\vec{V}_1(0) = \begin{bmatrix} 8.12 \\ 10 \\ -10 \\ -3.22 \\ -1.12 \end{bmatrix}$$

2.2.3 *Crossover*

To increase the potential diversity of the population, a crossover operation comes into play after generating the donor vector through mutation. The DE family of algorithms can use two kinds of crossover schemes - *exponential* and *binomial* [6, 7]. The donor vector exchanges its *body parts* i.e. components with the target vector $\vec{X}_i(t)$ under this operation to form the *trial* vector $\vec{U}_i(t) = [u_{i,1}(t), u_{i,2}(t), \ldots, u_{i,D}(t)]^T$. In exponential crossover, we first choose an integer n randomly among the numbers [0, D-1]. This integer acts as a starting point in the target vector, from where the crossover or exchange of components with the donor vector starts. We also choose another integer L from the interval [1, D]. L denotes the number of components; the donor vector actually contributes to the target. After a choice of n and L the trial vector:

$$u_{i,j}(t) = \begin{cases} v_{i,j}(t), \text{ for } j = \langle n \rangle_D, \langle n+1 \rangle_D, \ldots, \langle n+L-1 \rangle_D \\ x_{i,j}(t), \text{ for all other } j \in [0, D-1] \end{cases} \quad (2.4)$$

where the angular brackets $\langle \ \rangle_D$ denote a modulo function with modulus D. The integer L is drawn from [1,2,…, D] according to the following lines of pseudo code.

```
L = 0;
do
{
   L=L+1;
} while (rand (0, 1) < CR) AND (L<D));
```

Hence in effect Probability ($L \geq \upsilon$) = $(CR)^{\upsilon-1}$ for any $\upsilon > 0$. '*CR*' is called *crossover rate* and it appears as a control parameter of DE just like *F*. For each donor vector, a new set of *n* and *L* must be chosen randomly as shown above.

Example 2.3

This example illustrates the exponential crossover scheme of DE. From the last example, we see that, target vector $\vec{X}_1(0)$ and the corresponding donor vector $\vec{V}_1(0)$ are respectively given by,

$$\vec{X}_1(0) = \begin{bmatrix} 3.82 \\ 4.78 \\ -9.34 \\ 5.36 \\ -3.77 \end{bmatrix} \qquad \vec{V}_1(0) = \begin{bmatrix} 8.12 \\ 10 \\ -10 \\ -3.22 \\ -1.12 \end{bmatrix}$$

Suppose $n = 2$ and $L=3$ for this specific example. Then the components of $\vec{X}_1(0)$ that get replaced by the corresponding components of $\vec{V}_1(0)$ are, from $x_{1,3}(0)$ to $x_{1,5}(0)$. The newly formed offspring vector finally takes the following form:

$$\vec{U}_1(0) = \begin{bmatrix} 3.82 \\ 4.78 \\ -10 \\ -3.22 \\ 1.12 \end{bmatrix}$$

Figure 2.5 provides a visual feel of the process.

On the other hand, binomial crossover is performed on each of the *D* variables whenever a randomly picked number between 0 and 1 is less than or equal to the *CR* value. In this case the number of parameters inherited from the donor has a (nearly) binomial distribution. The scheme may be outlined as:

$$u_{i,j,G} = \begin{cases} v_{i,j,G}, & \text{if } (rand_{i,j}(0,1) \leq CR \text{ or } j = j_{rand}) \\ x_{i,j,G}, & \text{otherwise,} \end{cases} \tag{2.5}$$

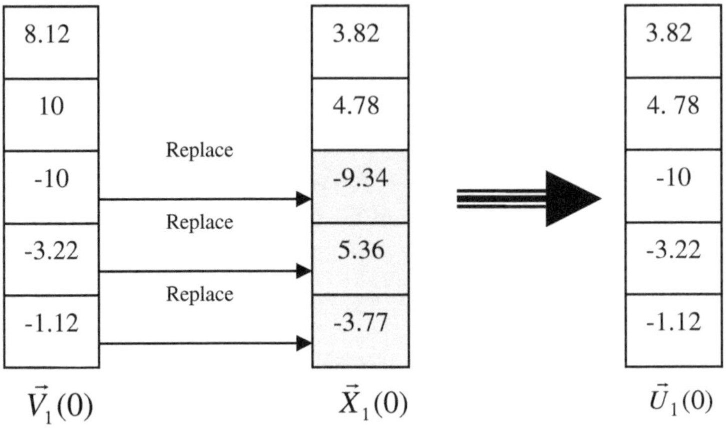

Fig. 2.5 Illustrating the exponential crossover in DE

where $rand_{i,j}(0,1) \in [0,1]$ is a uniformly distributed random number, which is called anew for each *j*-th component of the *i*-th parameter vector. $j_{rand} \in [1,2,...., D]$ is a randomly chosen index, which ensures that $\vec{U}_{i,G}$ gets at least one component from $\vec{V}_{i,G}$. It is instantiated once for each vector in one generation. We note that for this additional demand, *CR* is only approximating the true probability p_{Cr} that a component of the trial vector will be inherited from the donor. Also, one may observe that in a two-dimensional search space, three possible trial vectors may result from uniformly crossing a mutant/donor vector $\vec{V}_i(t)$ with the target vector $\vec{X}_i(t)$. These trial vectors are

i) $\vec{U}_i(t) = \vec{V}_i(t)$ such that both the components of $\vec{U}_i(t)$ are inherited from $\vec{V}_i(t)$.
ii) $\vec{U}_i'(t)$, in which the first component (j = 1) comes from $\vec{V}_i(t)$ and the second one (j = 2) from $\vec{X}_i(t)$.
iii) $\vec{U}_i''(t)$, in which the first component (j = 1) comes from $\vec{X}_i(t)$ and the second one (j = 2) from $\vec{V}_i(t)$.

The possible trial vectors due to uniform crossover are illustrated in Figure 2.6.

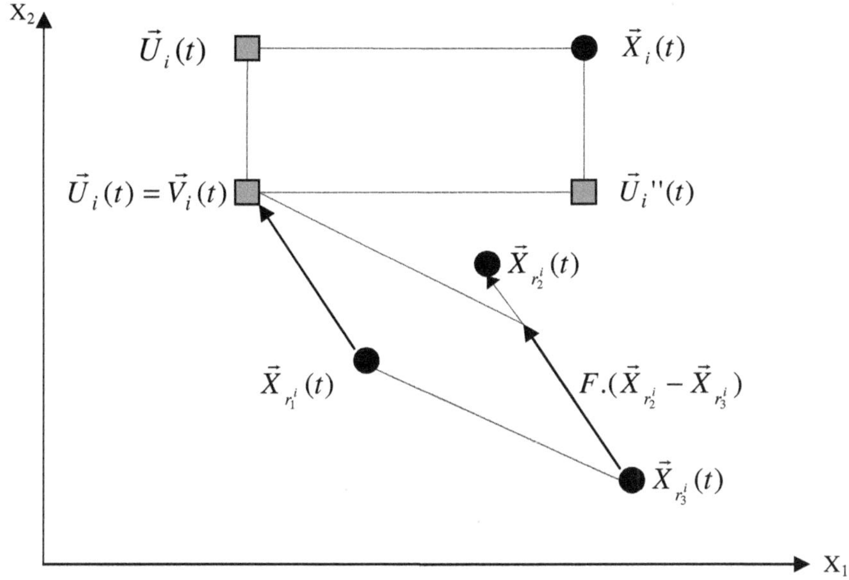

Fig. 2.6 Different possible trial vectors formed due to uniform/binomial crossover between the target and the mutant vectors in two-dimensional search space

Example 2.4

This example illustrates the 'binomial' crossover process. Suppose $CR = 0.8$. Now again refer to the target vector $\vec{X}_1(0)$ and the donor vector $\vec{V}_1(0)$ of Example 2.2. Suppose for $\vec{X}_1(0)$, the random number $j_{rand} = 3$. For each component of $\vec{X}_1(0)$ one random number $rand_{1,j}$ (0, 1) is generated between 0 and 1. If the value of the said random number is greater than 0.8 (i.e. value of *CR*), or $j = 3$ then the corresponding component of $\vec{X}_1(0)$ is replaced by $\vec{V}_1(0)$ else it is left unaltered. The process is summarized in the following table. Note from

Table 2.1 The binomial crossover process in Example 2.3

Components of $\vec{X}_1(0)$	Components of $\vec{V}_1(0)$	$rand_{1,j}(0, 1)$	j_{rand}	*CR*	Components of $\vec{U}_1(0)$
3.82	8.12	0.85			8.12
4.78	10	0.73			10
-9.34	-10	0.89	3	0.8	-10
5.36	-3.22	0.12			-3.22
-3.77	-1.12	0.63			-1.12

table 2.1 that although $rand_{1,3}(0, 1) = 0.58 > CR$, as $j = 3 = j_{rand}$ holds, we have $u_1^3(0) = v_1^3(0) = -10$

2.2.4 Selection

The last stage of a DE-iteration is the 'selection' i.e. deciding who between the target vector $\vec{X}_i(t)$ and the newly formed trial vector $\vec{U}_i(t)$ will survive to the next generation. The decision whether original $\vec{X}_i(t)$ will be retained in the population or will be replaced by $\vec{U}_i(t)$ in the next time step t+1 is entirely dependent upon the 'survival of the fittest' concept. If the trial vector yields a better fitness value it will replace the target vector in the next time step. Here by *better fitness value* we mean a lower value of the objective function in case of a minimization problem, and a higher value of the same if it is a maximization problem. The selection operation may be outlined as:

$$\vec{X}_i(t+1) = \begin{cases} \vec{U}_i(t) & \text{if } f(\vec{U}_i(t)) \le f(\vec{X}_i(t)) \\ \vec{X}_i(t) & \text{if } f(\vec{U}_i(t)) > f(\vec{X}_i(t)) \end{cases} \tag{2.6}$$

where $f(\vec{X})$ is the function to be minimized. Since the selection process employs a binary decision, i.e. any one between the target vector and its offspring survives the population size remains fixed through out generations. The fitness of the population members either improves over generations or remains unchanged, but never deteriorates.

Example 2.5

This example illustrates the selection mechanism of the classical DE algorithm. The objective function we are trying to minimize is the five dimensional sphere function given by,

$$f(\vec{X}) = \sum_{i=1}^{5} x_i^2$$

Now as found in Example 2.4, the first population member $\vec{X}_1(0)$ and its corresponding offspring vector $\vec{U}_1(0)$ are given by,

$$\vec{X}_1(0) = \begin{bmatrix} 3.82 \\ 4.78 \\ -9.34 \\ 5.36 \\ -3.77 \end{bmatrix} \qquad \vec{U}_1(0) = \begin{bmatrix} 3.82 \\ 4.78 \\ -10 \\ -3.22 \\ 1.12 \end{bmatrix}$$

The fitness (or objective function value) of the parent genome $\vec{X}_1(0)$ is calculated as,

$$f(\vec{X}_1(0)) = (3.82)^2 + (4.78)^2 + (-9.34)^2 + (5.36)^2 + (-3.77)^2$$
$$= 167.6189$$

Similarly, objective function value for the offspring genome is given by,

$$f(\vec{U}_1(0)) = (3.82)^2 + (4.78)^2 + (-10)^2 + (-3.22)^2 + (1.12)^2$$
$$= 149.2976$$

clearly we see, $f(\vec{U}_1(0)) < f(\vec{X}_1(0))$. Hence at $t = 1$, the target vector $\vec{X}_1(0)$ is replaced by the trial $\vec{U}_1(0)$. The vector $\vec{X}_1(1)$ looks like,

$$\vec{X}_1(1) = \begin{bmatrix} 3.82 \\ 4.78 \\ -10 \\ -3.22 \\ 1.12 \end{bmatrix}$$

2.2.5 *Summary of DE Iteration*

An iteration of the classical DE algorithm consists of the four basic steps – initialization of a population of search variable vectors, mutation, crossover or recombination and finally selection. After having illustrated these stages, we now formally present the whole of the algorithm in a pseudo-code below.

Pseudo-code for the DE algorithm

Step1. Set the generation number $t = 0$ and randomly initialize a population of NP individuals $P_t = \{\vec{X}_1(t), \ldots\ldots, \vec{X}_{NP}(t)\}$ with $\vec{X}_i(t) = [x_{i,1}(t), x_{i,2}(t), x_{3,i}(t), \ldots\ldots, x_{i,D}(t)]$ and each individual uniformly distributed in the range $[\vec{X}_{\min}, \vec{X}_{\max}]$, where $\vec{X}_{\min} = \{x_{\min,1}, x_{\min,2}, \ldots, x_{\min,D}\}$ and $\vec{X}_{\max} = \{x_{\max,1}, x_{\max,2}, \ldots, x_{\max,D}\}$ with $i = [1,2,\ldots., NP]$.

Step 2. WHILE stopping criterion is not satisfied

DO

FOR $i = 1$ to NP //do for each individual sequentially

Step 2.1 ***Mutation Step***

Generate a donor vector $\vec{V}_i(t) = \{v_{i,1}(t), \ldots\ldots., v_{i,D}(t)\}$ corresponding to the i-th target vector $\vec{X}_i(t)$ via one of the mutation scheme of DE (equations (2.3)).

Step 2.2 ***Crossover Step***

Generate a trial vector $\vec{U}_i(t) = \{u_{i,1}(t), ..., u_{i,D}(t)\}$ for the i-th target vector $\vec{X}_i(t)$ through exponential crossover (equation (2.4)) or binomial crossover (equation (2.5))

Step 2.3 ***Selection Step***

Evaluate the trial vector $\vec{U}_i(t)$

IF $f(\vec{U}_i(t)) \le f(\vec{X}_i(t))$, THEN $\vec{X}_i(t+1) = \vec{U}_i(t)$,

$$f(\vec{X}_i(t+1)) = f(\vec{U}_i(t))$$

IF $f(\vec{U}_i(t)) < f(\vec{X}_{best}(t))$, THEN $\vec{X}_{best}(t) = \vec{U}_i(t)$,

$$f(\vec{X}_{best}(t)) = f(\vec{U}_i(t))$$

END IF

END IF

ELSE $\vec{X}_i(t+1) = \vec{X}_i(t)$, $f(\vec{X}_i(t+1)) = f(\vec{X}_i(t))$

END FOR

Step 2.4 Increase the iteration count $t = t + 1$

END WHILE

The parameters used in the algorithm namely scale factor '*F*' and crossover rate '*CR*' should be submitted before to invoke the main computational part of the algorithm – the while loop. The terminating condition can be defined in two ways:

i) by a fixed number of iterations t_{max}, with a suitably large value of t_{max} depending upon the complexity of the objective function and alternatively,
ii) when best fitness of the population does not change appreciably over successive iterations.

The following example demonstrates a simple DE algorithm used to minimize the two-dimensional sphere function.

Example 2.6

This example illustrates the complete search on the fitness landscape of a two dimensional sphere function by a simple DE. Sphere is perhaps one of the simplest two dimensional functions and has been chosen to provide easy visual depiction of the search process. The function is given by,

$$f(\vec{X}) = x_1^2 + x_2^2$$

As can be easily perceived, the function has only one global minima $f^* = 0$ at $\vec{X}^* = [0, 0]^T$. We start with a randomly initialized population of five vectors in the search range [-10, 10]. Initially, these vectors are given by:

$$\vec{X}_1(0) = [5, -9]^T$$
$$\vec{X}_2(0) = [6, 1]^T$$
$$\vec{X}_3(0) = [-3, 5]^T$$
$$\vec{X}_4(0) = [-7, 4]^T$$
$$\vec{X}_5(0) = [6, 7]^T$$

Figure 2.7 illustrates the initial orientation of the search variable vectors in the two dimensional $x_1 - x_2$ space. The concentric circular lines are the *constant cost contours* of the function i.e. locus in the $x_1 - x_2$ plane along which $f(\vec{X}) = x_1^2 + x_2^2 =$ constant. Now following the mutation and recombination schemes as presented in expressions (2.4) and (2.6), we form five donor vectors and then create five offspring vectors for time $t = 1$. Next we apply the selection method described by (2.7) and evolve the entire population at time $t = 1$. These steps are summarized in Table 2.2.

Table 2.2 Evolution of the population from $t = 0$ to $t = 1$ in example 2.6

Population at t = 0	Fitness at t = 0	Donor vector at t =0	Offspring Vector at t =0	Fitness of offspring at t = 1	Evolved population at t = 1
$\vec{X}_1(0) =$ [2,-1]	5	$\vec{X}_1(0)$ =[-0.4,10.4]	$\vec{U}_1(0)$ =[-0.4,-1]	1.16	$\vec{X}_1(1)$ =[-0.4,-1]
$\vec{X}_2(0) =$ [6, 1]	37	$\vec{V}_2(0)$ =[1.2, -0.2]	$\vec{U}_2(0)$ =[1.2, 1]	2.44	$\vec{X}_2(1)$ =[1.2, 1]
$\vec{X}_3(0) =$ [-3, 5]	34	$\vec{V}_3(0)$ =[-4.4, -0.2]	$\vec{U}_3(0)$ =[-4.4, -0.2]	19.4	$\vec{X}_3(1)$ =[-4.4, -0.2]
$\vec{X}_4(0) =$ [-2, 6]	40	$\vec{V}_4(0)$ =[9.2, -4.2]	$\vec{U}_4(0)$ =[9.2, 6]	120.64	$\vec{X}_4(1)$ =[-2, 6]
$\vec{X}_5(0) =$ [6, 7]	85	$\vec{V}_5(0)$ =[5.2, 0.2]	$\vec{U}_5(0)$ =[6, 0.2]	36.04	$\vec{X}_5(1)$ =[6, 0.2]

In Figures 2.7 to 2.10, we have shown the orientation of the gradually evolving search variable vectors or genomes in two-dimensional variable space. The * marks in each figure indicates the tips of the search variable vectors. It can be easily observed that all the five vectors tend to converge towards the global minima at $t = 6$ i.e. only within 6 iterations of the DE algorithm. The best solution vector found at $t = 6$ is [0.1, -0.05] and it yields a cost function value $f(0.1,-0.05) = (0.1)^2 + (-0.05)^2 = 0.0125$, which is very close to the global minimum $f^* = 0$.

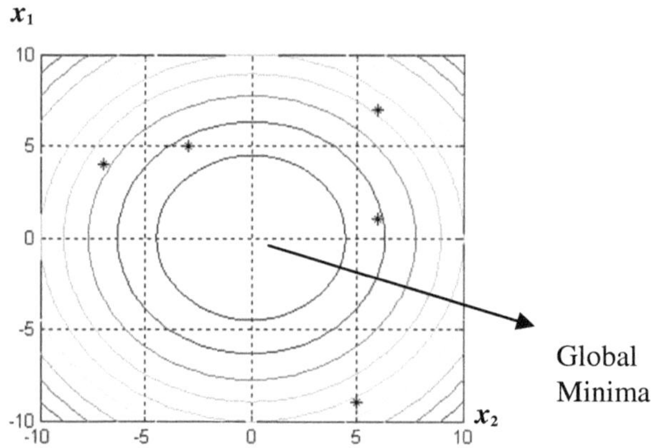

Fig. 2.7 Orientation of the initial solutions in the two dimensional search spaces. Each * mark denotes the tip of a search variable vector.

Fig. 2.8 Orientation of the solutions at time $t = 1$

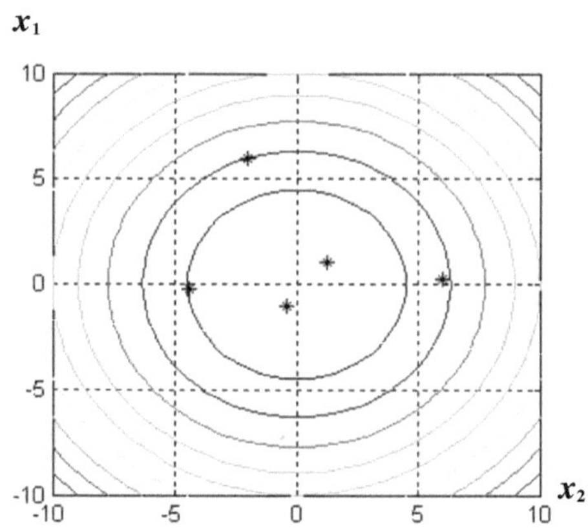

Fig. 2.9 Solutions at time $t = 4$

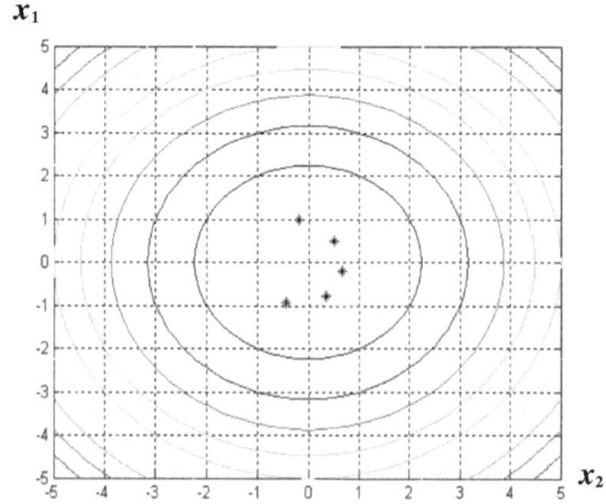

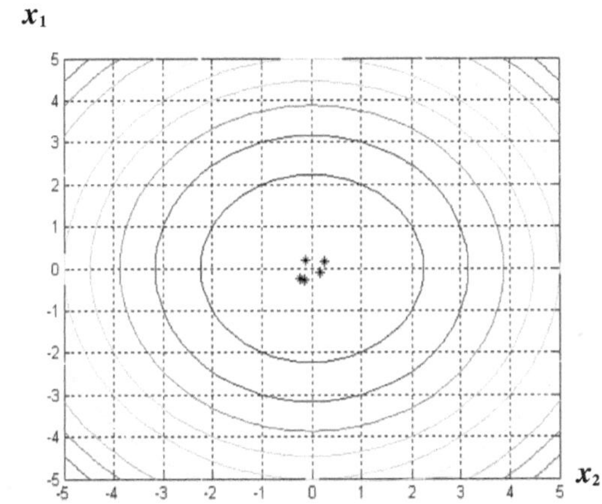

Fig. 2.10 Trial Solutions at time $t = 6$

2.3 The Complete Differential Evolution Algorithm Family of Storn and Price

Actually it is the process of mutation which demarcates one DE scheme from another. In the previous section, we have illustrated the basic steps of a simple DE. The mutation scheme used in equation (2.3) uses a randomly selected vector $\vec{X}_{r1}$ and only one weighted difference vector $F.(\vec{X}_{r2} - \vec{X}_{r3})$ is used to perturb it. Hence, in literature, the particular mutation scheme given by equation (2.3) is referred to as DE/rand/1. We can now have an idea of how the different DE schemes are named. The general convention is DE/x/y. DE stands for Differential Evolution, x represents a string denoting the type of the vector to be perturbed (whether it is randomly selected or it is the best vector in the population with respect to fitness) and y is the number of difference vectors considered for perturbation of x. The other four different mutation schemes, suggested by Storn and Price [13, 14] are summarized below.

1) Scheme DE/target-to-best/1

DE/target-to-best/1 (also known as DE/current-to-best/1 in some earlier papers on DE [14]) follows the same procedure as that of the simple DE scheme illustrated earlier, the only difference being that, now the donor vector, used to perturb each population member, is created using any two randomly selected member of the population as well as the best vector of the current generation (i.e. the vector yielding best suited objective function value at $t = t$). This can be expressed for the i-th donor vector at time $t = t+1$ as,

$$\vec{V}_i(t) = \vec{X}_i(t) + F.(\vec{X}_{best}(t) - \vec{X}_i(t)) + F.(\vec{X}_{r_1^i}(t) - \vec{X}_{r_2^i}(t)), \quad (2.7)$$

where $\vec{X}_i(t)$ is the target vector and $\vec{X}_{best}(t)$ is the best member of the population regarding fitness at current time step $t = t$.

2) Scheme DE/ best/1

In this scheme everything is identical to DE/rand/1 except the fact that the trial vector is formed as,

$$\vec{V}_i(t) = \vec{X}_{best}(t) + F.(\vec{X}_{r_1^i}(t) - \vec{X}_{r_2^i}(t)). \tag{2.8}$$

here the vector to be perturbed is the best vector of the current population and the perturbation is caused by using a single difference vector.

3) Scheme DE/ best/2

Under this method, the donor vector is formed by using two difference vectors as shown below,

$$\vec{V}_i(t) = \vec{X}_{best}(t) + F.(\vec{X}_{r_1^i}(t) - \vec{X}_{r_2^i}(t)) + F.(\vec{X}_{r_3^i}(t) - \vec{X}_{r_4^i}(t)). \tag{2.9}$$

Owing to the central limit theorem the random variations in the parameter vector seems to shift slightly into the Gaussian direction which seems to be beneficial for many functions.

4) Scheme DE/ rand/2

Here the vector to be perturbed is selected randomly and two weighted difference vectors are added to the same to produce the donor vector. Thus for each target vector, a totality of five other distinct vectors are selected from the rest of the population. The process can be expressed in the form of an equation as,

$$\vec{V}_i(t) = \vec{X}_{r_1^i}(t) + F_1.(\vec{X}_{r_2^i}(t) - \vec{X}_{r_3^i}(t)) + F_2.(\vec{X}_{r_4^i}(t) - \vec{X}_{r_5^i}(t)). \tag{2.10}$$

Here F_1 and F_2 are two weighing factors selected in the range from 0 to 1. To reduce the number of parameters we may choose $F_1 = F_2 = F$. We would like to mention here that The indices $r_1^i, r_2^i, r_3^i, r_4^i$, and r_5^i are mutually exclusive integers randomly chosen from the range [1, *NP*], which are also different from the index *i*. These indices are randomly generated once for each mutant vector.

Storn and Price [13] suggested total ten different working strategies of DE and some guidelines in applying these strategies to any given problem. These strategies were derived from the five different DE mutation schemes outlined above. Each mutation strategy was combined with either the 'exponential' type crossover or the 'binomial' type crossover. This yielded a total of 5× 2 = 10 DE strategies, which are listed below.

1. DE/best/1/exp
2. DE/rand/1/exp
3. DE/rand-to-best/1/exp
4. DE/best/2/exp
5. DE/rand/2/exp
6. DE/best/1/bin
7. DE/rand/1/bin
8. DE/rand-to-best/1/bin
9. DE/best/2/bin
10. DE/rand/2/bin

The general convention used above is again DE/x/y/z, where DE stands for Differential Evolution, x represents a string denoting the vector to be perturbed, y is the number of difference vectors considered for perturbation of x, and z stands for the type of crossover being used (exp: exponential; bin: binomial).

2.4 Control Parameters of the Differential Evolution

There are three main control parameters of the DE algorithm: the mutation scale factor *F*, the crossover constant *CR*, the population size *NP*. In this Section, we focus on the effect of each of these parameters on the performance of the DE as well as the state-of-the-art methods for tuning these parameters.

A good volume of research work has been undertaken so far to improve the ultimate performance of the DE by tuning its control parameters. Storn and Price in [6] have indicated that a reasonable value for *NP* could be chosen between 5*D* and 10*D* (*D* being the dimensionality of the problem), and a good initial choice of *F* was 0.5. The effective value range of *F* is usually between 0.4 and 1.

Gamperle *et al.* [15] evaluated different parameter settings for DE on the Sphere, Rosenbrock's, and Rastrigin's functions (please see the appendix for an introduction to the benchmark functions used to test DE and its variants). Their experimental results revealed that the global optimum searching capability and the convergence speed are very sensitive to the choice of control parameters *NP, F* and *CR*. Furthermore, a plausible choice of the population size *NP* is between 3*D* and 8*D*, the scaling factor $F = 0.6$ and the crossover rate *CR* is between [0.3, 0.9]. Recently, the authors in [16] claim that typically $0.4 < F < 0.95$ with $F = 0.9$ being a good first choice. *CR* always lies in (0, 0.2) when the function is separable, while in (0.9, 1) when the function's parameters are dependent.

As can be perceived from the literature, several claims and counter-claims were reported concerning the rules for choosing the control parameters, confusing the engineers who try to solve real-world optimization problems with the DE. Further, most of these claims lack sufficient experimental justifications. Therefore researchers naturally consider some techniques such as self-adaptation to avoid manual tuning of the parameters of DE. Usually self-adaptation is applied to tune the control parameters *F* and *CR*. Liu and Lampinen [17] introduced Fuzzy Adaptive Differential Evolution (FADE) using fuzzy logic controllers whose inputs incorporate the relative function values and individuals of successive generations to adapt the parameters for the mutation and crossover operation. Based on the experimental results over a set of benchmark functions, the FADE algorithm outperformed the conventional DE. In this context, Qin and Suganthan [18] proposed a proposed a Self-adaptive DE (SaDE) algorithm, in which both the trial vector generation strategies and their associated parameters are gradually self-adapted by learning their previous experiences of generating promising solutions.

Zaharie [19] proposed a parameter adaptation strategy for DE (ADE) based on the idea of controlling the population diversity, and implemented a multi-population approach. Following the same line of thinking, Zaharie and Petcu [20] designed an adaptive Pareto DE algorithm for multi-objective optimization and also analyzed its parallel implementation. Abbass [21] self-adapted the crossover rate *CR* for multi-objective optimization problems, by encoding the value of *CR* into each individual, simultaneously evolved with other search variables. The scaling factor *F* was generated for each variable from a Gaussian distribution *N* (0, 1).

The upper limit of the scale factor *F* is empirically taken as 1. Although that does not necessarily mean that a solution is not possible with $F > 1$, however, until date, no benchmark function that was successfully optimized with DE required $F > 1$. Following a methodology proposed by H. –G. Beyer [22], Zaharie [23] derived a lower limit of *F* and the study [23] revealed that if *F* be sufficiently small, the population can converge even in the absence of selection pressure. With a few simplifying assumptions, Zaharie proved the following relation between the expected variance of the original population $P_{x,t}$ at time step $t = t$ and the variance of the trial population $P_{u,t}$:

$$E(Var(P_{u,t})) = \left(2.F^2.p_{Cr} - \frac{2.p_{Cr}}{NP} + \frac{p_{Cr}^2}{NP} + 1 \right).\mathrm{var}(P_{x,t}), \tag{2.11}$$

where, p_{Cr} is the probability of crossover (Zaharie neglected the j_{rand} part in equation (2.5) and therefore p_{Cr} is the absolute probability that a component of the target vector is exchanged with that of the donor vector). Consequently, the DE control parameter combinations that satisfy the equation:

$$2.F^2 - \frac{2}{NP} + \frac{p_{Cr}}{NP} = 0 \tag{2.12}$$

may be considered as critical since they result in a population whose variance remains constant except for random fluctuations. Thus when the selection step is absent, according to equation (2.12), *F* will display a critical value F_{crit} such that the population variance decreases when $F < F_{crit}$ and increases if $F > F_{crit}$. Solving equation (2.13) clearly

$$F_{crit} = \sqrt{\frac{\left(1 - {p_{Cr}}/{2}\right)}{NP}} \tag{2.13}$$

Zaharie experimentally confirmed that F_{crit} establishes a lower limit on the value of *F* in the sense that smaller values will induce convergence even on a level of objective function landscape.

Omran *et al.* [24] introduced a self-adaptive scaling factor parameter *F*. They generated the value of *CR* for each individual from a normal distribution *N* (0.5, 0.15). This approach (called 'SDE') was tested on four benchmark functions and performed better than other versions of DE. Besides adapting the control parameters *F* or *CR*, some researcher also adapted the population size. Teo [25] proposed DE with Self Adapting Populations (DESAP), based on self-adaptive Pareto DE proposed by Abbass [21]. Das *et al.* [26] introduced two schemes for the adaptation of scale factor F in DE. In the first scheme (called DERANDSF: DE with Random Scale Factor), F was made to vary randomly between 0.5 and 1in successive iterations. They also suggested decreasing F linearly from 1.0 to 0.5 in their second scheme (called DETVSF: DE with Time varying Scale Factor). This encourages the individuals to sample diverse zones of the search space during the early stages of the search. During the later stages a decaying scale factor helps to adjust the movements of trial solutions finely so that they can explore the interior of a relatively small space in which the suspected global optimum lies.

2.5 Important Variants of the Differential Evolution Algorithm

Since its advent in 1995, DE has been attracting the attention of the researchers from diverse domains of knowledge, all over the world. This has resulted into the wealth of variants of the basic DE algorithm, some of which are application specific while some are generalized for global numerical optimization. Without forgetting the No Free Lunch (NFL) theorem [27], one may note that each of these modified DE algorithms has its own advantages and disadvantages and none is suitable for tackling all kinds of optimization problems that appear in real world. The interested readers are redirected to appropriate references for details of the discussed algorithms wherever necessary.

2.5.1 Differential Evolution Using Trigonometric Mutation

Recently, Fan and Lampinen [28] proposed a trigonometric mutation operator for DE to speed up its performance. To implement the scheme, for each target vector, three distinct vectors are randomly selected from the DE population. Suppose for the i-th target vector $\vec{X}_i(t)$, the selected population members are $\vec{X}_{r1}(t)$, $\vec{X}_{r2}(t)$ and $\vec{X}_{r3}(t)$. The indices r_1, r_2 and r_3 are mutually exclusive integers randomly chosen from the range [1, NP], which are also different from the index i. Now three weighing coefficients are formed according to the following equations:

$$p' = \left|f(\vec{X}_{r1})\right| + \left|f(\vec{X}_{r2})\right| + \left|f(\vec{X}_{r3})\right| \tag{2.14}$$

$$p_1 = \left|f(\vec{X}_{r1})\right| \Big/ p' \tag{2.15}$$

$$p_2 = \left|f(\vec{X}_{r2})\right| \Big/ p' \tag{2.16}$$

$$p_3 = \left|f(\vec{X}_{r3})\right| \Big/ p' \tag{2.17}$$

where $f()$ is the function to be minimized. Let rand (0, 1) be a uniformly distributed random number in (0, 1) and r be the trigonometric mutation rate in the same interval (0, 1). The trigonometric mutation scheme may now be expressed as,

$$\vec{V}_i(t+1) = (\vec{X}_{r1} + \vec{X}_{r2} + \vec{X}_{r3})/3 + (p_2 - p_1).(\vec{X}_{r1} - \vec{X}_{r2}) + (p_3 - p_2).(\vec{X}_{r2} - \vec{X}_{r3}) + (p_1 - p_3).(\vec{X}_{r3} - \vec{X}_{r1}) \quad \text{if } rand(0,1) < \mathrm{r}$$

$$\vec{V}_i(t+1) = \vec{X}_{r1} + F.(\vec{X}_{r2} - \vec{X}_{r3}) \quad \text{else.} \tag{2.18}$$

Thus we find that the scheme proposed by Fan *et al.* used trigonometric mutation with a probability of r and the mutation scheme of DE/rand/1 with a probability of (1- r).

2.5.2 Differential Evolution Using Arithmetic Recombination

The binomial crossover scheme usually employed in most of the DE variants creates new combinations of parameters; it leaves the parameter values themselves unchanged. As mentioned earlier, binomial crossover is in spirit same as the discrete recombination used in conjunction with many EAs. However, in *continuous* or *arithmetic* recombination, the individual components of the trial vector are expressed as a linear combination of the components from mutant/donor vector and the target vector. The common form of the arithmetic recombination between two vectors $\vec{X}_{r1}(t)$ and $\vec{X}_{r2}(t)$ adopted by most of the EAs [29, 30] may be put as:

$$\vec{W}_i(t) = \vec{X}_{r1}(t) + k_i.(\vec{X}_{r1}(t) - \vec{X}_{r2}(t)) \tag{2.19}$$

The *coefficient of combination* k_i can either be a constant or a random variable (e.g. rand (0, 1)) [30]. Generally speaking, if this coefficient is sampled anew for each vector then the resulting process is known as line recombination. However, if the combination coefficient is elected randomly anew for each component of the vectors to be crossed, then the process is known as intermediate recombination and may be formalized for the *j*-th component of the recombinants as:

$$w_{i,j}(t) = x_{r1,j}(t) + k_j.(\vec{X}_{r1,j}(t) - \vec{X}_{r2,j}(t)) \tag{2.20}$$

In Figure 2.11, we have schematically shown the regions searched by discrete, line and arithmetic recombination between donor vector $\vec{V}_i(t)$ and the target vector $\vec{X}_i(t)$ when the coefficient of combination is a uniformly distributed random number between 0 and 1. The two recombinant vectors occupy the opposite corners of a hypercube whose remaining corners are the trial vectors

$\vec{U}_i^{/}(t)$ and $\vec{U}_i^{//}(t)$ created by discrete recombination. Line recombination, as its name suggests, searches along the axis connecting the recombinant vectors, while the intermediate recombination explores the entire D-dimensional volume contained within the hypercube.

As can be perceived from Figure 2.12, both the discrete as well as intermediate recombination is not a rotationally invariant process. If the coordinate system rotates through an angle, the corners of the hypercube are relocated, which in turn redefines the area searched by the intermediate recombination. On the other hand, the line recombination is rotationally invariant.

Since a good global optimizer needs to be rotationally invariant, to overcome the limitation of the discrete recombination usually employed in DE, a new trial vector generation strategy 'DE/current-to-rand/1' is proposed in [14], which replaces the crossover operator prescribed in equation (2.6) with the rotationally invariant arithmetic line recombination operator to generate the trial vector $\vec{U}_i(t)$ by linearly combining the target vector $\vec{X}_i(t)$ and the corresponding donor vector $\vec{V}_i(t)$ as follows:

$$\vec{U}_i(t) = \vec{X}_i(t) + k_i.(\vec{V}_i(t) - \vec{X}_i(t)) \tag{2.21}$$

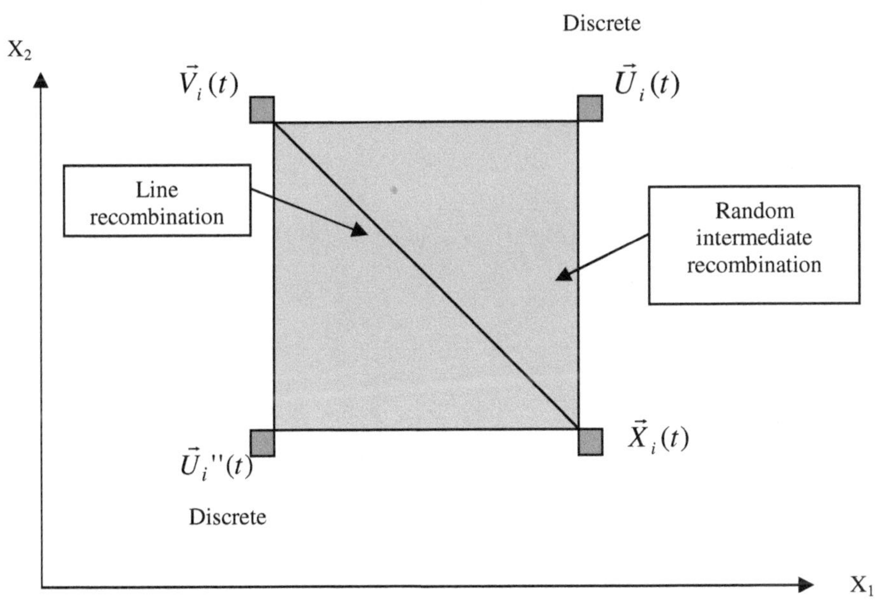

Fig. 2.11 Domains of the different recombinant vectors generated using discrete, line and random intermediate recombination

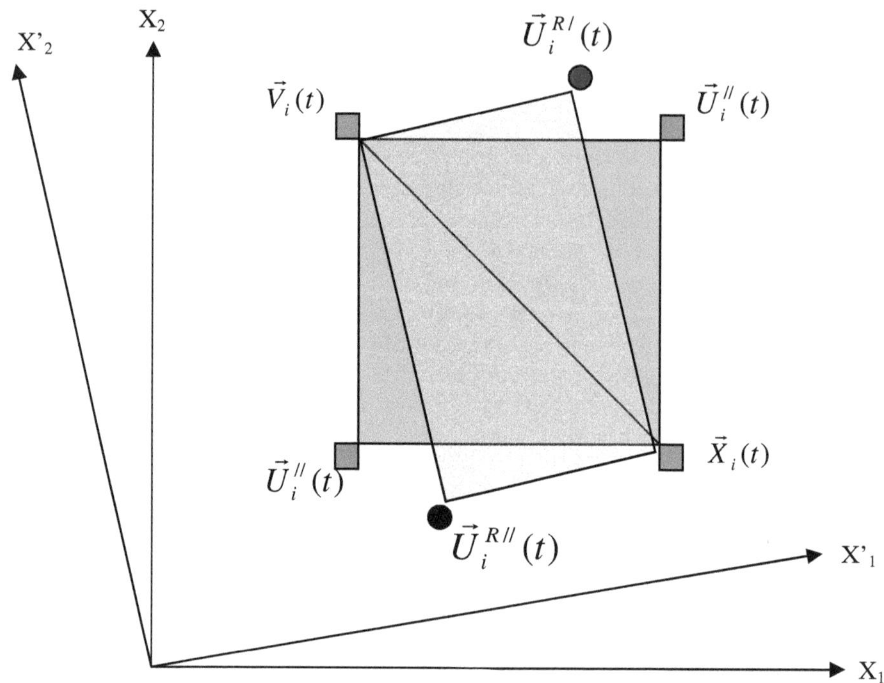

Fig. 2.12 Change of the trial vectors generated through the discrete and random intermediate recombination due to rotation of the coordinate system. $\vec{U}_i^{R/}(t)$ and $\vec{U}_i^{R//}(t)$ indicate the new trial vectors due to discrete recombination in rotated coordinate system.

Now incorporating equation (2.3) in (2.21) we have:

$$\vec{U}_i(t) = \vec{X}_i(t) + k_i.(\vec{X}_{r_1}(t) + F.(\vec{X}_{r_2}(t) - \vec{X}_{r_3}(t)) - \vec{X}_i(t)), \tag{2.22}$$

which further simplifies to:

$$\vec{U}_i(t) = \vec{X}_i(t) + k_i.(\vec{X}_{r_1}(t) - \vec{X}_i(t)) + F'.(\vec{X}_{r_2}(t) - \vec{X}_{r_3}(t)), \tag{2.23}$$

where k_i is the combination coefficient, which has been proven [18] to be effective when it is chosen with a uniform random distribution from [0, 1] and $F' = k_i.F$ is a new constant here.

2.5.3 *Self Adaptive Differential Evolution*

Selecting an optimal set of control parameter values is, very often, a problem-specific task for DE. The trial and error method used for tuning the control parameters is time consuming, less reliable and requires multiple runs on the given problem. To circumvent this problem, recently in 2006, Brest *et al.* proposed a self adaptation scheme for the DE control parameters [31]. They encoded control parameters *F* and *CR* into the individual and adjusted introducing two new parameters τ_1 and τ_2. The self adaptive encoding of the control parameters is depicted in Figure 2.13.

Fig. 2.13 Self adaptive encoding scheme

$\vec{X}_1(t)$	$F_1(t)$	$CR_1(t)$
$\vec{X}_1(t)$	$F_2(t)$	$CR_2(t)$
………	…….	……
$\vec{X}_{NP}(t)$	$F_{NP}(t)$	$CR_{NP}(t)$

In their algorithm (called 'SADE'), a set of *F* and *CR* values was assigned to each individual in the population, augmenting the dimensions of each vector. The better values of these encoded control parameters lead to better individuals that in turn, are more likely to survive and produce offspring and, thus, propagate these better parameter values. The new control parameters for the next generation are computed as follows:

$$\left.\begin{aligned} F_i(t+1) &= F_l + rand_1 * F_u && \text{if } rand_2 < \tau_1 \\ &= F_i(t) && \text{else.} \end{aligned}\right\} \tag{2.24}$$

$$\left.\begin{aligned} CR_i(t+1) &= rand_3 && \text{if } rand_4 < \tau_2 \\ &= CR_i(t) && \text{else.} \end{aligned}\right\} \tag{2.25}$$

where $rand_j, j \in \{1,2,3,4\}$ are uniform random values in the interval [0, 1]. τ_1 and τ_2 may be interpreted as the probabilities of adjusting the factors *F* and *CR*. In [34], Brest *et al.* have used $\tau_1 = \tau_2 = 0.1$. As $F_l = 0.1$ and $F_u = 0.9$, the new *F* takes a value from [0.1, 0.9] while the new *CR* takes a value from [0, 1]. As the $F_i(t+1)$ and $CR_i(t+1)$ values are obtained before the mutation is performed, they influence the mutation, crossover, and selection operations of the new vector $\vec{X}_i(t+1)$. Since the rule for the adaptation of control parameters *F* and *CR* remained very simple, the new DE variant does not increase the time complexity, in comparison to the original DE algorithm. The SADE was found to perform well with respect to several other DE variants as well as some standard evolutionary algorithms in a statistically significant manner, over a test-suit of 21 well known numerical benchmark functions [31].

2.5.4 The DE/rand/1/Either-Or Algorithm

DE/rand/1/either-or is a state-of-the-art DE variant described by Price *et al.* [13, page 118]. In this algorithm, the trial vectors that are pure mutants occur with a probability p_F and those that are pure recombinants occur with a probability $1 - p_F$. The scheme for trial vector generation may be outlined as:

$$\left.\begin{aligned} U_i(t) &= \vec{X}_{r1}(t) + F.(\vec{X}_{r2}(t) - \vec{X}_{r3}(t)) \quad \text{if } rand_i(0,1) < p_F \\ &= \vec{X}_{r0,G} + k.(\vec{X}_{r1} + \vec{X}_{r2} - 2.\vec{X}_{r0}) \quad \text{Otherwise} \end{aligned}\right\} \tag{2.26}$$

Price *et al.* recommended $k = 0.5.(F+1)$ as a good choice for the parameter *K* for a given *F*. The DE/rand/1/either-or algorithm provides a simple way to implement the dual axis search in the *k*-*F* plane (*k* indicating the combination coefficient of the arithmetic crossover and *F* being the scale factor). The scheme provides efficient solutions for functions that are best minimized by either mutation-only ($p_F = 1$) or recombination only ($p_F = 0$), as well as generic functions that can be solved by randomly interleaving both the operations ($0 < p_F < 1$).

2.5.5 The Opposition-Based Differential Evolution

The concept of *opposition-based learning* was introduced by Tizhoosh [32] and its applications were introduced in [32-34]. Rahnamayan *et al.* [35] have recently proposed an *Opposition-based DE* (ODE) for faster global search and optimization. The algorithm also finds important applications to the noisy optimization problems [36]. The conventional DE was enhanced by utilizing *opposition number* based optimization concept in three levels, namely, population initialization, generation jumping, and local improvement of the population's best member. Before elaborating on the ODE algorithm, we need to define the *opposite numbers* [32].

Definition 2.1: Let *x* be a real number defined in the closed interval $[a,b]$, i. e. $x \in [a,b]$. Then the opposite number $\breve{x}$ of *x* may be defined as:

$$\breve{x} = a + b - x \tag{2.27}$$

The definition may be extended to higher dimensions in the following way.

Definition 2.2: Let $P(x_1, x_2, ..., x_D)$ be a point in *D*-dimensional hyperspace where $\vec{X} = [x_1, x_2, ..., x_D] \in \Re^D$ and $x_i \in [a_i, b_i]\ \forall i \in \{1,2,...,D\}$. Then the opposite point of *P* may be defined as $P(\breve{x}_1, \breve{x}_2, ..., \breve{x}_D)$ such that

$$\breve{x}_i = a_i + b_i - x_i \tag{2.28}$$

The ODE changes the classical DE using the concept of opposite numbers at three different stages. Below we briefly outline these changes.

1) Opposition Based Population Initialization

In most of the existing versions of the DE, random number generation is the only choice to create initial population. But as demonstrated by Tizhoosh [32], concept of opposition-based optimization can help in obtaining fitter starting points in the search space even when there is no a priori knowledge about solutions. Rahnamayan *et al.*[35] proposed the following scheme:

(a) Generating uniformly distributed random population $P(NP)$, *NP* is the population size

(b) Calculate the opposite population $OP(NP)$. The *k*-th opposite individual corresponding to the *k*-th parameter vector of $P(NP)$ may be given as (following equation (2.27),

$OP_{k,j} = a_{k,j} + b_{k,j} - P_{k,j}$, where $k = 1,2,....,NP$ and $j = 1,2,....,D$

$a_{k,j}$ and $b_{k,j}$ denote the interval boundaries of *j*-th parameter of the *k*-th vector i. e. $x_{k,j} \in [a_{k,j}, b_{k,j}]$.

(c) Select *NP* fittest individuals from set the $\{P(NP), OP(NP)\}$ as initial population.

2) Opposition Based Generation Jumping

Generation jumping means after each iteration, depending upon a predetermined probability *Jr*, instead of generating new population by evolutionary process, the opposite population may be calculated (if $rand(0,1) < Jr$) and the *NP* fittest individuals are selected from the current population and the corresponding opposite population (exactly similar to what was performed for opposition-based population initialization) depending on some. The authors suggest taking a small value of *Jr* such that approximately $Jr \in (0,0.04)$.

3) Opposition Based Best Individual Jumping

This stage improves the best candidate solution (the fittest member) in the current population by applying small perturbations and through the following steps:

(a) Create difference-offspring of the best individual in the current population by:

$$new\_best = best + F'.(\vec{X}_{r1} - \vec{X}_{r2}) \tag{2.29}$$

where *r1* and *r2* are mutually different random integer indices selected from {1, 2, ..., n}. *F'* is a real constant which determines amplification of the added differential vector. *F'* should be set to a small number $F' \in [0,0.02]$ because we need a small/local exploration to improve the current best member. In contrast, a large value for F' can reduce the chance to obtain a better candidate.

(b) Calculate opposite of offspring created in (a) using equation (2.29), call *op_newbest*,

(c) Replace the current best member by the fittest member of the set $\{$*best, newbest, op_newbest*$\}$.

It should be mentioned that to calculate the opposite individuals for generation jumping and also for the best individual jumping (step (3)), the opposite of each variable is calculated dynamically. It means, the maximum and minimum values

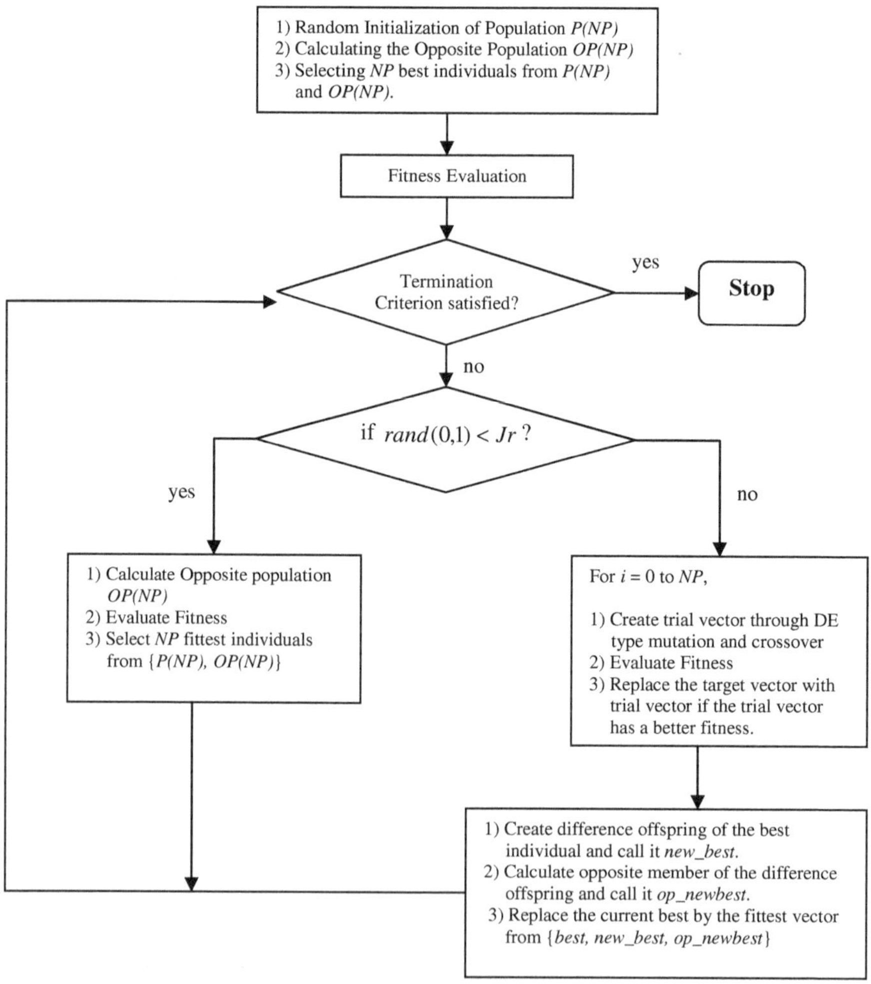

Fig. 2.14 Pseudo-code of the ODE Algorithm

of each variable in *current population* ($[a_j^p, b_j^p]$) are used to calculate opposite point instead of using variables' predefined interval boundaries i.e.

$$OP_{k,j} = a_{k,j}^p + b_{k,j}^p - P_{k,j}, \text{ where } k = 1,2,....,NP \text{ and } j = 1,2,....,D$$

This dynamic behavior of the opposite point calculation increases the chance to find fitter opposite points. A flow chart representation of the ODE algorithm has been provided in Figure 2.14. The experiments conducted in [35] confirmed that the proposed ODE algorithm performs better than the DE in terms of convergence speed over noisy and noise-free functions.

2.5.6 The Binary Differential Evolution Algorithm

Recently Pampara *et al.* [37] proposed a new DE variant which can operate in binary problem spaces without deviating from the basic search mechanism of the classical DE. The algorithm was named by its authors as the *Angle Modulated DE (AMDE)* as it employs a trigonometric function as a bit string generator. The angle modulation scheme draws inspiration from the domain of telecommunication engineering [38]. The trigonometric generating function used in the angle modulation function is a composite sinusoidal function which may be given as:

$$g(x) = \sin(2\pi(x-a)\times b\times \cos A)) + d\,, \tag{2.30}$$

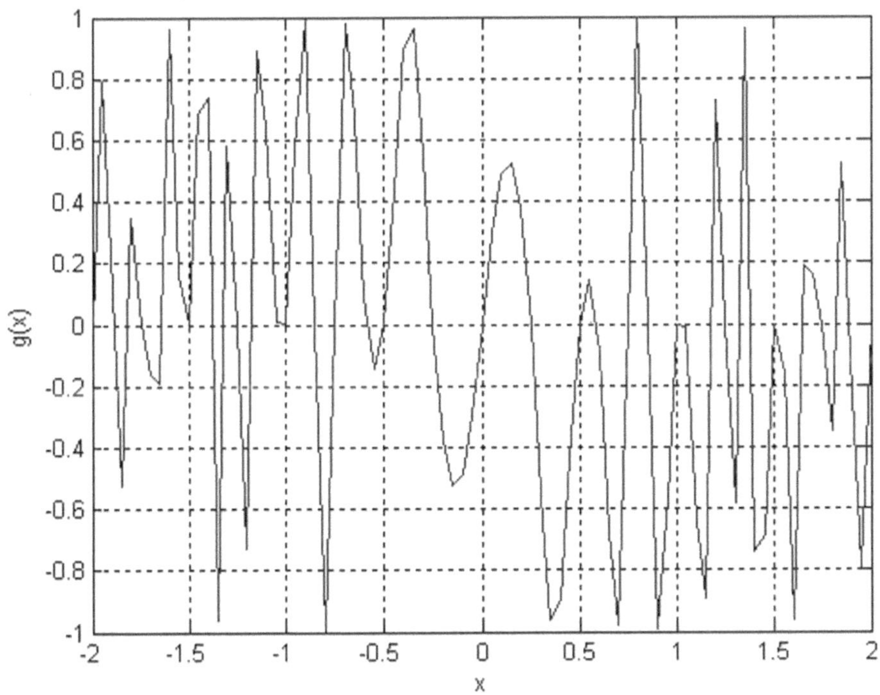

Fig. 2.15 Bit string generating function (a =0, b = 1, c = 1, and d = 0)

where $A = 2\pi \times c \times (x - a)$ and x is a single element from a set of evenly separated intervals determined by the required number of bits that need to be generated. For example, a selection of 10 evenly spaced points between 0 and 10 is needed in order to generate 10 bit values. Figure 2.15 graphically represents the bit string generating function with default values a =0, b = 1, c = 1 and d = 0.

The optimization algorithm is then applied to evolve a 4-dimensional tuple (a, b, c, d). This tuple represents the coefficient values for equation (2.30). The optimization process evolves the tuple parameters instead of actually evolving the bit string. After an iteration of the optimization algorithm, the tuple parameters are substituted back into equation (2.30). The resulting generating function is sampled at evenly spaced intervals and for each interval a bit value is recorded. The generated bit vector represents the potential solution within the binary problem space of the original problem. The generation of the bit string vector values is a simple procedure. The sample points are fed into the generating function and the output of equation (2.30) is evaluated. If the resulting output value at the sample point is positive, a bit value of 1 is recorded, otherwise a bit value of 0 is recorded. The pseudo code of the AMDE algorithm has been provided below:

Angle Modulated Differential Evolution Algorithm

Initialize a population and set control parameter values
repeat
Select required individuals for reproduction scheme
Produce an offspring individual
Evaluate the fitness of the offspring individual by generating bit string and passing to fitness function
if $f_{AMDE}(offspring) \le f_{AMDE}(parent)$ **then**
Replace parent individual with offspring
else
Retain parent individual
end if
until Stopping condition is met

where $f_{AMDE}()$ is the fitness function of the AMDE algorithm. The MADE algorithm was found to outperform the binary PSO (binPSO) algorithm by the authors.

2.5.7 *Differential Evolution with Adaptive Local Search*

Noman and Iba [39] proposed a crossover based adaptive Local Search (LS) operation to improve the performance of the classical DE. Typically in LS, every candidate solution has more than one neighbour solution; the choice of which one to move to is taken using only information about the solutions in the neighbourhood of the current one, hence the name local search. If the choice of the neighbouring solution is done by taking the one locally maximizing the criterion, the metaheuristic takes the name *hill-climbing*. The authors in [39] propose an LS, whose length of the search can be adjusted adaptively using a hill-climbing heuristic.

The incorporation of a crossover-based local search (XLS) with adaptive length (adaptive length XLS, shortened as AHCXLS) in DE resulted into a DE-variant called by the authors: DEahcSPX, where SPX is the simplex-based crossover scheme proposed by Tsutsui *et al.* for real-coded GAs [40]. Below we provide the pseudo-codes for AHXLS, DEahcSPX and the SPX algorithms.

AHCXLS (*I*, n_p)

Step 1: P[1] = *I* ;
Step 2: Repeat from $i = 2$ to n_p times
Step 3: P[i] = select random individuals from the population.
Step 4: End Repeat
Step 5: **C = *Crossover***(P)
Step 6: If C is better than P[1],
P[1] = C;
Step 7: Else
Return (P[1]).
Step 8: Go to Step 5.

DEahcSPX

Step 1: Generate an Initial Population ***P(t)***
Step 2: ***Evaluate*** the fitness of ***P(t)***
Step 3: *B* = ***Best_Index*** (***P(t)***)
Step 4: ***P(t)***.[*B*] = AHCXLS (***P(t)***.[*B*], n_p)
Step 5: for each individual *I* in ***P(t)***
Reproduce an offspring *J* from *I*.
Step 6: $\boldsymbol{P(t+1)} = \boldsymbol{P(t+1)} \cup Select(I, J)$
Step 7: Ste $t = t+1$
Step 8: Repeat Steps 3 to 8 until termination criteria is met.

SPX

Step 1: Choose n_p parents $\vec{X}_i(t)$, $i = 1,2,....,n_p$ according to the generational model used and calculate their center of mass $\vec{O}$ as:

$$\vec{O} = \frac{1}{n_p} \sum_{i=1}^{n_p} \vec{X}_i(t) . \tag{2.31}$$

Step 2: Generate random number $\vec{r}_i$ as: $\vec{r}_i = u^{\frac{1}{i+1}}, i = 1,2,...,n_p - 1$ where u is a uniform random number and $u \in [0,1]$.

Step 3: Calculate $\vec{Y}_i$ and $\vec{C}_i$ as:

$$\vec{Y}_i = \vec{O} + \varepsilon.(\vec{X}_i(t) - \vec{O}),\ i = 1,2,....n_p \tag{2.32}$$

$$\vec{C}_i = \begin{cases} 0 & \text{for } i = 1, \\ r_{i-1}(\vec{Y}_{i-1} - \vec{Y}_i + C_{i-1}) & \text{for } i = 2,....,n_p. \end{cases} \tag{2.33}$$

where $\varepsilon = 1.0$ is the expansion rate, a control parameter of SPX.

Step 4: Generate and offspring $\vec{C}$ as:

$$\vec{C} = \vec{Y}_{n_p} + \vec{C}_{n_p} \tag{2.34}$$

The experimental results reported by Noman and Iba [39] indicate that the DEahcSPX could outperform the classical DE (more precisely the DE/rand/1/bin variant) in terms of convergence speed over a set of carefully chosen numerical benchmarks [41, 31]. The overall performance of the adaptive LS scheme was reported better than the other crossover-based LS strategies and the overall performance of the newly proposed DE algorithm was shown to be superior to or at least comparable with some other Memetic Algorithms (MAs) [42] selected from literature. The proposed LS scheme was also found prospective for adaptive DE variants.

2.5.8 *Self-adaptive Differential Evolution (SaDE) with Strategy Adaptation*

In [43], Qin *et al.* have proposed a new self-adaptive variant of DE (SaDE), in which both trial vector generation strategies and their associated control parameter values are gradually self-adapted by learning from their previous experiences in generating promising solutions. Consequently, it is possible to determine a more suitable generation strategy along with its parameter settings can be determined adaptively to match different phases of the search process / evolution.

In SaDE, four effective trial vector generation strategies namely the DE/rand/1/bin, DE/rand-to-best/2/bin, DE/rand/2/bin and finally DE/current-to-rand/1 were chosen to constitute a strategy candidate pool. The first three DE-variants are equipped with binomial type crossover (a most popular kind of recombination in DE-community [13]) which is described in equation (2.5). Also note that the DE/rand/1, DE/rand/2 and the DE/current-to-rand/1 schemes have already been described in equations (2.3), (2.10) and (2.23) respectively. The DE/rand-to-best/2/bin scheme may be illustrated for the j-th component of the i-th target vector as:

$$u_{i,j} = \begin{cases} x_{i,j} + F.(x_{best,j} - x_{i,j}) + F.(x_{r1,j} - x_{r2,j}) + F.(x_{r3,j} - x_{r4,j}) & \text{if } rand[0,1) < CR \text{ or } j = j_{rand} \\ x_{j,i} & \text{otherwise,} \end{cases} \tag{2.35}$$

where the symbols bear their usual meanings.

In the SaDE algorithm, for each target vector in the current population, one trial vector generation strategy is selected from the candidate pool according to the probability learned from its success rate in generating improved solutions (that can survive to the next generation) within a certain number of previous generations, called the *Learning Period* (*LP*). The selected strategy is subsequently applied to the corresponding target vector to generate a trial vector. More specifically, at each generation, the probabilities of choosing each strategy in the candidate pool are summed to 1. These probabilities are initially equal (1/*K* for *K* total number of strategies in the pool) and are then gradually adapted during evolution, based on the *Success* and *Failure Memories* [43] over the previous *LP* generations. The adaptation of the probabilities take place in such a fashion that, the larger the success rate for the *k-th* strategy in the pool within the previous *LP* generations, the larger is the probability of applying it to generate the trial vectors at the current generation.

Now, as far as the control parameters of DE are concerned, SaDE leaves *NP* as a user-specified parameter since it highly replies on the complexity of a given problem. The parameter *F*, in SaDE, is approximated by a normal distribution with mean value 0.5 and standard deviation 0.3, denoted by *N* (0.5, 0.3). A set of *F* values are randomly sampled from such normal distribution and applied to each target vector in the current population. This way, SaDE attempts to maintain both exploitation (with small *F* values) and exploration (with large *F* values) power throughout the entire evolution process. Following suggestions in [14], the control parameter *K* in the strategy "DE/current-to-rand/1" is hereby randomly generated within [0, 1] so as to eliminate one additional parameter. SaDE gradually adjusts the range of *CR* values for a given problem according to previous *CR* values that have generated trial vectors successfully entering the next generation. Specifically, it is assumed that *CR* obeys a normal distribution with mean value CR_m and standard deviation *Std* =0.1, denoted by $N(CR_m, Std)$, where CR_m is initialized as 0.5. The *Std* should be set as a small value to guarantee that most *CR* values generated by N (CR_m, *Std*) are between [0, 1], even when CR_m is near 0 or 1. Hence, the value of *Std* is set as 0.1.

The performance of SaDE was compared with the conventional DE and 3 adaptive DE-variants over a suite of 26 bound constrained numerical optimization problems, and the authors reported that, SaDE was more effective in obtaining better quality solutions, which are more stable with the relatively smaller standard deviation, and had higher success rates.

2.5.9 DE with Neighborhood-Based Mutation

2.5.9.1 The DE/target-to-best/1 - A Few Drawbacks

Most of the population-based search algorithms try to balance between two contradictory aspects of their performance: *exploration* and *exploitation*. The first one means the ability of the algorithm to 'explore' or search every region of the feasible search volume, while the second one denotes the ability to converge to the near optimal solutions as quickly as possible. The DE variant, known as DE/target-to-best/1 (equation (2.7)) uses the best vector of the population to generate donor vectors. By 'best' we mean the vector that corresponds to the best fitness (e.g., the lowest objective function value for a minimization problem) in the entire population at a particular generation. The scheme promotes exploitation since all the vectors/genomes are attracted towards the same best position (pointed to by the 'best' vector) on the fitness landscape through iterations, thereby converging faster to that point. But as a result of such exploitative tendency, in many cases, the population may lose its global exploration abilities within a relatively small number of generations, thereafter getting trapped to some locally optimal point on the search space.

In addition, DE employs a greedy selection strategy (the better between the target and the trial vectors is selected) and uses a fixed scale factor F (typically in [0.4, 1]). Thus if the difference vector $\vec{X}_{r_1}(t) - \vec{X}_{r_2}(t)$, used for perturbation is small (this is usually the case when the vectors come very close to each other and the population converges to a small domain), the vectors may not be able to explore any better region of the search space, thereby finding it difficult to escape large plateaus or suboptimal peaks/valleys. Mezura-Montes *et al.,* while comparing the different variants of DE for global optimization in [44], have noted that DE/target-to-best/1 shows a poor performance and remains inefficient in exploring the search space especially for multi-modal functions. The same conclusions were reached at by Price *et al.* in [13, page 156].

2.5.9.2 Motivations for the Neighborhood-Based Mutation

A proper trade–off between exploration and exploitation is necessary for the efficient and effective operation of a population-based stochastic search technique like DE, PSO etc. The DE/target-to-best/1, in its present form, favors exploitation only, since all the vectors are attracted by the same best position found so far by the entire population, thereby, converging faster towards the same point.

In this context Das *et al.* [45] propose two kinds of neighborhood models for DE. The first one is called the *local neighborhood model*, where each vector is mutated using the best position found so far in a small neighborhood of it and *not* in the entire population. On the other hand, the second one, referred as the *global mutation model*, takes into account the globally best vector $\vec{X}_{best}(t)$ of the entire population at current generation $t = t$ for mutating a population member. Note that DE/target-to-best/1 employs only the global mutation strategy.

A vector's neighborhood is the set of other parameter vectors that it is connected to; it considers their experience when updating its position. The graph of inter-connections is called the neighborhood structure. Generally, neighborhood connections are independent of the positions pointed by the vectors. In the *local* model, whenever a parameter vector points to a good region of the search space, it only directly influences its immediate neighbors. Its second degree neighbors will only be influenced after those directly connected to them become highly successful themselves. Thus, there is a delay in the information spread through the population regarding the best position of each neighborhood. Therefore, the attraction to specific points is weaker, which prevents the population from getting trapped in local minima. We would like to mention here that vectors belonging to a *local neighborhood* are not necessarily local in the sense of their geographical nearness or similar fitness values. As will be seen in the next section, the overlapping neighborhoods have been created in DE according to the order of the indices of the population members, following the neighborhood models in PSO.

Finally we combine the local and the global model using a *weight factor* that appears as a new parameter in the algorithm. The weight factor may be tuned in many different ways. In what follows we describe these issues in sufficient details. Please note that the neighborhoods of different vectors were chosen randomly and not according to their fitness values or geographical locations on the fitness landscape, following the PSO philosophy [46]. This preserves the diversity of the vectors belonging to the same neighborhood.

2.5.9.3 The Local and Global Neighborhood-Based Mutations in DE

Suppose we have a DE population $P(t) = [\vec{X}_1(t), \vec{X}_2(t), \ldots, \vec{X}_{NP}(t)]$ where each $\vec{X}_i(t)$ $(i = 1,2,\ldots,NP)$ is a *D*-dimensional parameter vector. The vector indices are sorted only randomly (as obtained during initialization) in order to preserve the diversity of each neighborhood. Now for every vector $\vec{X}_i(t)$ we define a neighborhood of radius *k* (where *k* is a non-zero integer from 0 to $(NP-1)/2$ as the neighborhood size must be smaller than the population size i.e. $2k+1 \le NP$), consisting of vectors $\vec{X}_{i-k}(t), \ldots, \vec{X}_i(t), \ldots, \vec{X}_{i+k}(t)$. We assume the vectors to be organized on a ring topology with respect to their indices, such that vectors $\vec{X}_{NP}(t)$ and $\vec{X}_2(t)$ are the two immediate neighbors of vector $\vec{X}_1(t)$. The concept of local neighborhood is schematically illustrated in Figure 2.16. Please note that the neighborhood topology is static and has been defined on the set of indices of the vectors. Although various neighborhood topologies (like star, wheel, pyramid, 4-clusters, and circular) have been proposed in literature for the PSO algorithms [46], after some initial experimentation over numerical benchmarks, we find that in case of DE (where the population size is usually much larger than in the case of PSO) the circular or ring topology provides best performance as compared to other salient neighborhood structures.

For each member of the population a local donor vector is created by employing the best (fittest) vector in the neighborhood of that member and any two other vectors chosen from the same neighborhood. The model may be expressed as:

$$\vec{L}_i(t) = \vec{X}_i(t) + \alpha \cdot (\vec{X}_{n\_best_i}(t) - \vec{X}_i(t)) + \beta \cdot (\vec{X}_p(t) - \vec{X}_q(t)), \qquad (2.36)$$

where the subscript n_best_i indicates the best vector in the neighborhood of $\vec{X}_i(t)$ and $p, q \in [i-k, i+k]$ with $p \ne q \ne i$. Similarly the global donor vector is created as:

$$\vec{g}_i(t) = \vec{X}_i(t) + \alpha \cdot (\vec{X}_{g\_best}(t) - \vec{X}_i(t)) + \beta \cdot (\vec{X}_{r_1}(t) - \vec{X}_{r_2}(t)), \qquad (2.37)$$

where the subscript *g_best* indicates the best vector in the entire population at iteration $t=t$ and $r_1, r_2 \in [1, NP]$ with $r_1 \neq r_2 \neq i$. α and β are the scaling factors.

Note that in equations (2.36) and (2.37), the first perturbation term on the R.H.S (the one multiplied by α) is an arithmetical recombination operation, while the second term (the one multiplied by β) is the differential mutation. Thus in both the global and local mutation models, we basically generate mutated recombinants and not pure mutants.

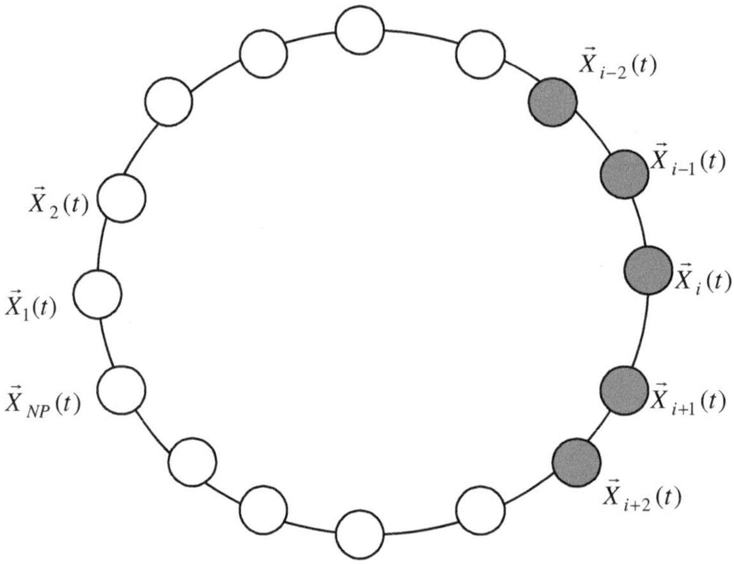

Fig. 2.16 The ring topology of neighborhood in DE. The dark spheres indicate a neighborhood of radius 2 of the *i*-th population member where $i = 9$

Now we combine the local and global donor vectors using a scalar weight $w \in (0,1)$ to form the actual donor vector of the proposed algorithm:

$$\vec{V}_i(t) = w.\vec{g}_i(t) + (1-w).\vec{L}_i(t)\,. \qquad (2.38)$$

Clearly if $w = 1$ and in addition $\alpha = \beta = F$, the donor vector generation scheme in equation (2.38) reduces to that of DE/target-to-best/1. Hence the latter may be considered as a special case of this more general strategy involving both global and local neighborhood of each vector synergistically. From now onwards, we shall refer to this version as DEGL (DE with Global and Local neighborhoods). Rest of the algorithm is exactly similar to DE/rand/1/bin. DEGL uses a binomial crossover scheme.

Please note that in each generation, the vectors belonging to a DE population are perturbed sequentially. If a target vector $\vec{X}_i(t)$ is replaced by the corresponding trial vector $\vec{U}_{i,G}$, the neighborhood-best $\vec{X}_{n\_best_i,G}$ and the globally best vector $\vec{X}_{g\_best,G}$ may also be updated by $\vec{U}_{i,G}$, provided the latter yields a lower value of the objective function.

2.5.9.4 Control Parameters in DEGL

DEGL introduces four new parameters α, β, w, and the neighborhood radius k. Among them α and β are playing the same role as the constant F in equation (2.7). Thus, in order to reduce the number of parameters further, we take $\alpha = \beta = F$. The most crucial parameter in DEGL is perhaps the weight factor w, which controls the balance between the exploration and exploitation capabilities. Small values of w (close to 0) in (16) favor the local neighborhood component, thereby resulting in better exploration. On the other hand, large values (close to 1) favor the global variant component, promoting exploitation. Therefore, values of w around the middle point, 0.5, of the range [0, 1] result in the most balanced DEGL versions. However, such balanced versions do not take full advantage of any special structure of the problem at hand (e.g., uni-modality, convexity etc.). In such cases, weight factors that are biased towards 0 or 1 may exhibit better performance. Moreover, on-line adaptation of w during the execution of the algorithm can enhance its performance. Optimal values of the weight factor always depend on the problem at hand. We considered three different schemes for the selection and adaptation of w to gain intuition regarding DEGL's performance and we describe them in the following paragraphs.

1) Increasing Weight Factor: All vectors have the same weight factor which is initialized at 0 and is increased up to 1 during the execution of the algorithm. Thus, exploration is favored in the first stages of the algorithm's execution (since $w = 0$ corresponds to the local neighborhood model) and exploitation is promoted at the final stages, where w assumes higher values. Let t denote the current time step (iteration number), $w(t)$ be the weight factor at time t, and t_{max} denotes the maximum number of time steps. Then we considered two different increasing schedules in our study:

A) Linear Increment: w is linearly increased from 0 to 1.

$$w(t) = \frac{t}{t_{\max}} \tag{2.39}$$

B) Exponential Increment: The weight factor increases from 0 to 1 in an exponential fashion as follows:

$$w(t) = \exp\left(\frac{t}{t_{\max}} . \ln(2)\right) - 1 \tag{2.40}$$

This scheme results in slow transition from exploration to exploitation in the early stages of the algorithm's execution, but it exhibits faster transition in the later stages.

2) Random Weight Factor: In this scheme the weight factor of each vector is made to vary as a uniformly distributed random number in (0, 1) i. e. $w_i(t) \sim rand(0,1)$. Such a choice may decrease the convergence speed (by introducing more diversity).

3) Self Adaptive Weight Factor: In this scheme, each vector has its own weight factor. The factor is incorporated in the vector as an additional variable, augmenting the dimension of the problem. Thus, a generation now consists of vectors $\vec{a}_i(t) = \{\vec{X}_i(t), \vec{S}_i(t)\}$ where $\vec{S}_i(t) = \{w_i(t)\}$ and $w_i(t)$ is the weight factor for vector $\vec{X}_i(t)$. During the initialization phase of DE, $w_i(t)$ is randomly initialized in (0.0, 1.0). Next, while evolving a vector $\vec{a}_i(t)$, at first local and global mutant vectors $\vec{L}_i(t)$ and $\vec{g}_i(t)$ are formed for $\vec{X}_i(t)$ following equations (14) and (15). The sub-vector $\vec{S}$ undergoes global mutation only and weight factors perturbing $\vec{S}$ come from the same population members $\vec{a}_{r_1}(t)$ and $\vec{a}_{r_2}(t)$, which were also used to form $\vec{g}_i(t)$. The mutation of $w_i(t)$ leads to the formation of a new trial weight factor $w_i^{/}(t)$ according to the following equation:

$$w_i^{/}(t) = w_i(t) + F.(w_{g\_best}(t) - w_i(t)) + F.(w_{r_1}(t) - w_{r_2}(t)), \qquad (2.41)$$

where, $w_{g\_best}(t)$ is the weight factor associated with the best parameter vector $\vec{X}_{g\_best}(t)$. Value of the newly formed $w_i^{/}(t)$ is restricted to the range [0.05, 0.95] in the following way:

$$\text{if } w_i^{/}(t) > 0.95, \; w_i^{/}(t) = 0.95.$$

$$\text{else if } w_i^{/}(t) < 0.05, \; w_i^{/}(t) = 0.05$$

$w_i^{/}(t)$ is then used to combine $\vec{L}_i(t)$ and $\vec{g}_i(t)$ according to equation (2.38) and this leads to the formation of new donor parameter vector $\vec{V}_i(t)$. The donor vector thus formed exchanges its components with $\vec{X}_i(t)$ following the binomial crossover and results into the trial vector $\vec{U}_i(t)$. Note that the weight factor does not undergo crossover. Now the newly formed weight factor is promoted to the next generation only if $\vec{U}_i(t)$ yields an equal or lower objective function value as compared to $\vec{X}_i(t)$, i. e.

$$\vec{a}_i(t+1) = \{\vec{X}_i(t+1) = \vec{U}_i(t), \vec{S}_i(t+1) = \{w_i^/(t)\}\}, \text{ if } f(\vec{U}_i(t)) \le f(\vec{X}_i(t))$$
$$\vec{a}_i(t+1) = \{\vec{X}_i(t+1) = \vec{X}_i(t), \vec{S}_i(t+1) = \{w_i(t)\}\}, \text{ otherwise} \qquad (2.42)$$

The process is repeated sequentially for each vector in a generation. Note that the weight factors associated with the neighborhood-best and globally best vectors are not updated each time, when a trial vector replaces the corresponding target. The weight factor for a parameter vector is changed only once according to equations (2.39) and (2.40) in each generation. According to the self-adaptation scheme, the dynamics of DEGL is allowed to determine the optimal $w_i(t)$ for each vector, individually, capturing any special structure of the problem at hand. In what follows this scheme will be denoted as DEGL/SAW (DEGL with Self Adaptive Weight)

Finally we would like to point out that a proper selection of the neighborhood size affects the trade–off between exploration and exploitation. However, there are no general rules regarding the selection of neighborhood size, and it is usually based on the experience of the user.

2.5.9.5 Runtime Complexity of DEGL – A Discussion

Runtime-complexity analysis of the population-based stochastic search techniques like DE, GA etc. is a critical issue by its own right. Following the works of Zielinski *et al.* [47] we note that the average runtime of a standard DE algorithm usually depends on its stopping criterion. While computing the run-time complexity, we usually take into account the fundamental floating-point arithmetic and logical operations performed by an algorithm [48]. We may neglect very simple operations like copy/assignment etc. as these are merely data-transfer operations between the ALU and/or CPU registers and hardly require any complex digital circuitry like adder, comparator etc. [48, 49]. Now, in each generation of DE, a loop over *NP* is conducted, containing a loop over *D*. Since the mutation and crossover operations are performed at the component level for each DE vector, the number of fundamental operations in DE/rand/1/bin is proportional to the total number of loops conducted until the termination of the algorithm. Thus, if the algorithm is stopped after a fixed number of iterations t_{max}, then the runtime complexity is $O(NP \cdot D \cdot t_{\max})$.

In case of DE/target-to-best/1, computational complexity of finding the globally best vector depends only on comparing the objective function value against the single best vector's objective function value. Note that the best objective function evaluation value must be upgraded for each newly generated trial vector, if it replaces the target vector. Now that means in the worst possible case (when the target vector is always replaced by the trial vector), this is done $NP \cdot t_{\max}$ times. Thus, the overall runtime remains. $O(\max(NP \cdot t_{\max}, NP \cdot D \cdot t_{\max})) = O(NP \cdot D \cdot t_{\max})$.

In DEGL, besides the globally best vector, we have to take into account the best vector of each neighborhood as well. Each individual vector is endowed with a small memory, which can keep track of the best vector in its neighborhood and the corresponding objective function value. At the very onset, once all the vectors are initialized, a search is performed to detect the neighborhood-best for each individual. Note that this search is performed only once at $t = 0$. In subsequent generations, these locally best vectors only need to be updated in the memory of the neighboring vectors. This is just like the updating phase of the globally best vector in DE/target-to-best/1. Now let us try to estimate the cost of the initial search. Note that the neighborhoods in DEGL are actually overlapping in nature (on the index-graph) and this is clearly illustrated in Figure 2.17. Any two adjacent vectors (w.r.t. their indices) will have $2k+1+1-2=2k$ number of common neighbors.

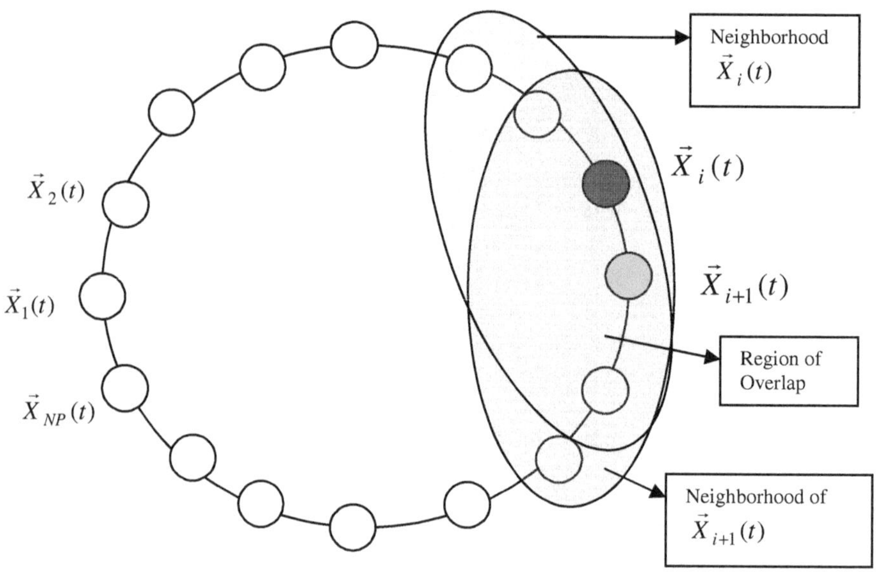

Fig. 2.17 The overlapping of neighborhoods in DEGL

Suppose $N_k(\vec{X}_i(t))$ indicates the set of vectors belonging to the immediate neighborhood of radius k for the vector $\vec{X}_i(t)$. Then evidently the cardinality of both the sets $N_k(\vec{X}_i(t)) \cap N_k^c(\vec{X}_{i+1}(t))$ and $N_k^c(\vec{X}_i(t)) \cap N_k(\vec{X}_{i+1}(t))$ is exactly 1 (where N_k^c stands for complement of the set N_k). We observe that $\vec{X}_{i-k}(t) \in N_k(\vec{X}_i(t)) \cap N_k^c(\vec{X}_{i+1}(t))$ and $\vec{X}_{i+k+1}(t) \in N_k^c(\vec{X}_i(t)) \cap N_k(\vec{X}_{i+1}(t))$. Now we start by detecting the best vector of the neighborhood of any population member, say $\vec{X}_i(t)$ and call it $\vec{X}_{n\_best_i}(t)$. This is equivalent to finding the lowest entry from an array of $2k+1$ numbers (objective function values) and requires $2k$ number of comparisons. Next, to calculate the best vector in the neighborhood of $\vec{X}_{i+1}(t)$, if $\vec{X}_{n\_best_i}(t) \neq \vec{X}_{i-k}(t)$ then we simply need to compare the objective function values of $\vec{X}_{i+k+1}(t)$ and $\vec{X}_{n\_best_i}(t)$ in order to determine $\vec{X}_{n\_best_{i+1}}(t)$. This requires only 1 comparison. But if unfortunately $\vec{X}_{n\_best_i}(t) = \vec{X}_{i-k}(t)$, we shall have to find the neighborhood best of $\vec{X}_{i+1}(t)$ by taking its $2k$ neighbors into account and this requires $O(k)$ runtime. Hence in worst possible case (when always the current neighborhood's best vector is excluded from the serially next vector's neighborhood) searching the best vectors of all the neighborhoods is completed in $O(NP \cdot k)$ time.

Once the search for all neighborhood-bests is finished, in subsequent generations, best vector in the neighborhood of $\vec{X}_i(t)$ is updated only if a newly generated trial vector $\vec{U}_i(t)$ replaces the target vector $\vec{X}_i(t)$ and in addition to that $f(\vec{U}_i(t)) < f(\vec{X}_{n\_best_i}(t))$. It is possible that $\vec{X}_{n\_best_i}(t)$ differs from $\vec{X}_{n\_best_{i+1}}(t)$ i.e. two vectors, adjacent on the index graph, may have distinct neighborhood-best vectors. This happens when the best vector in the neighborhood of $\vec{X}_{i+1}(t)$ is $\vec{X}_{i+k+1}(t)$. Under this condition, it is possible that $\vec{U}_i(t)$ is better than $\vec{X}_{n\_best_i}(t)$ but not better than $\vec{X}_{n\_best_{i+1}}(t)$. Hence in order to update the best vectors in the memories of all the neighbors of $\vec{X}_i(t)$ (when $f(\vec{U}_i(t)) < f(\vec{X}_{n\_best_i}(t))$ is satisfied) we have to compare the objective function values of $\vec{U}_i(t)$ and the neighborhood-bests in the memories of $2k$ neighbors of $\vec{X}_i(t)$. Thus in worst possible case, updating of all the local best vectors in the memories of the neighbors of each vector requires $O(NP \cdot k)$ comparisons in each generation. Evidently, over $t_{\max}$ generations, the number of additional comparisons necessary is $O(NP \cdot k \cdot t_{\max})$. This implies that, the worst case complexity of DEGL is actually $O(\max(NP \cdot k \cdot t_{\max}, NP \cdot D \cdot t_{\max}))$. Now, the asymptotic order of complexity for DEGL remains $O(NP \cdot k \cdot t_{\max})$ if $k \leq D$. Please note that this condition is usually satisfied when DEGL is applied to the optimization of higher dimensional functions. For example, the usual population size for DE is $NP = 10D$. If the neighborhood size is approximately 10% of the population size (which, as can be seen later, provides reasonably good results with

DEGL), we have $2k+1=(0.1)\cdot NP = D \Rightarrow k=\frac{D-1}{2}$ with $D>1$. Clearly, in this case we have $k \le D$. Simple algebraic manipulations suggest that this condition holds true if the neighborhood size is below 20% of the population size NP and $D>1$. Hence, we can say that under such conditions, $O(\max(NP\cdot k\cdot t_{\max}, NP\cdot D\cdot t_{\max}))=O(NP\cdot D\cdot t_{\max})$ and then DEGL does not impose any serious burden on the runtime complexity of the existing DE variants.

2.5.9.6 Comparative Performance of DEGL

This section is intended to provide a glimpse of the superior performance of DEGL over four popular numerical benchmarks [50], in comparison to some state-of-the-art variants of DE.

1. *Quartic Function i.e. Noise:*

$$f_7(\vec{X}) = (\sum_{i=1}^{D} i.x_i^4) + rand\ [0,1)$$

$$-1.28 \le x_i \le 1.28,\ \min(f_1) = f_1(1,\ldots.,1) = 0$$

2.*Generalized Rastrigin's Function:*

$$f_2(\vec{X}) = \sum_{i=1}^{D} [x_i^2 - 10\cos(2\pi x_i) + 10],$$

$$-5.12 \le x_i \le 5.12,\ \min(f_2) = f_2(0,\ldots.,0) = 0$$

3. *Ackley's Function:*

$$f_3(\vec{X}) = -20\exp\left(-0.2\sqrt{\frac{1}{D}\sum_{i=1}^{D} x_i^2}\right) - \exp\left(\frac{1}{D}\sum_{i=1}^{D}\cos 2\pi x_i\right) + 20 + e$$

$$-32 \le x_i \le 32,\ \min(f_3) = f_3(0,\ldots.,0) = 0$$

4. *Generalized Griewank Function:*

$$f_4(\vec{X}) = \frac{1}{4000}\sum_{i=1}^{D} x_i^2 - \prod_{i=1}^{D}\cos(\frac{x_i}{\sqrt{i}}) + 1$$

$$-600 \le x_i \le 600,\ \min(f_4) = f_4(0,\ldots.,0) = 0$$

Simulations were carried out to obtain a comparative performance analysis of DEGL/SAW (that appears to be the best performing algorithm among all the variants of DEGL) with respect to: (a) DE/rand/1/bin [13] (b) DE/target-to-best/1/bin [13] (c) DE/rand/1/either-or [13] (d) SADE [31] and (e) NSDE [51].

Among the competitors, the first two belong to the classical DE family of Storn and Price. The DE/rand/1/bin algorithm was chosen because of its wide popularity in solving numerical optimization or engineering problems [13]. The NSDE algorithm was proposed by Yang *et al.* [51] as a hybridization of DE with the Neighborhood Search (NS), which appears as a main strategy underpinning Evolutionary Programming (EP) [52]. The resulting algorithm, known as NSDE, performs mutation by adding a normally distributed random value to each target-vector component.

The mean and the standard deviation (within parentheses) of the best-of-run values for 50 independent runs of each of the six algorithms are presented in Table 2.3 for the 25 and 100 dimensional functions f_1 to f_4. Since all the algorithms start with the same initial population over each problem instance, we used paired t-tests to compare the means of the results produced by best and the second best algorithms (with respect to their final accuracies). The t-tests are quite popular among researchers in evolutionary computing and they are fairly robust to violations of a Gaussian distribution with large number of samples like 50 [53]. In the 10-th columns of Table 2.3 we report the statistical significance level of the difference of the means of best two algorithms. Note that here '+' indicates the t value of 49 degrees of freedom is significant at a 0.05 level of significance by two-tailed test, '.' means the difference of means is not statistically significant and 'NA' stands for Not Applicable, covering cases for which two or more algorithms achieve the best accuracy results.

Table 2.3 Average and the standard deviation of the best-of-run solution for 50 independent runs and the success rate tested on f_1 to f_4

Function	Dim	Max FEs	Mean Best Value (Standard Deviation) DE/rand/1/bin	DE/target-to-best/1/bin	DE/rand/1/either-or	SADE [27]	NSDE [30]	DEGL/SAW	Statistical Significance
f_1	25	5×10^5	4.9391e-03 (5.92e-04)	9.0982e-03 (2.08e-04)	6.9207e-04 (4.26e-06)	3.7552e-02 (9.02e-03)	4.3482e-03 (6.50e-04)	**1.0549e-07 (2.33e-06)**	+
	100	5×10^6	2.8731e-02 (2.33e-02)	3.3921e-02 (3.32e-02)	4.3332e-03 (5.76e-02)	5.9281e-02 (4.31e-03)	9.8263e-02 (2.90e-03)	**6.9921e-06 (4.56e-05)**	+
f_2	25	5×10^5	1.0453e-03 (8.04e-02)	9.5278e-01 (4.72e-01)	1.7109e-23 (2.726e-24)	6.7381e-24 (3.728e-21)	4.8392e-21 (8.872e-20)	**5.8492e-25 (5.333e-27)**	.
	100	5×10^6	2.1121e-02 (4.86e-03)	6.76249e-01 (4.27e-01)	**8.4719e-23 (9.36e-22)**	5.8824e-21 (4.83e-20)	5.5732e-05 (5.93e-04)	1.7728e-22 (3.88e-20)	.
f_3	25	5×10^5	4.1902e-08 (3.36e-08)	9.8035e-03 (6.80e-03)	6.9437e-15 (4.86e-15)	7.8343e-15 (2.85e-15)	5.9749e-10 (3.2231e-04)	**5.9825e-23 (1.00e-22)**	+
	100	5×10^6	7.6687e-05 (6.767e-05)	6.76249e-01 (4.237e-01)	6.9398e-13 (4.852e-13)	3.0665e-12 (5.125e-13)	4.1232e-05 (7.496e-06)	**8.52742e-17 (1.365e-15)**	+
f_4	25	5×10^5	6.8318e-22 (3.837e-25)	7.94504e-07 (8.03e-08)	3.0905e-34 (7.462e-34)	1.8274e-28 (7.682e-29)	7.9318e-26 (3.774e-28)	**2.9931e-36 (4.736e-35)**	+
	100	5×10^6	2.1962e-10 (8.45e-11)	5.27284e-05 (4.63e-07)	3.2928e-12 (2.77e-13)	8.9569e-13 (1.02e-14)	5.0392e-10 (4.29e-08)	**4.11464e-15 (6.02e-16)**	+

Convergence characteristics of the six competitive algorithms over two of these benchmark functions have been presented in Figure 2.18. For more details of the experimental results please see [45].

The results presented here adequately illustrate that the proposed DEGL algorithm can meet or beat its competitor DE-variants in a statistically meaningful way over numerical benchmark functions.

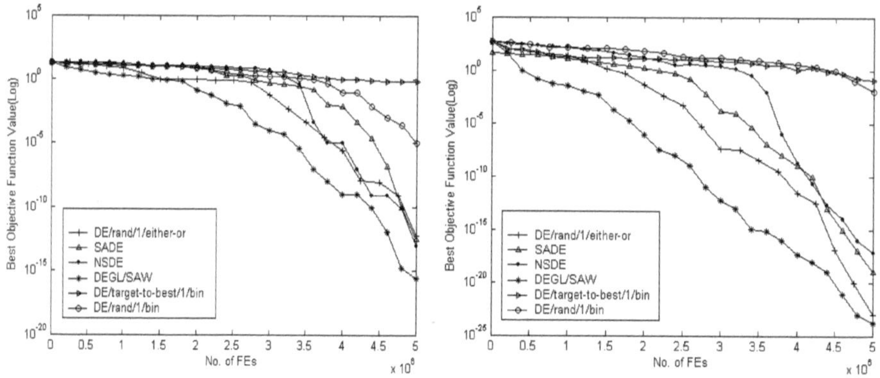

(a) Generalized Ackley's Function (f_3)

(b) Generalized Rastrigin's Function (f_2)

Fig. 2.18 Progress towards the optimum solution for median run of six algorithms over two difficult test functions

2.6 Hybridization of Differential Evolution with Other Stochastic Search Techniques

Over the past few years, DE has been successfully hybridized with a few other stochastic search algorithms like the PSO (Particle Swarm Optimization) [54], BFOA (Bacterial Foraging Optimization Algorithm) [55], and SA (Simulated annealing) [56]. The first synergy between DE and PSO was reported by Zang and Xie [57] called DEPSO. The method proposed in [57] alternates the original PSO algorithm and the DE operator at the odd iterations and at the even iterations. DEPSO achieved better convergence results than both the original algorithms over certain constrained optimization problems. Das *et al.* [58] modified the selection mechanism of the classical DE family by using the concepts of SA called Annealed Differential Evolution (AnDE). In SA the probability of accepting the inferior solutions is dynamically altered with iterations. Besides using a SA based selection scheme, AnDE introduced a center of mass based mutation strategy, in which the trial vectors are stochastically attracted towards the mean vector of the current population, instead of the best one as done in the case of DE/current-to-best/bin and DE/best/2 schemes. Biswas *et al.* [59] proposed a synergistic coupling of DE and a recently developed swarm intelligence algorithm, well-known as the Bacterial Foraging Optimization Algorithm (BFOA) [55]. Foraging can be modeled as an optimization process where an animal seeks to maximize energy per unit time spent for foraging. BFOA emulates the foraging behavior of a group of *Escherichia coli* bacteria living in our intestine. The bacterial group undergoes four principal processes: chemotaxis, swarming, reproduction, and elimination-dispersal. The computational chemotaxis of BFOA, which may also be viewed as a stochastic gradient search, has been coupled with DE type mutation and crossing over of the optimization agents. This leads to the new hybrid algorithm called CDE (Chemotactic Differential Evolution) [59]. CDE was shown to overcome the problems of slow and premature convergence of both the classical DE and BFOA over several benchmark functions as well as real world optimization problems.

Das *et al.* [60] proposed a hybridization of PSO and DE called Particle Swarm Optimization with Differentially Perturbed Velocity (PSO-DV). PSO-DV introduces a differential operator (borrowed from DE) in the velocity-update scheme of PSO. The operator is invoked on the position vectors of two randomly chosen particles (population-members), not on their individual best positions. Further, unlike the PSO scheme, a particle is actually shifted to a new location only if the new location yields a better fitness value, i.e., a selection strategy has been incorporated into the swarm dynamics. In the proposed algorithm, for each particle i in the swarm two other distinct particles, say j and k ($i \neq j \neq k$), are selected randomly. The difference between their positional coordinates is taken as a difference vector:

$$\vec{\delta} = \vec{X}_k - \vec{X}_j \tag{2.43}$$

Then the d-th velocity component ($1 < d < D$) of the target particle i is updated as:

$$\begin{aligned} v_{i,d}(t) &= \omega.v_{i,d}(t-1) + \beta.\delta_d + \varphi_2.rand2_{i,d}(0,1).(p_d^g - x_{i,d}(t-1)) \\ & \qquad\qquad\qquad\qquad \text{if } rand_d(0,1) \leq CR \\ &= v_{i,d}(t-1) \qquad\qquad\quad \text{otherwise,} \end{aligned} \tag{2.44}$$

where CR is the crossover probability, δ_d is the d-th component of the difference vector defined earlier, and β is a scale factor in [0, 1]. In essence the cognitive part of the velocity update formula in PSO is replaced with the vector differential operator to produce some additional exploration capability. Clearly, for $CR \leq 1$, some of the velocity components will retain their old values. Now, a new trial location $\vec{Tr}_i$ is created for the particle by adding the updated velocity to the previous position $\vec{X}_i$:

$$\vec{Tr}_i = \vec{X}_i(t-1) + \vec{V}_i(t) \tag{2.45}$$

The particle is placed at this new location only if the coordinates of the location yield a better fitness value. Thus if we are seeking the minimum of the objective function $f(\vec{X})$, then the target particle is relocated as follows:

$$\begin{aligned} \vec{X}_i(t) &= \vec{Tr}_i \qquad\qquad \text{if } f(\vec{Tr}_i) \leq f(\vec{X}_i(t)) \\ \vec{X}_i(t) &= \vec{X}_i(t-1) \qquad \text{otherwise} \end{aligned} \tag{2.46}$$

Therefore, every time its velocity changes, the particle either moves to a better position in the search space or sticks to its previous location. The current location of the particle is thus the best location it has ever found. In other words, unlike the classical PSO, in the present scheme, $p_{i,d}^l$ always equals $x_{i,d}$. So the cognitive part involving $\left|p_{i,d}^{l} - x_{i,d}\right|$ is automatically eliminated in our algorithm. If a particle gets stagnant at any point in the search space (i.e., if its location does not change for a predetermined number of iterations), then the particle is shifted by a random mutation (explained below) to a new location. This technique helps escape local minima and also keeps the swarm "moving":

If $((\vec{X}_i(t-1) = \vec{X}_i(t) = \vec{X}_i(t+1) = ... = \vec{X}_i(t+N-1))$ and $(f(\vec{X}_i(t+N) \neq f^*)$
then
for (r = 1 to D)

$$x_{i,r}(t+N) = x_{\min,j} + rand_r(0,1).(x_{\max,r} - x_{\min,r}) \qquad (2.47)$$

where f^* is the global minimum of the fitness function, N is the maximum number of iterations up to which stagnation can be tolerated and ($x_{\max,r}$, $x_{\min,r}$) define the permissible bounds of the search space in the r-th dimension. The scheme is conceptually outlined with a humanoid particle over the fitness landscape of the sphere function, in Figure 2.19.

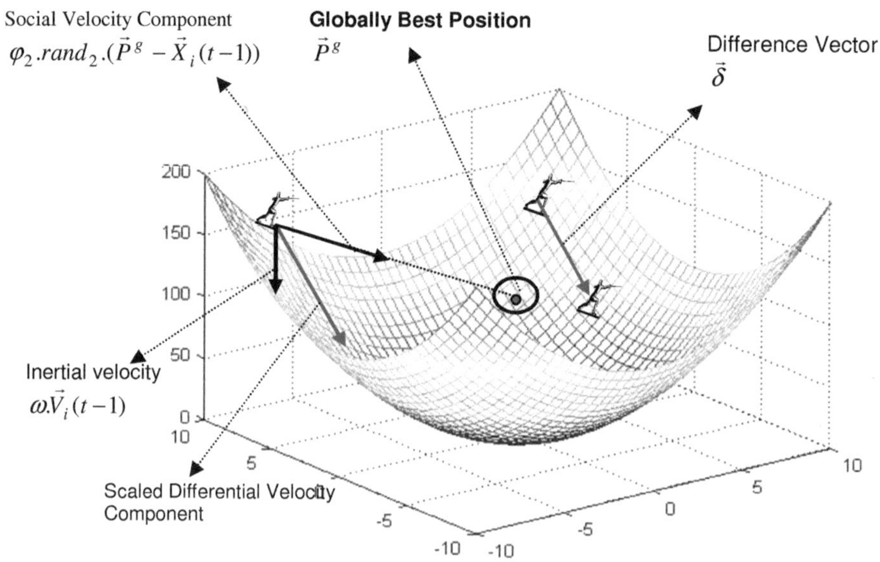

Fig. 2.19 Illustrating PSO-DV on a two dimensional function surface

2.7 Conclusions

This chapter presented the foundations and the present state-of-the-art on the DE family of algorithms in a comprehensive manner. The classical DE algorithm may be viewed as a simple GA comprising of four basic steps: *initialization*, *mutation*, *crossover*, and *selection*. Each of these steps was outlined with hand-traced examples based on a simple cost function, at the beginning of the chapter. Storn and Price suggested five different mutation strategies for DE, each of which can be used with any one of two different crossover schemes known as *exponential* and *binomial* crossover. This results into 10 distinct members of the classical DE family, which were discussed in details in the Chapter. DE has three major control parameters: the scale factor *F*, the crossover rate *CR* and the population size *NP*. Effect of these parameters on the performance of DE was addressed and some methods for choosing suitable values of these parameters were also discussed in the Chapter. The article also briefly reviewed nine prominent variants of DE, including DE with trigonometric mutation, DE with arithmetic recombination, DE/rand/1/either-or, self-adaptive DE, opposition-based DE, binary DE, DE with adaptive local search, Self-adaptive DE with strategy adaptation, and finally DE with a neighborhood-based mutation. An interesting hybrid algorithm resulting from the synergy of DE and particle swarm optimization was outlined as well.

References

[1] Back, T., Fogel, D.B., Michalewicz, Z.: Handbook of Evolutionary Computation. IOP and Oxford University Press, Bristol (1997)

[2] Fogel, D.B.: Evolutionary Computation. IEEE Press, Piscataway (1995)

[3] Michalewicz, Z.: Genetic Algorithms + Data Structures = Evolution Programs. Springer, Berlin (1992)

[4] Holland, J.H.: Adaptation in Natural and Artificial Systems. University of Michigan Press, Ann Arbor (1975)

[5] Goldberg, D.E.: Genetic algorithms in search. In: Optimization and Machine Learning. Addison-Wesley, Reading (1989)

[6] Storn, R., Price, K.: Differential Evolution - a simple and efficient adaptive scheme for global optimization over continuous spaces, Technical Report TR-95-012, ICSI (1995), `http://http.icsi.berkeley.edu/~storn/litera.html`

[7] Storn, R., Price, K.: Differential evolution – A simple and efficient heuristic for global optimization over continuous spaces. Journal of Global Optimization 11(4), 341–359 (1997)

[8] Nelder, J.A., Mead, R.: A simplex method for function minimization. Computer Journal 7, 308–313 (1965)

[9] Avriel, M.: Nonlinear Programming: Analysis and Methods. Dover Publishing (2003)

[10] Price, W.L.: Global optimization by controlled random search. Computer Journal 20(4), 367–370 (1977)

[11] Fogel, L.J., Owens, A.J., Walsh, M.J.: Artificial Intelligence through Simulated Evolution. John Wiley, Chichester (1966)

[12] Eiben, A.E., Smith, J.E.: Introduction to Evolutionary Computing. Springer, Heidelberg (2003)

[13] Price, K., Storn, R., Lampinen, J.: Differential Evolution - A Practical Approach to Global Optimization. Springer, Berlin (2005)

[14] Price, K.V.: An introduction to differential evolution. In: Corne, D., Dorigo, M., Glover, V. (eds.) New Ideas in Optimization, pp. 79–108. Mc Graw-Hill, UK (1999)

[15] Gamperle, R., Muller, S.D., Koumoutsakos, A.: Parameter study for differential evolution. In: WSEAS NNA-FSFS-EC 2002, Interlaken, Switzerland, Feburary 11-15 (2002)
[16] Ronkkonen, J., Kukkonen, S., Price, K.V.: Real parameter optimization with differential evolution. In: The 2005 IEEE Congress on Evolutionary Computation (CEC 2005), vol. 1, pp. 506–513. IEEE Press, Los Alamitos (2005)
[17] Liu, J., Lampinen, J.: A Fuzzy adaptive differential evolution algorithm. Soft computing- A Fusion of Foundations, Methodologies and Applications 9(6), 448–462 (2005)
[18] Qin, A.K., Suganthan, P.N.: Self-adaptive differential evolution algorithm for numerical optimization. In: IEEE Congress on Evolutionary Computation, pp. 1785–1791 (2005)
[19] Zaharie, D.: Control of population diversity and adaptation in differential evolution algorithms. In: Matousek, D., Osmera, P. (eds.) Proc. of MENDEL 2003 9th International Conference on Soft Computing, Brno, Czech Republic, pp. 41–46 (June 2003)
[20] Zaharie, D., Petcu, D.: Adaptive pareto differential evolution and its parallelization. In: Wyrzykowski, R., Dongarra, J., Paprzycki, M., Waśniewski, J. (eds.) PPAM 2004. LNCS, vol. 3019, pp. 261–268. Springer, Heidelberg (2004)
[21] Abbass, H.: The Self-Adaptive pareto differential evolution algorithm., In: Proceedings of the 2002 Congress on Evolutionary Computation, pp. 831–836 (2002)
[22] Beyer, H.G.: On the dynamics of EAs without selection. In: Banzaf, W., Reeves, C. (eds.) Foundations of genetic algorithms, pp. 5–26. Morgan Kaufmann, San Mateo (1999)
[23] Zaharie, D.: Critical Values for the Control Parameters of Differential Evolution Algorithms. In: Matousek, R., Osmera, P. (eds.) Proc. of Mendel 2002, 8th International Conference on Soft Computing, Brno, Czech Republic, pp. 62–67 (2002)
[24] Omran, M., Salman, A., Engelbrecht, A.P.: Self-adaptive differential evolution. In: Hao, Y., Liu, J., Wang, Y.-P., Cheung, Y.-m., Yin, H., Jiao, L., Ma, J., Jiao, Y.-C. (eds.) CIS 2005. LNCS, vol. 3801, pp. 192–199. Springer, Heidelberg (2005)
[25] Teo, J.: Exploring dynamic self-adaptive populations in differential evolution. Soft Computing - A Fusion of Foundations, Methodologies and Applications (2006)
[26] Das, S., Konar, A., Chakraborty, U.K.: Two improved differential evolution schemes for faster global search. In: ACM-SIGEVO Proceedings of GECCO, Washington D.C., pp. 991–998 (June 2005)
[27] Wolpert, D.H., Macready, W.G.: No free lunch theorems for optimization. IEEE Transactions on Evolutionary Computation 1, 67 (1997)
[28] Fan, H.-Y., Lampinen, J.: A trigonometric mutation operation to differential evolution. International Journal of Global Optimization 27(1), 105–129 (2003)
[29] Ashlock, D.: Evolutionary Computation for Modeling and Optimization. Springer, Heidelberg (2006)
[30] Eiben, A.E., Smith, J.E.: Introduction to Evolutionary Computing. Springer, Heidelberg (2003)
[31] Brest, J., Greiner, S., Bošković, B., Mernik, M., Žumer, V.: Self-adapting Control parameters in differential evolution: a comparative study on numerical benchmark problems. IEEE Transactions on Evolutionary Computation 10(6), 646–657 (2006)

[32] Tizhoosh, H.R.: Opposition-Based Learning: A New Scheme for Machine Intelligence. In: Int. Conf. on Computational Intelligence for Modeling Control and Automation - CIMCA 2005, Vienna, Austria, vol. I, pp. 695–701 (2005)

[33] Tizhoosh, H.R.: Reinforcement learning based on actions and opposite actions. In: ICGST International Conference on Artificial Intelligence and Machine Learning (AIML 2005), Cairo, Egypt (2005)

[34] Tizhoosh, H.R.: Opposition-based reinforcement learning. Journal of Advanced Computational Intelligence and Intelligent Informatics 10(3) (2006)

[35] Rahnamayan, S., Tizhoosh, H.R., Salama, M.M.A.: Opposition-based differential evolution. IEEE Transactions on Evolutionary Computation 12(1), 64–79 (2008)

[36] Rahnamayan, S., Tizhoosh, H.R., Salama, M.M.A.: Opposition-based differential evolution for optimization of noisy problems. In: Proc. 2006 IEEE Congress on Evolutionary Computation (CEC 2006), Vancouver, pp. 1865–1872 (July 2006)

[37] Pampara, G., Engelbrecht, A.P., Franken, N.: Binary differential evolution. In: IEEE Congress on Evolutionary Computation. CEC 2006 (2006)

[38] Proakis, J.G., Salehi, M.: Communication System Engineering, 2nd edn. Prentice Hall Publishers, Englewood Cliffs (2002)

[39] Noman, N., Iba, H.: Accelerating Differential Evolution Using an Adaptive Local Search. IEEE Transactions on Evolutionary Computation 12(1), 107–125 (2008)

[40] Tsutsui, S., Yamamura, M., Higuchi, T.: Multi-parent recombination with simplex crossover in real coded genetic algorithms. In: Proc. Genetic Evol. Comput. Conf (GECCO 1999), pp. 657–664 (July 1999)

[41] Suganthan, P.N., Hansen, N., Liang, J.J., Deb, K., Chen, Y.-P., Auger, A., Tiwari, S.: Problem definitions and evaluation criteria for the CEC 2005 special session on real-parameter optimization. Technical Report, Nanyang Technological University, Singapore, and KanGAL Report #2005005, IIT Kanpur, India (May 2005)

[42] Ong, Y.-S., Keane, A.J.: Meta-lamarckian learning in memetic algorithms. IEEE Transactions on Evolutionary Computation 8(2), 99–110 (2004)

[43] Qin, A.K., Huang, V.L., Suganthan, P.N.: Differential evolution algorithm with strategy adaptation for global numerical optimization. IEEE Transactions on Evolutionary Computations (2009), doi:10.1109/TEVC.2008.927706

[44] Mezura-Montes, E., Velázquez-Reyes, J., Coello, C.A.C.: A comparative study of differential evolution variants for global optimization. In: Genetic and Evolutionary Computation Conference (GECCO 2006), pp. 485–492 (2006)

[45] Das, S., Abraham, A., Chakraborty, U.K., Konar, A.: Differential Evolution Using a Neighborhood based Mutation Operator. IEEE Transactions on Evolutionary Computation (accepted, 2008)

[46] Mendes, R., Kennedy, J.: The fully informed particle swarm: simpler, maybe better. IEEE Transactions of Evolutionary Computation 8(3) (2004)

[47] Zielinski, K., Peters, D., Laur, R.: Run time analysis regarding stopping criteria for differential evolution and particle swarm optimization. In: Proc. of the 1st International Conference on Experiments/Process/System Modelling/ Simulation/ Optimization, Athens, Greece (2005)

[48] Aho, A.V., Hopcroft, J.E., Ullman, J.D.: Data Structures and Algorithms. Addison-Wesley, Reading (1983)

[49] Cormen, T.H., Leiserson, C.E., Rivest, R.L.: Introduction to Algorithms, 1st edn. MIT Press and McGraw-Hill (1990)

[50] Yao, X., Liu, Y., Lin, G.: Evolutionary programming made faster. IEEE Transactions on Evolutionary Computation 3(2), 82–102 (1999)
[51] Yang, Z., He, J., Yao, X.: Making a Difference to Differential Evolution. In: Michalewicz, Z., Siarry, P. (eds.) Advances in Metaheuristics for Hard Optimization, pp. 415–432. Springer, Heidelberg (2007)
[52] Michalewicz, Z., Fogel, D.B.: How to Solve It: Modern Heuristics. Springer, Berlin (1999)
[53] Flury, B.: A First Course in Multivariate Statistics, vol. 28. Springer, Heidelberg (1997)
[54] Kennedy, J., Eberhart, R.C.: Swarm Intelligence. Morgan Kaufmann, San Francisco (2001)
[55] Passino, K.M.: Biomimicry of bacterial foraging for distributed optimization and control. IEEE Control Systems Magazine, 52–67 (2002)
[56] Kirkpatrik, S., Gelatt, C., Vecchi, M.: Optimization by Simulated Annealing. Science 220, 671–680 (1983)
[57] Zhang, W.-J., Xie, X.-F.: DEPSO: Hybrid particle swarm with differential evolution operator. In: Proc. IEEE Int. Conf. Syst., Man, Cybern., pp. 3816–3821 (2003)
[58] Das, S., Konar, A., Chakraborty, U.K.: Annealed Differential Evolution. In: IEEE Congress in Evolutionary Computation, CEC 2007. IEEE press, USA (2007)
[59] Biswas, A., Dasgupta, S., Das, S., Abraham, A.: A Synergy of Differential Evolution and Bacterial Foraging Algorithm for Global Optimization. Neural Network World 17(6), 607–626 (2007)
[60] Das, S., Konar, A., Chakraborty, U.K.: Improving particle swarm optimization with differentially perturbed velocity. In: Proc. Genetic Evol. Comput. Conf. (GECCO), pp. 177–184 (June 2005)

Chapter 3
Modeling and Analysis of the Population-Dynamics of Differential Evolution Algorithm

Abstract. Theoretical analysis of the dynamics of evolutionary algorithms is believed to be very important to understand the search behavior of evolutionary algorithms and to develop more efficient algorithms. In this chapter, we illustrate the dynamics of a canonical Differential Evolution (DE) algorithm with DE/rand/1 type mutation and binomial crossover. The chapter proposes a simple mathematical model of the underlying evolutionary dynamics of a one-dimensional DE-population. The model shows that the fundamental dynamics of each agent (parameter vector) in DE employs the gradient-descent type search strategy, with a learning rate parameter that depends on control parameters like scale factor F and crossover rate CR of DE. The stability and convergence-behavior of the proposed dynamics is analyzed in the light of Lyapunov's stability theorems. The mathematical model developed in this Chapter, provides important insights into the search mechanism of DE in a near neighborhood of an isolated optimum. Empirical studies over simple objective functions are conducted in order to validate the theoretical analysis.

3.1 Introduction

Theoretical analysis of evolutionary algorithms has received increasing attention in the recent years [1]. A few examples of interesting topics are, among many others, convergence analysis [2, 3], dynamics of evolution strategies [4], genetic algorithms [5, 6], and analysis of average computational time [7]. However, the dynamics of EAs during optimization and the roles of each genetic operator are still unclear and stand as a significant research problem at its own right. The analysis of dynamics of EAs is very helpful not only to understand working mechanism of EAs [8] but also to improve performance of EAs and to propose new algorithms [9] because the solution of an optimizer is the result of the dynamics of EAs.

Since its inception in 1995, a good volume of research has been undertaken in order to improve the performance of the DE algorithm over complex and multi-modal fitness landscapes. There exists a plethora of works concerning the empirical study of parameter selection and tuning process in DE [10-18] and its

S. Das, A. Abraham, and A. Konar: Metaheuristic Clustering, SCI 178, pp. 111–135.
springerlink.com

application to optimization problems [19, 20]. Little research has, however, been undertaken to model the underlined search dynamics of DE, which would enable us to understand how and why DE manages to find the optima of many difficult numerical functions so fast. Some significant work in this direction was reported in [21, 22, and 13] by Zaharie, where she theoretically analyzed the influence of the variation operators and their parameters on the expected population variance. In [21], Zaharie showed that the expected population variance (after applying mutation and crossover or recombination) of DE is greater than that of the ES algorithm analyzed in [23]. This finding could explain to some extent the excellent performance of DE on certain test functions. The works of Zaharie, however, did not focus on modeling DE as a dynamical system and analyzing its stability and convergence properties from there.

In this chapter, we provide a simple mathematical model of DE/rand/1/bin scheme (which is the most popular variant of DE family [19]). The model attempts to find out an expected *velocity* of each parameter vector towards the optimum over successive generations [24, 25]. It also tries to relate the search mechanism of DE with that of the classical gradient descent search technique [26]. An earlier attempt to hybridize DE and GA with the gradient descent techniques can be found in [27]. Our model, however, indicates that the fundamental equation governing the expected velocity of the search agents over a continuous fitness landscape in DE has itself got a striking resemblance with that of the steepest descent search. The term analogous to the learning rate in steepest descent, for DE, becomes a function of control parameters like F and CR. Our model points out that DE employs some kind of estimation of the gradient (*not* any analytical expression of the gradient itself though) in order to direct its search towards the optima. Based on the proposed model, the stability and convergence of the DE-vectors in a small neighborhood centered on an isolated optimum, has been investigated with the Lyapunov stability theorems [28, 29]. The Lyapunov's theorems are widely used in nonlinear system analysis to determine the necessary conditions for stability of a dynamical system. The theoretical results, presented in this context, indicate that the crossover rate CR mainly governs the time taken by a single search-agent to converge to an arbitrarily small neighborhood around the optimum. A few simple experimental results have also been provided in order to support the theoretical claims made in the paper. In the appendix we provide an equivalent mathematical model for the DE/current-to-rand/1 scheme which uses arithmetic recombination operator so that the trial vectors may remain rotationally invariant [19].

3.2 The Mathematical Model of the Population-Dynamics in DE

In this section, we outline a simple mathematical model of the dynamics of the parameter vectors of a DE population over a one-dimensional fitness landscape. The analysis is based on simple probability theory and differential calculus .It provides an insight into the search process developed through DE type mutation and crossover operators. Although there are a number of different DE mutation strategies, we confine ourselves to the most popular one among them – DE/rand/1/bin scheme for the sake of simplicity of expressions that allow one to visualize how different operators in DE contribute to the dynamics of the population. The analysis can be easily extended to other DE type mutations.

3.2.1 Assumptions

Suppose $f(x)$, function of single variable x, is to be optimized using the DE Algorithm. Let $\{x_1, x_2, ...x_{NP}\}$ be a set of trial solutions forming the population subjected to DE search where NP denotes the population size. In order to validate our analysis, we make certain assumptions, which are listed below:

i) The objective function $f(x)$ is assumed to be of class C^2 (please note that a function f is said to be of class C^k if the derivatives $f^1, f^2, ..., f^{(k)}$ exist and are continuous [30]). Also let $f(x)$ be Lipschitz continuous [30], that is given any two points x and $y \in \Re$, f satisfies the Lipschitz condition $|f(x) - f(y)| \leq L.|x - y|$ with the value of Lipschitz constant $L \leq 1$ in the region of the fitness landscape (i.e. $f(x)$ is actually a contraction mapping), where our analysis applies. Moreover the objective function is unimodal in the region of interest.
ii) The population of NP trial solutions is limited within a small region i.e. individual trial solutions are located very close to each other. According to [21], this is usually the case during the later stages of the search, when the parameter vectors concentrate in a compact cluster around the global optimum, and especially when the scaling factor F is set at 0.5.

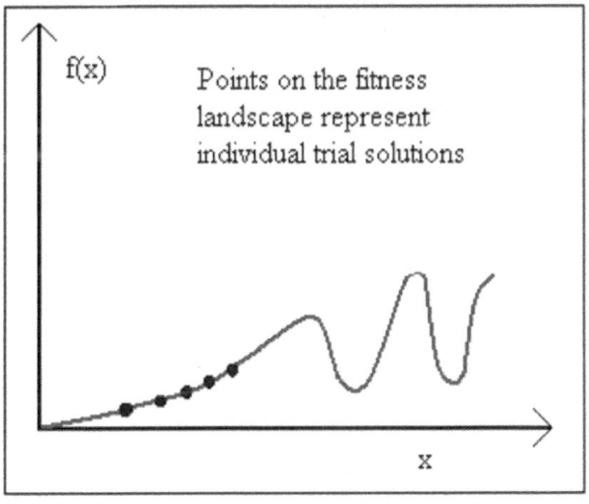

Fig. 3.1 A DE population dispersed on a one-dimensional arbitrary fitness landscape

iii) Dynamics is modeled assuming the vectors as search-agents moving in continuous time.

Figure 3.1 depicts a favorable portion of a one- dimensional arbitrary objective function for our analysis.

3.2.2 Modeling Different Steps of DE

Let x_m be the m-th individual of the population, where $m = 1,2,..., NP$. It is used as a target vector in a particular DE iteration. During an iteration of DE, it undergoes three steps: mutation, crossover, and selection. Each step is modeled individually and finally they are merged to get a generalized expression. In the following analysis, upper case letter denotes random variables.

Three trial solutions are chosen at random from the population. Let X_{r1}, X_{r2}, X_{r3} are three trial solutions (random variables) picked up randomly from population. Here, we assume trial solutions are drawn with replacement. i. e. each trial solution chosen at a particular draw is returned to the population before next draw. This assumption makes X_{r1}, X_{r2}, X_{r3} independent of each other.

This means $P(X_{ri} = x_l \mid X_{rj} = x_k) = P(X_{ri} = x_l)$

$$\Rightarrow P(X_{ri} = x_l \cap X_{rj} = x_k) = P(X_{ri} = x_l)P(X_{rj} = x_k)$$

Where, $i, j = 1,2,3$ and $k, l = 1(1)NP$ and $i \neq j$

Difference of X_{r2}, X_{r3} is scaled by a factor F and then X_{r1} is added with the scaled difference. Let V_m be the generated donor vector.

$$\therefore V_m = X_{r1} + F(X_{r2} - X_{r3})$$

For the one-dimensional analysis we omit the restriction that at least one component of the trial vector must come from the donor. Hence in this case *CR* equals the true probability of the event that $U_m = V_m$. Equipped with these assumptions we may assert the following theorems:

Theorem 3.1: The expected value of a trial vector U_m corresponding to the target vector x_m is given by

$$E(U_m) = (1 - CR)x_m + CRx_{av} \tag{3.1}$$

and the expected value of U_m^2 is then given by,

$$E(U_m{}^2) = (1 - CR)x_m{}^2 + CR(2F^2 + 1)Var(x) + CRx_{av}{}^2 \tag{3.2}$$

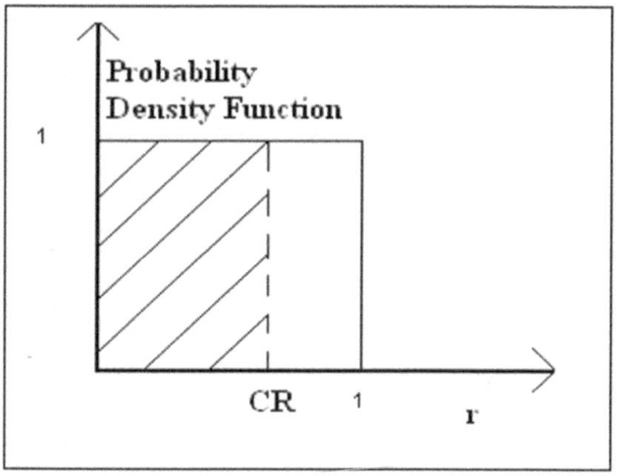

Fig. 3.2 Probability density function of r

where x_{av} is the mean of the population i.e. $x_{av} = \frac{1}{NP}\sum_{i=1}^{NP} x_i$ and $Var(x)$ is the variance of the target population.

Proof: From Figure 3.2, probability of the event, $r \le CR = P(r \le CR) =$ Area of the shaded region.

$$= 1 \times CR = CR$$

Now, $r \le CR$ and $r > CR$ are mutually exclusive and exhaustive events.

$\therefore P(r > CR) = 1 - P(r \le CR) = 1 - CR$

$\therefore E(U_m) = P(r > CR)x_m +$

$$\sum_{i=1}^{NP}\sum_{j=1}^{NP}\sum_{k=1}^{NP}[P\{(r \le CR) \cap ((X_{r1} = x_i) \cap (X_{r2} = x_j) \cap (X_{r3} = x_k))\}\{x_i + F(x_j - x_k)\}]$$

Now, we have assumed that mutation and crossover are independent of each other i.e. r is independent of X_{r1}, X_{r2}, X_{r3}.

$$P\{(r \le CR) \cap ((X_{r1} = x_i) \cap (X_{r2} = x_j) \cap (X_{r3} = x_k))$$
$$= P(r \le CR)P((X_{r1} = x_i) \cap (X_{r2} = x_j) \cap (X_{r3} = x_k))$$

$$\therefore E(U_m) = P(r > CR)x_m + P(r \le CR)\sum_{i=1}^{NP}\sum_{j=1}^{NP}\sum_{k=1}^{NP} P\{(X_{r1} = x_i) \cap (X_{r2} = x_j) \cap (X_{r3} = x_k)\}[x_i + F(x_j - x_k)]$$

X_{r1}, X_{r2}, X_{r3} are independent random variables.

Hence, $P\{(X_{r1} = x_i) \cap (X_{r2} = x_j) \cap (X_{r3} = x_k)\} = P(X_{r1} = x_i)P(X_{r2} = x_j)P(X_{r3} = x_k)$

$$\therefore E(U_m) = P(r > CR)x_m + P(r \le CR)\sum_{i=1}^{NP}\sum_{j=1}^{NP}\sum_{k=1}^{NP} P(X_{r1} = x_i)P(X_{r2} = x_j)P(X_{r3} = x_k)[x_i + F(x_j - x_k)]$$

Now, $P(X_{r1} = x_i) = P(X_{r2} = x_j) = P(X_{r3} = x_k) = \dfrac{1}{NP}$

$$\therefore E(U_m) = P(r > CR)x_m + P(r \le CR)\sum_{i=1}^{NP}\sum_{j=1}^{NP}\sum_{k=1}^{NP}\frac{1}{NP^3}[x_i + F(x_j - x_k)]$$

$$\therefore E(U_m) = (1-CR)x_m + CR\frac{1}{NP^3}\sum_{i=1}^{NP}\sum_{j=1}^{NP}\sum_{k=1}^{NP}[x_i + F(x_j - x_k)]$$

$$\therefore E(U_m) = (1-CR)x_m + CR\frac{1}{NP^3}[\sum_{i=1}^{NP}\sum_{j=1}^{NP}\sum_{k=1}^{NP}x_i + F(\sum_{i=1}^{NP}\sum_{j=1}^{NP}\sum_{k=1}^{NP}x_j - \sum_{i=1}^{NP}\sum_{j=1}^{NP}\sum_{k=1}^{NP}x_k)]$$

$$\therefore E(U_m) = (1-CR)x_m + CR\frac{1}{NP^3}[\sum_{i=1}^{NP}\sum_{j=1}^{NP}\sum_{k=1}^{NP}x_i]$$

$$\therefore E(U_m) = (1-CR)x_m + CR\frac{1}{NP}\sum_{i=1}^{NP}x_i$$

$$\therefore E(U_m) = (1-CR)x_m + CRx_{av}$$

Now, similar to the previous one,

$$\therefore E(U_m^2) = P(r > CR)x_m^2 + \sum_{i=1}^{NP}\sum_{j=1}^{NP}\sum_{k=1}^{NP}[P\{(r \le CR)\cap((X_{r1} = x_i)\cap(X_{r2} = x_j)\cap(X_{r3} = x_k))\}\{x_i + F(x_j - x_k)\}^2]$$

Proceeding in the same manner,

$$\therefore E(U_m^2) = (1-CR)x_m^2 + CR\frac{1}{NP^3}\sum_{i=1}^{NP}\sum_{j=1}^{NP}\sum_{k=1}^{NP}\{x_i + F(x_j - x_k)\}^2$$

$$\Rightarrow E(U_m^2) = (1-CR)x_m^2 + CR\frac{1}{NP^3}\sum_{i=1}^{NP}\sum_{j=1}^{NP}\sum_{k=1}^{NP}\{x_i^2 + F^2(x_j - x_k)^2 + 2Fx_i(x_j - x_k)\}$$

$$\Rightarrow E(U_m^2) = (1-CR)x_m^2 + CR[(2F^2+1)\frac{1}{NP}\sum_{i=1}^{NP}x_i^2 - 2F^2(\frac{1}{NP}\sum_{i=1}^{NP}x_i)^2]$$

$$\because \sum_{i=1}^{NP}\sum_{j=1}^{NP}\sum_{k=1}^{NP}x_i x_j = \sum_{i=1}^{NP}\sum_{j=1}^{NP}\sum_{k=1}^{NP}x_i x_k \text{ and}$$

$$\because \sum_{i=1}^{NP}\sum_{j=1}^{NP}\sum_{k=1}^{NP}x_i^2 = \sum_{i=1}^{NP}\sum_{j=1}^{NP}\sum_{k=1}^{NP}x_j^2 = \sum_{i=1}^{NP}\sum_{j=1}^{NP}\sum_{k=1}^{NP}x_k^2$$

also $$\because \sum_{i=1}^{NP}\sum_{j=1}^{NP}\sum_{k=1}^{NP}x_i x_j = NP\sum_{i=1}^{NP}x_i\sum_{j=1}^{NP}x_j = NP\left(\sum_{i=1}^{NP}x_i\right)^2$$

and $\because \sum_{i=1}^{NP}\sum_{j=1}^{NP}\sum_{k=1}^{NP} x_i x_j = \sum_{i=1}^{NP}\sum_{j=1}^{NP}\sum_{k=1}^{NP} x_j x_k = \sum_{i=1}^{NP}\sum_{j=1}^{NP}\sum_{k=1}^{NP} x_k x_i = NP\left(\sum_{i=1}^{NP} x_i\right)^2$

So,

$$E(U_m^{\ 2}) = (1-CR)x_m^{\ 2} + CR(2F^2+1)[\frac{1}{NP}\sum_{i=1}^{NP} x_i^{\ 2} - (\frac{1}{NP}\sum_{i=1}^{NP} x_i)^2] + CR(\frac{1}{NP}\sum_{i=1}^{NP} x_i)^2$$

$$\Rightarrow E(U_m^{\ 2}) = (1-CR)x_m^{\ 2} + CR(2F^2+1)Var(x) + CRx_{av}^{\ 2}$$

Where, $Var(x) = \frac{1}{NP}\sum_{i=1}^{NP} x_i^{\ 2} - (\frac{1}{NP}\sum_{i=1}^{NP} x_i)^2$ and $x_{av} = \frac{1}{NP}\sum_{i=1}^{NP} x_i$, and hence the proof.

Remark: Note that if $CR = 0$, $E(U_m) = x_m$ and $E(U_m^2) = x_m^2$, i.e. the expected value of the trial vector remains same as that of the target vector. This is intuitively simple as for $CR = 0$, the trial vector can only inherit from the target but nothing from the donor/mutant vector. Again if $CR = 1$, $E(U_m) = x_{av}$ and $E(U_m^2) = (2F^2+1).Var(x) + x_{av}^2$. Clearly if $0 < CR < 1$, the expected value of the trial vector lies in between x_m and x_{av}.

Theorem 3.2: If the DE population may be modeled as a continuous-time, dynamic system, then the expectation value of the velocity of an individual point on the fitness landscape may be given as:

$$E(\frac{dx_m}{dt}) = -\frac{k}{8}CR\{(2F^2+1)Var(x) + (x_{av} - x_m)^2\}f'(x_m) + \frac{1}{2}CR(x_{av} - x_m) \tag{3.3}$$

Proof: Let us assume that mutation and crossover occur in unit time to give rise to off offspring. In selection x_m is replaced by U_m if the objective function value for $x = U_m$ is less than or equal to that for $x = x_m$.This decision-making is performed using Heaviside's unit step function [31] , which is defined as follows:

$$\begin{aligned} u(p) &= 1 \quad \text{if} \quad p \geq 0 \\ &= 0 \quad \text{otherwise.} \end{aligned}$$

Now, let at time t position of m - th trial solution be x_m and at $t + \Delta t$ it is changed to $x_m + \Delta x_m$

Then, $\frac{\Delta x_m}{\Delta t} = u[\frac{f(x_m) - f(x_m + \Delta x_m)}{\Delta t}](U_m - x_m)$

$$\Rightarrow \frac{\Delta x_m}{\Delta t} = u[\frac{f(x_m) - f(x_m + \Delta x_m)}{\Delta x_m} \frac{\Delta x_m}{\Delta t}](U_m - x_m)$$

$$\Rightarrow \underset{\Delta t \to 0}{Lt} \frac{\Delta x_m}{\Delta t} = \underset{\Delta t \to 0}{Lt}\, u[\frac{f(x_m) - f(x_m + \Delta x_m)}{\Delta x_m} \frac{\Delta x_m}{\Delta t}](U_m - x_m)$$

$$\Rightarrow \underset{\Delta t \to 0}{Lt} \frac{\Delta x_m}{\Delta t} = \underset{\Delta t \to 0}{Lt}\, u[-\frac{f(x_m + \Delta x_m) - f(x_m)}{\Delta x_m} \frac{\Delta x_m}{\Delta t}](U_m - x_m)$$

$$\therefore \frac{dx_m}{dt} = u[-f'(x_m)\frac{dx_m}{dt}](U_m - x_m) \tag{3.4}$$

Now, we have to replace unit step function by logistic function to carry out the analysis. Ideally,

$$u(p) = \underset{k \to \infty}{lt} \frac{1}{1 + e^{-kp}}.$$

Let us take a moderate value of k for analysis. The approximation of unit-step function with logistic function has been depicted in Figure 3.3. $u(p) \approx \frac{1}{1+e^{-kp}}$. Now, if p is very small. Then, $e^{-kp} \approx 1 - kp$ [neglecting higher order terms]

$$\therefore u(p) \approx \frac{1}{1 + e^{-kp}} \approx \frac{1}{2 - kp} = \frac{1}{2}(1 - \frac{kp}{2})^{-1}$$

Again assuming p to be very small and neglecting higher order terms in expansion of $(1 - \frac{kp}{2})^{-1}$ we obtain,

$$u(p) \approx \frac{1}{2} + \frac{k}{4} p \tag{3.5}$$

Now, the population has a small divergence. $\therefore U_m - x_m$ is not very large and hence $\left|\frac{dx_m}{dt}\right|$ is small.

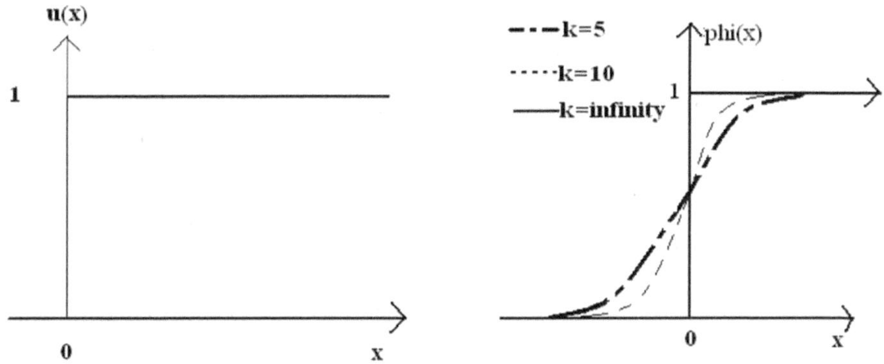

Fig. 3.3 The unit step and the logistic functions

Also we have assumed that fitness landscape has a moderate slope i.e. $|f'(x_m)|$ is also small, which in turn suggests that $\left|f'(x_m)\frac{dx_m}{dt}\right|$ is small. Thus from equations (3.4) and (3.5) we get,

$$\frac{dx_m}{dt} = [\frac{1}{2} - \frac{k}{4} f'(x_m)\frac{dx_m}{dt}](U_m - x_m)$$

$$\Rightarrow \frac{dx_m}{dt} = \frac{\frac{1}{2}(U_m - x_m)}{1 + \frac{k}{4} f'(x_m)(U_m - x_m)} \tag{3.6}$$

Now, $\left|\frac{k}{4} f'(x_m)(U_m - x_m)\right|$ is small

$$\therefore [1 + \frac{k}{4} f'(x_m)(U_m - x_m)]^{-1} \approx 1 - \frac{k}{4} f'(x_m)(U_m - x_m).$$

From (3.6) we get,

$$\frac{dx_m}{dt} = -\frac{k}{8}(U_m - x_m)^2 f'(x_m) + \frac{U_m - x_m}{2} \tag{3.7}$$

Now U_m is a random variable. $\therefore \frac{dx_m}{dt}$, which is a function of U_m, is also a random variable.

Let us try to compute its expected value.

$$E(\frac{dx_m}{dt}) = -\frac{k}{8} f'(x_m) E(U_m - x_m)^2 + \frac{1}{2} E(U_m - x_m)$$

$$\Rightarrow E(\frac{dx_m}{dt}) = -\frac{k}{8} f'(x_m)[E(U_m^{\ 2}) + x_m^{\ 2} - 2x_m E(U_m)] + \frac{1}{2}[E(U_m) - x_m] \tag{3.8}$$

Substituting values of $E(U_m), E(U_m^{\ 2})$ from equation (3.1) and (3.2) to equation (3.8) we get,

$$E(\frac{dx_m}{dt}) = -\frac{k}{8} CR\{(2F^2+1)Var(x) + (x_{av} - x_m)^2\} f'(x_m) + \frac{1}{2} CR(x_{av} - x_m)$$

and hence the proof.

Theorem 3.4: Let x_{av} denote the centroid (mean of all points) of the current population and $x_{av} = \frac{1}{NP}\sum_{m=1}^{NP} x_m$. Also let us denote $\varepsilon_m = x_{av} - x_m$ = deviation of individual from average. Then expected velocity of the centroid of the population may be given by,

$$E(\frac{dx_{av}}{dt}) = -\frac{k}{8} CR(2F^2+1)Var(x) f_{av}' - \frac{k}{8} CR(\frac{1}{N}\sum_{m=1}^{NP} \varepsilon_m^{\ 2} f'(x_{av} + \varepsilon_m)) \tag{3.9}$$

Proof: Now, $x_{av} = \frac{1}{NP}\sum_{i=1}^{NP} x_i = \frac{1}{NP}\sum_{m=1}^{NP} x_m$

$$\Rightarrow \frac{dx_{av}}{dt} = \frac{d}{dt}(\frac{1}{NP}\sum_{m=1}^{NP} x_m) = \frac{1}{NP}\sum_{m=1}^{NP}\frac{dx_m}{dt}$$

$$\Rightarrow E(\frac{dx_{av}}{dt}) = E(\frac{1}{NP}\sum_{m=1}^{NP}\frac{dx_m}{dt}) = \frac{1}{NP}\sum_{m=1}^{NP} E(\frac{dx_m}{dt})$$

$$\Rightarrow E(\frac{dx_{av}}{dt}) = \frac{1}{NP}(-\frac{k}{8}CR\sum_{m=1}^{NP}\{(2F^2+1)Var(x) + (x_{av} - x_m)^2\} f'(x_m) + \frac{1}{2}CR\sum_{m=1}^{NP}(x_{av} - x_m))$$

Now, $\sum_{m=1}^{NP}(x_{av} - x_m) = 0$

$$\therefore E(\frac{dx_{av}}{dt}) = \frac{1}{NP}(-\frac{k}{8}CR.\sum_{m=1}^{NP}\{(2F^2+1)Var(x) + (x_{av} - x_m)^2\}f^{'}(x_m)$$

Let us denote $\varepsilon_m = x_{av} - x_m$ = deviation of individual from average.

$$\therefore E(\frac{dx_{av}}{dt}) = -\frac{k}{8}CR(2F^2+1)Var(x)(\frac{1}{NP}\sum_{m=1}^{NP}f^{'}(x_m)) - \frac{k}{8}CR(\frac{1}{NP}\sum_{m=1}^{NP}\varepsilon_m{}^2 f^{'}(x_m))$$

$$\Rightarrow E(\frac{dx_{av}}{dt}) = -\frac{k}{8}CR(2F^2+1)Var(x)\left(\frac{1}{NP}\sum_{m=1}^{NP}f^{'}(x_m)\right) - \frac{k}{8}CR(\frac{1}{NP}\sum_{m=1}^{NP}\varepsilon_m{}^2 f^{'}(x_{av}+\varepsilon_m))$$

$$\therefore E(\frac{dx_{av}}{dt}) = -\frac{k}{8}CR(2F^2+1)Var(x)f_{av}{}^{'} - \frac{k}{8}CR(\frac{1}{N}\sum_{m=1}^{NP}\varepsilon_m{}^2 f^{'}(x_{av}+\varepsilon_m)) \tag{3.10}$$

Where, $f^{'}{}_{av} = \frac{1}{NP}\sum_{m=1}^{NP}f^{'}(x_m)$ = average of the gradients for trial solution points on fitness landscape. This completes the proof.

Remark: From theorem 3.3, we may write,

$$E(\frac{dx_m}{dt}) = -\alpha_{DE}f^{'}(x_m) + \beta_{DE} \tag{3.11}$$

Where, $\alpha_{DE} = -\frac{k}{8}CR\{(2F^2+1)Var(x) + (x_{av} - x_m)^2\}$ and

$\beta_{DE} = \frac{1}{2}CR(x_{av} - x_m)$

The classical gradient descent search algorithm is given by the following dynamics (continuous) in single dimension [25]:

$$\frac{d\theta}{dt} = -\alpha.G + \beta \tag{3.12}$$

where α is the learning rate and β is the momentum.

The resemblance of equations (3.11) and (3.12) is not difficult to recognize and this suggests that, the dynamics of actual DE uses some kind of estimation for the gradient of the objective function. In equation (3.9), $-\alpha_{DE}f^{'}(x_m)$ term on the R.H.S. is responsible for moving along the direction of the negative gradient,

whereas β_{DE} represents a component of velocity of a trial solution towards the mean vector (center of mass) of the population.
Evidently very near to an optimum, when $f'(x_m) \to 0$,

$$E(\frac{dx_m}{dt}) \approx \beta_{DE} = \frac{1}{2}CR(x_{av} - x_m) \tag{3.13}$$

If the population is not much scattered, $(x_{av} - x_m)$ is very small.

$$\therefore E(\frac{dx_m}{dt}) \to 0$$

So, it converges towards optima. Under these circumstances, the population members are confined within a small region of the search space and hence $Var(x) \to 0$, $x_{av} - x_m \to 0$ and also $\varepsilon_m \to 0$. Thus, from (3.13) and (3.10) we get $E(\frac{dx_m}{dt}) \to 0$ and $E(\frac{dx_{av}}{dt}) \to 0$.

3.3 A State Space Formulation of the DE Population

To study stability of DE algorithm we first model it as an autonomous control system. Here each population member x_m is a state variable of the control system. We try to get state equations of the system. Then we proceed for stability tests. From (3.3) we get,

$$E(\frac{dx_m}{dt}) = -\frac{k}{8}CR\{(2F^2+1)Var(x)+(x_{av}-x_m)^2\}f'(x_i)+\frac{1}{2}CR(x_{av}-x_m),$$

for $m = 1,2,..., NP$.

We have assumed the population of trial vectors is located very close to optima. So $f'(x_m)$ is negligibly small. Hence the equation can be written as,

$$E(\frac{dx_m}{dt}) = \frac{1}{2}CR(x_{av} - x_m), \qquad \text{for } i = 1,2,..., NP$$

$$E(\frac{dx_m}{dt}) = \frac{1}{2}CR(\frac{1}{NP}\sum_{j=1}^{NP} x_j - x_m), \quad \text{for } i = 1,2,..., NP \tag{3.14}$$

Actually equation (3.14) represents NP number of simultaneous equations. Next, we represent them using matrix notation. From (3.14) we get,

$$\begin{bmatrix} E(\frac{dx_1}{dt}) \\ E(\frac{dx_2}{dt}) \\ \cdots\cdots \\ \cdots\cdots \\ E(\frac{dx_{NP}}{dt}) \end{bmatrix} = \frac{1}{2}CR \begin{bmatrix} \frac{1}{NP}-1 & \frac{1}{NP} & \cdots\cdots & \frac{1}{NP} \\ \frac{1}{NP} & \frac{1}{NP}-1 & \cdots\cdots & \frac{1}{NP} \\ \cdots\cdots & \cdots\cdots & \cdots\cdots & \cdots\cdots \\ \cdots\cdots & \cdots\cdots & \cdots\cdots & \cdots\cdots \\ \frac{1}{NP} & \frac{1}{NP} & \cdots\cdots & \frac{1}{NP}-1 \end{bmatrix} \begin{bmatrix} x_1 \\ x_2 \\ \cdots \\ \cdots \\ \cdots \\ \cdots \\ x_{NP} \end{bmatrix} \tag{3.15}$$

The above matrix equation is of the form $\left[E\left(\frac{d\vec{x}}{dt}\right)\right] = A[\vec{x}]$, Where

$\vec{x} = [x_1, x_2, ..., x_{NP}]^T$ is the set of state variables and:

$$A = \frac{1}{2}CR \begin{bmatrix} \frac{1}{NP}-1 & \frac{1}{NP} & \cdots\cdots & \frac{1}{NP} \\ \frac{1}{NP} & \frac{1}{NP}-1 & \cdots\cdots & \frac{1}{NP} \\ \cdots\cdots & \cdots\cdots & \cdots\cdots & \cdots\cdots \\ \cdots\cdots & \cdots\cdots & \cdots\cdots & \cdots\cdots \\ \frac{1}{NP} & \frac{1}{NP} & \cdots\cdots & \frac{1}{NP}-1 \end{bmatrix}$$

We know that eigen-values of matrix A are the poles of the system. Eigen-values are those values of λ for which $\det[\lambda I - A] = 0$ is satisfied, where I is the identity matrix of order NP.

$$\det[\lambda I - A] = 0 \Rightarrow \det \begin{bmatrix} \frac{2}{CR}\lambda - \frac{1}{NP}+1 & -\frac{1}{NP} & \cdots\cdots & -\frac{1}{NP} \\ -\frac{1}{NP} & \frac{2}{CR}\lambda - \frac{1}{NP}+1 & \cdots\cdots & -\frac{1}{NP} \\ \cdots\cdots & \cdots\cdots & \cdots\cdots & \cdots\cdots \\ \cdots\cdots & \cdots\cdots & \cdots\cdots & \cdots\cdots \\ -\frac{1}{NP} & -\frac{1}{NP} & \cdots\cdots & \frac{2}{CR}\lambda - \frac{1}{NP}+1 \end{bmatrix} = 0$$

After doing simple algebraic operations on the rows of the determinant in LHS of the above equation, we get,

$$\lambda(\lambda + \frac{CR}{2})^{NP-1} = 0 \tag{3.16}$$

Note that (3.16) is the characteristic equation of matrix A. From (3.16) we get tha eigen values as $\lambda = 0, -\frac{CR}{2}, -\frac{CR}{2}, \ldots\ldots, -\frac{CR}{2}$.

These values of λ are the system poles. Since one pole is at zero, the system is not asymptotically stable and must have a DC component. In the following section, we check whether the system is stable in the sense of Lyapunov.

3.4 Lyapunov Stability Analysis of the DE Population

In this section we analyze the stability of the population-dynamics represented by equation (3.16) using the concept of Lyapunov stability theorems [28]. We begin this treatment by explaining some basic concepts and their interpretations from the standard literature on nonlinear control theory [28, 29].

Definition 3.1

A point $\vec{x} = \vec{x}_e$ is called an *equilibrium state,* if the dynamics of the system is given by

$$\frac{d\vec{x}}{dt} = f(\vec{x}(t))$$

becomes zero at $\vec{x} = \vec{x}_e$ for any t i.e. $f(\vec{x}_e(t)) = 0$. The equilibrium state is also called equilibrium (stable) point in D-dimensional hyperspace, when the state $\vec{x}_e$ has D-components.

Definition 3.2

A scalar function $V(\vec{x})$ is said to be *positive definite* with respect to the point $\vec{x}_e$ in the region $\|\vec{x} - \vec{x}_e\| \le K$, if $V(\vec{x}) > 0$ at all points of the region except at $\vec{x}_e$ where it is zero.

Definition 3.3

A scalar function $V(\vec{x})$ is said to be *negative definite* if $-V(\vec{x})$ is positive definite.

Definition 3.4

A dynamics $\frac{d\vec{x}}{dt} = f(\vec{x}(t))$ is *asymptotically stable* at the equilibrium point $\vec{x}_e$, if

a) it is stable in the sense of Lyapunov, i.e., for any neighborhood $S(\varepsilon)$ surrounding $\vec{x}_e$ ($S(\varepsilon)$ contains points $\vec{x}$ for which $\|\vec{x}-\vec{x}_e\| \le \varepsilon$) where there is a region $S(\delta)$($S(\delta)$ contains points $\vec{x}$ for which $\|\vec{x}-\vec{x}_e\| \le \delta$), $\delta < \varepsilon$, such that trajectories of the dynamics starting within $S(\delta)$ do not leave $S(\varepsilon)$ as time $t \to \infty$ and

b) the trajectory starting within $S(\delta)$ converges to the origin as time t ap proaches infinity.

The sufficient condition for stability of a dynamics can be obtained from the Lyapunov's theorem, presented below.

Lyapunov's stability theorem [28, 32]

Given a scalar function $V(\vec{x})$ and some real number $\varepsilon > 0$, such that for all $\vec{x}$ in the region $\|\vec{x}-\vec{x}_e\| \le \varepsilon$ the following conditions hold:

1) $V(\vec{x}_e) = 0$

2) $V(\vec{x}) > 0$ for $\vec{x} \neq \vec{x}_e$, i.e. $V(\vec{x})$ is positive definite.

3) $V(\vec{x})$ has continuous first partial derivatives with respect to all components of $\vec{x}$.

Then the equilibrium state $\vec{x}_e$ of the system $\dfrac{d\vec{x}}{dt} = f(\vec{x}(t))$ is

a) *asymptotically stable* if $\dfrac{dV}{dt} < 0$, i.e. $\dfrac{dV}{dt}$ is negative definite, and

b) *asymptotically stable* in the large if $\dfrac{dV}{dt} < 0$ for $\vec{x} \neq \vec{x}_e$, and in addition, $V(\vec{x}) \to \infty$ as $\|\vec{x}-\vec{x}_e\| \to \infty$.

Remark: Lyapunov stability analysis is based on the idea that if the total energy in the system continually decreases, then the system will asymptotically reach the zero energy state associated with an equilibrium point of the system. A system is said to be asymptotically stable if all the states approach the equilibrium state with time.

Theorem 3.6: The system defined in equations (3.14) and (3.15) satisfies Lyapunov's stability criterion.

***Proof*:** We are assuming the population is located very close to optima. Hence value of the gradient is negligibly small. So equation (3.14) holds true in such a region.

$$E(\frac{dx_m}{dt}) = \frac{1}{2} CR (\frac{1}{NP} \sum_{j=1}^{NP} x_j - x_m) \,,$$

for $m = 1,2,..., NP$

The condition for an equilibrium point is $E(\frac{dx_m}{dt}) = 0$, for $m = 1,2..., NP$ [according to definition 1.1]. We consider the case where the DE population is confined within a small neighborhood of an isolated optimum and over the entire population value of the gradient is very small. In this case, the preferred equilibrium point should be the optimum itself. This ensures that with time there is no change in values of state variables i.e. positions of the population members after they hit the optimum. Now from (3.14),

$$E(\frac{dx_m}{dt}) = 0$$

$$\Rightarrow \frac{1}{2} CR(\frac{1}{NP} \sum_{j=1}^{NP} x_j - x_m) = 0 \text{ , for } i = 1,2..., NP \,.$$

$$\Rightarrow x_i = \frac{1}{NP} \sum_{j=1}^{NP} x_j = x_{av} \text{ , for } i = 1,2.NP \,.$$

This is possible only if all of the state variables are equal in value i.e. $x_1 = x_2 = ... = x_{NP} = x_e$, where x_e is the equilibrium position. At this point we would like to mention that as the search progresses; the population-members in DE get to the better portions of the search space or remain constant owing to its greedy selection strategy. In case of a smooth, unimodal fitness landscape, the solution vectors generally crowd into a small neighborhood surrounding the optimum. Thus during the later stages of search, the equilibrium point x_e basically is identical to the optimum, once reaching at which point, population members are expected not to change any further and thus this point should satisfy the condition $x_1 = x_2 = ... = x_{NP}$ as well. This section examines the stability of the solution vectors very near to such an optimum point of the search space. Figure 3.4 shows a fitness landscape and an equilibrium position at the optimum. Next, we define Lyapunov's Energy function V as,

$$V(x_1, x_2,.......x_{NP}, t) = \sum_{i=1}^{NP} (x_i - x_{av})^2 \tag{3.17}$$

Clearly $V = 0$, if $x_1 = x_2 = ... = x_{NP} = x_e$

> 0, otherwise.

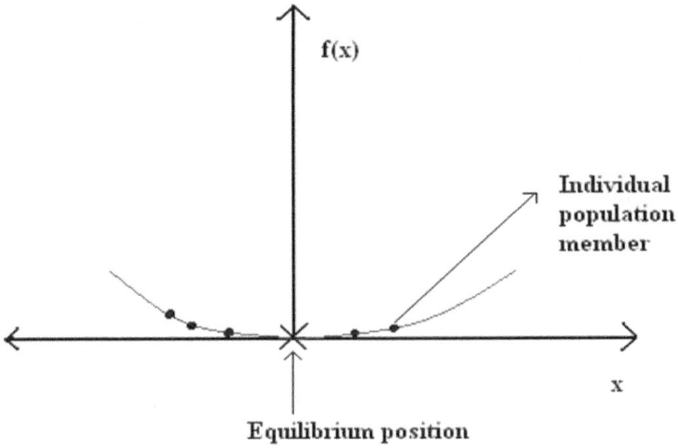

Fig. 3.4 State variables along with equilibrium position

Energy function is always positive except the equilibrium, where it becomes zero. So, energy function is a positive definite with respect to equilibrium [from definition 3.2]. It is to be noted that $V = NP(Var(x))$.

Clearly V has continuous partial derivative.

Differentiating both sides of (3.17) with respect to time, we get

$$\frac{dV}{dt} = 2\sum_{i=1}^{NP}(x_m - x_{av})(\frac{dx_m}{dt} - \frac{dx_{av}}{dt})$$

$$E\left(\frac{dV}{dt}\right) = 2\sum_{i=1}^{NP}(x_m - x_{av})(E\left(\frac{dx_m}{dt}\right) - E\left(\frac{dx_{av}}{dt}\right)) \tag{3.18}$$

From (3.14) we get,

$$E(\frac{dx_m}{dt}) = \frac{1}{2}CR(x_{av} - x_m) \text{ and}$$

$$E(\frac{dx_{av}}{dt}) = E(\frac{d}{dt}(\frac{1}{NP}\sum_{i=1}^{NP}x_i)) = \frac{1}{NP}E\left(\sum_{i=1}^{NP}\frac{dx_i}{dt}\right) = 0.$$

Putting expectation values in (3.18),

$$E\left(\frac{dV}{dt}\right) = -CR\sum_{i=1}^{NP}(x_i - x_{av})^2 \tag{3.19}$$

From equation (3.19) it is evident that $E\left(\frac{dV}{dt}\right)$ is 0 when $x_1 = x_2 = \ldots\ldots\ldots\ldots = x_{NP} = x_e$ and is negative otherwise. Hence $E\left(\frac{dV}{dt}\right)$ is a negative definite with respect to equilibrium point. Here V is positive definite and $E\left(\frac{dV}{dt}\right)$ is negative definite, satisfying Lyapunov's stability theorem. We can infer that the system is asymptotically stable and it converges to the optima.

Stability of DE system has been already ascertained. Now to gain further insight over convergence of the process we try to find out an estimate for convergence time.

Theorem 3.7: An estimate of the system time-constant can be $\frac{1}{CR}$

***Proof*:** Equation (3.19) can be written as

$$\frac{V}{-E\left(\frac{dV}{dt}\right)} = \frac{1}{CR}. \tag{3.20}$$

The term in the denominator of L.H.S of above expression is the expected or average value of time rate of change of energy function. Let the process be carried out repeatedly for same initial conditions and parameter values and an average of energy function is calculated for the runs and the average of the energy function be denoted by $\langle V \rangle$. Time rate of change of the average is also computed and let it be denoted as $\frac{d\langle V \rangle}{dt}$ .We assume that the runs of the algorithm are independent and probability associated with selecting a population member in any stage of the algorithm does not change with time i.e. the process is time invariant. In that case we may expect from (3.20)

$$\frac{\langle V \rangle}{-\frac{d\langle V \rangle}{dt}} = \frac{1}{CR}$$

$$\Rightarrow \langle V \rangle = V_0 \exp\left(-\frac{t}{1/CR}\right) \tag{3.21}$$

where V_0 is the initial value of energy function. We have seen that energy function decreases with time. We may define a time-constant for the system as the time interval in which the energy function reduces to $\frac{1}{e}$ part of its initial value. If we denote this time-constant by T, putting $\langle V \rangle = \frac{V_0}{e}$ and $t = T$ in (3.21)we get time-constant $T = \frac{1}{CR}$.

3.5 Computer Simulation Results

In this section we provide the phase plots ($v = \frac{dx}{dt}$ versus x plots) for DE/rand/1/bin, which supports the theoretical results derived in the earlier section. A population of 11 vectors is taken to optimize the single dimensional sphere function $f(x) = x^2$ using the DE Algorithm. In Figure 3.5 four phase-trajectories

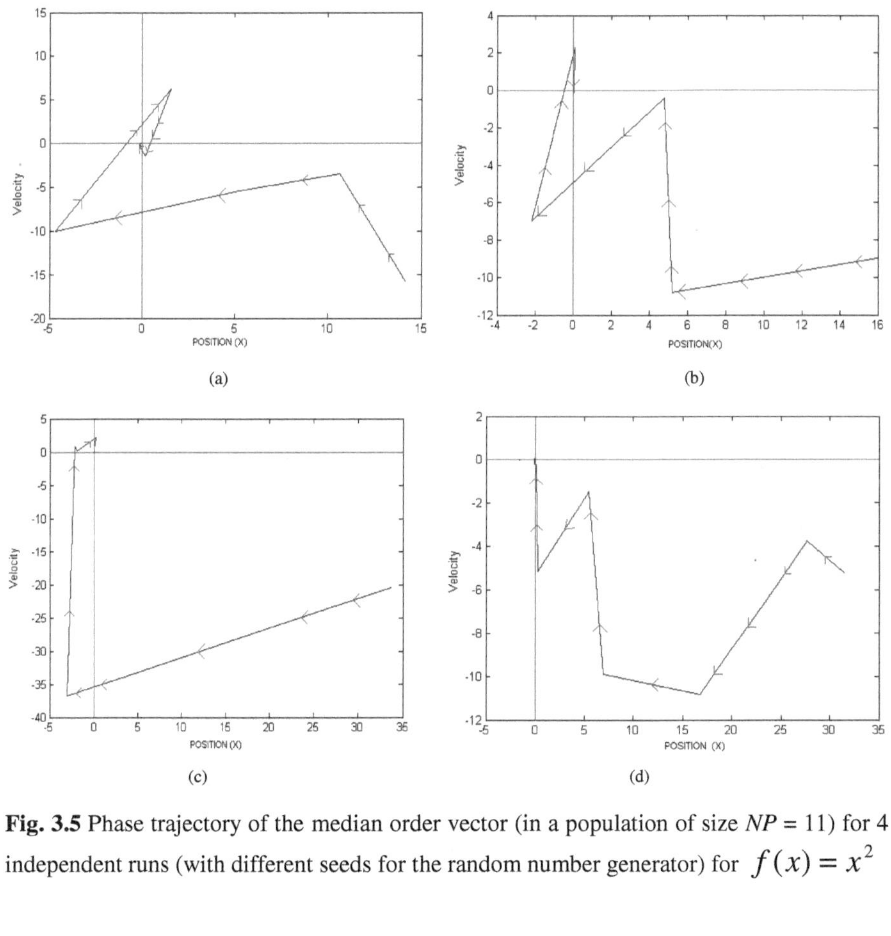

Fig. 3.5 Phase trajectory of the median order vector (in a population of size NP = 11) for 4 independent runs (with different seeds for the random number generator) for $f(x) = x^2$

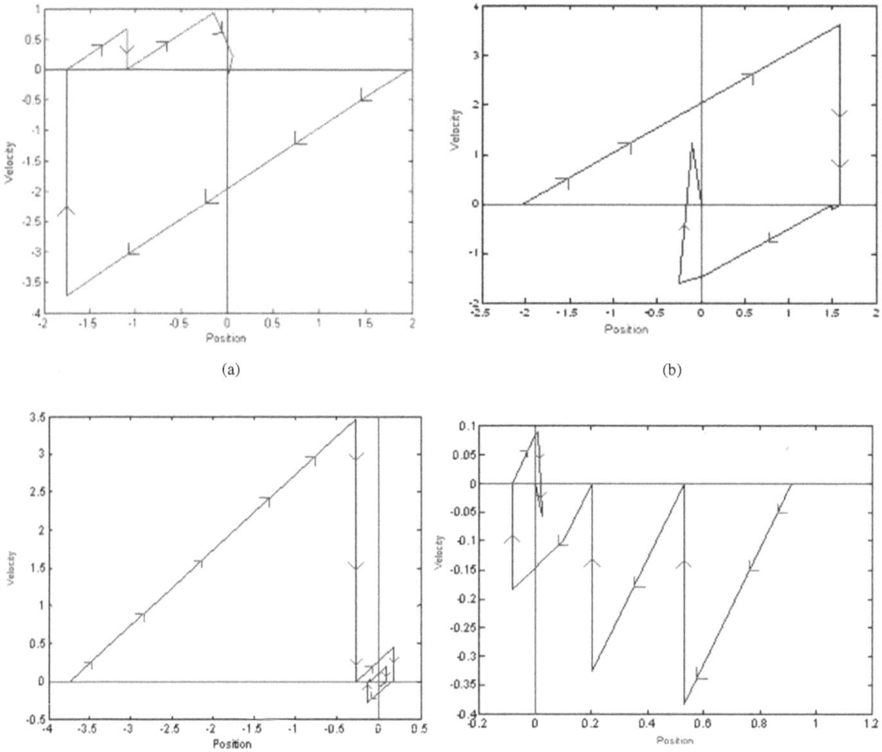

Fig. 3.6 Phase trajectory of the median order vector (in a population of size *NP* = 11) for 4 independent runs (with different seeds for the random number generator) for $f(x) = 1 - e^{-x^2}$

have been shown for the median vector (when the population is ranked according to the final fitness values of the vectors) over four independent runs (with different initial populations). These phase trajectories verifies our theoretical finding that near an optima, the expected velocity $E(\frac{dx}{dt})$ of individual member of population gradually approaches zero.

Similarly, we construct phase trajectories for objective function $f(x) = 1 - e^{-x^2}$. New set of phase trajectories is shown in Figure 3.6.

We have estimated convergence time from time rate of Lyapunov energy function in theorem 3.7. Now, according to (23) convergence time is inversely proportionate to crossover probability. In Figure 3.7 plots of time variations of Lyapunov's energy function is provided for various crossover probabilities (objective function used $f(x) = x^2$). From Figure 3.7 we observe as crossover probability increases convergence time gradually decreases. This matches with our theoretical finding of theorem 3.7. From Figure 3.7 we graphically determine convergence time, which is the time in which Lyapunov energy function diminishes to e-th (approx 2.71) fraction of its initial value. In Table 3.1 below we make a comparison between convergence time measured from Figure 3.6 and found from (3.20).

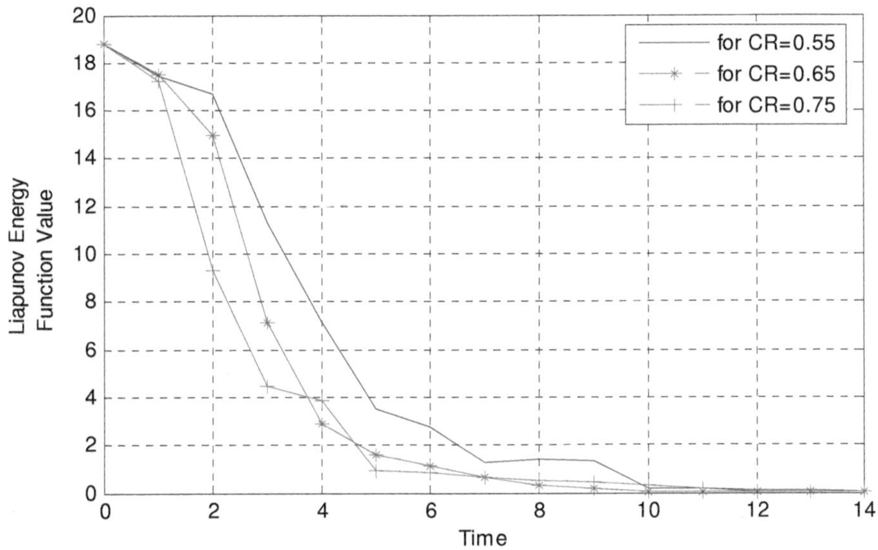

Fig. 3.7 Convergence characteristics for various values of crossover probability

Table 3.1 Comparison between calculated and experimental convergence time

Crossover probability	Convergence time (Expressed in number of iterations)	
	Measured graphically	Calculated theoretically
0.55	2.4	2.31
0.65	2.1	1.94
0.75	1.8	1.73

3.6 Conclusions

Differential Evolution (DE) has been regarded as a competitive form of Evolutionary Algorithm for function optimization problems in recent years. In this article we provide a simple analysis of the evolutionary dynamics undertaken by each of the population members in DE/rand/1/bin, which appears as one of the most popular and widely used variant of the DE. We apply simple statistical and calculus-based methods in order to derive a dynamical model of the DE-population that undergoes mutation, binomial crossover and selection. The selection mechanism in DE has been modeled by the well-known unit step function, which was subsequently approximated by continuous logistic function. One important finding of our analysis is the similarity of the fundamental differential equation governing the expected velocity of a search-agent in DE with that of the classical gradient descent search with momentum. This suggests that DE uses a stochastic estimate of the gradient of the objective function (which was assumed to be continuous in our analysis in order to keep the mathematics less rigorous) in order to locate the optima of the function. It is due to the gradient descent type search strategy, that DE converges much faster than algorithms like GA or Particle Swarm Optimization (PSO) over uni-modal benchmarks [33]. However, the actual algorithm does not take into account any analytical expression of the true function-gradient and due to the randomness introduced by mutation and crossover operations into the dynamics of an individual, can escape trapping in local optima in many cases.

Based on the mathematical model derived here, we also analyze the stability of a DE population, very near to an isolated optimum, which acts as the equilibrium point for the dynamics. Application of Lyapunov's stability theorems reveals that the near-equilibrium behavior of a DE population is inherently stable and free from any kind of oscillatory behaviors seen in other optimization algorithms like Bacterial Foraging Optimization (BFO) [34] or PSO [34]. Our analysis reveals that the control parameter *CR* governs the rate of convergence of a DE population to an optimum. Future research may focus on analyzing the stability of the DE dynamics based on a stochastic Lyapunov energy function approach [35].

Appendix

In this section, we carry out the previous analysis for 'DE/current-to-rand/1' previously described in section 2.5.2 of chapter 2. Besides previous assumptions described in section 3.1, we also assume $k_1 = k_2 = ... = k_{NP} = k_{crossover}$. This assumption is made to simplify the analysis. Similar to the derivations done in theorem 1 and 2, we calculate the following expectations.

$$E(U_m - x_m) = k_{crossover}(x_{av} - x_m) \tag{3.27}$$

$$E(U_m - x_m)^2 = k_{crossover}{}^2 \left\langle x^2 \right\rangle_{av} (1 + 2F^2) + (k_{crossover}{}^2 x_m{}^2 - 2k_{crossover}{}^2 x_m x_{av} - 2k_{crossover}{}^2 F^2 x_{av}{}^2) \tag{3.28}$$

Where, $\left\langle x^2 \right\rangle_{av} = \dfrac{1}{NP} \sum_{i=1}^{NP} x_i{}^2$

Selection step is exactly same in the two versions of algorithms. Theorem 3.3 also holds for this case. From theorem 3.3 we obtain expression for $E\left(\frac{dx_m}{dt}\right)$, which is as following,

$$E(\frac{dx_m}{dt}) = -\frac{k}{8} f^{'}(x_m)E(U_m - x_m)^2 + \frac{1}{2}E(U_m - x_m) \tag{3.29}$$

Substituting values from (3.27) and (3.28) we get,

$$E(\frac{dx_m}{dt}) = \alpha_{new} f^{'}(x_m) + \beta_{new} \tag{3.30}$$

where,

$$\alpha_{new} = -\frac{k}{8} k_{crossover}{}^2 \langle x^2 \rangle_{av} (1 + 2F^2) + (k_{crossover}{}^2 x_m{}^2 - 2k_{crossover}{}^2 x_m x_{av} - 2k_{crossover}{}^2 F^2 x_{av}{}^2$$

and $\beta_{new} = \frac{k_{crossover}}{2}(x_{av} - x_m)$.

Equation (3.30) shows that the fundamental dynamics of 'DE/current-to-rand/1' near an optimum also has a resemblance with the classical gradient descent algorithm. We carry out stability tests in a way exactly similar to that of done in section 4. We found that 'DE/current-to-rand/1' is also asymptotically stable, satisfying Lyapunov's criterion. In this case convergence time becomes $\frac{1}{k_{crossover}}$.

References

[1] Reeves, C.R., Rowe, J.E.: Genetic Algorithms – Principles and Perspectives: A Guide to GA Theory. Kluwer Academic Publishers, Dordrecht (2003)

[2] Rudolph, G.: Convergence analysis of canonical genetic algorithms. IEEE Transactions on Neural Networks 5(1), 96–101 (1994)

[3] Baeck, T.: Order statistics for convergence velocity analysis of simplified evolutionary algorithms. In: Foundations of Genetic Algorithms, pp. 91–102 (1994)

[4] Beyer, H.-G.: On the dynamics of EAs without selection. In: Proceedings of Foundations of Genetic Algorithms, 5 (FOGA-5), pp. 5–26 (1999)

[5] Vose, M.D.: The Simple Genetic Algorithm: Foundations and Theory. MIT Press, Cambridge (1999)

[6] Pruegel-Bennett, A.: Modeling genetic algorithm dynamics. In: Theoretical Aspects of Evolutionary Computing, pp. 59–85 (2001)

[7] He, J., Yao, X.: Towards an analytic framework for analyzing the computation time of evolutionary algorithms. Artificial Intelligence 145(1-2), 59–97 (2003)

[8] Okabe, T., Jin, Y., Sendhoff, B.: On the dynamics of evolutionary multi-objective optimization. In: Proceedings of Genetic and Evolutionary Computation Conference (GECCO 2002), pp. 247–255 (2002)
[9] Okabe, T., Jin, Y., Sendhoff, B.: Evolutionary multi-objective optimization with a hybrid representation. In: Proceedings of Congress on Evolutionary Computation (CEC 2003), pp. 2262–2269 (2003)
[10] Gamperle, R., Muller, S.D., Koumoutsakos, A.: Parameter study for differential evolution. In: WSEAS NNA-FSFS-EC 2002, Interlaken, Switzerland, February 11-15, 2002. WSEAS (2002)
[11] Ronkkonen, J., Kukkonen, S., Price, K.V.: Real parameter optimization with differential evolution. In: The 2005 IEEE Congress on Evolutionary Computation (CEC 2005), vol. 1, pp. 506–513. IEEE Press, Los Alamitos (2005)
[12] Liu, J., Lampinen, J.: A Fuzzy Adaptive Differential Evolution Algorithm. Soft computing- A Fusion of Foundations, Methodologies and Applications 9(6), 448–462 (2005)
[13] Zaharie, D.: Control of Population Diversity and Adaptation in Differential Evolution Algorithms. In: Matousek, D., Osmera, P. (eds.) Proc. of MENDEL 2003, 9th International Conference on Soft Computing, Brno, Czech Republic, pp. 41–46 (June 2003)
[14] Zaharie, D., Petcu, D.: Adaptive pareto differential evolution and its parallelization. In: Wyrzykowski, R., Dongarra, J., Paprzycki, M., Waśniewski, J. (eds.) PPAM 2004. LNCS, vol. 3019, pp. 261–268. Springer, Heidelberg (2004)
[15] Abbass, H.: The self-adaptive pareto differential evolution algorithm. In: Proceedings of the 2002 Congress on Evolutionary Computation, pp. 831–836 (2002)
[16] Omran, M.G.H., Salman, A., Engelbrecht, A.P.: Self-adaptive differential evolution. In: Hao, Y., Liu, J., Wang, Y.-P., Cheung, Y.-m., Yin, H., Jiao, L., Ma, J., Jiao, Y.-C. (eds.) CIS 2005. LNCS (LNAI), vol. 3801, pp. 192–199. Springer, Heidelberg (2005)
[17] Brest, J., Greiner, S., Boskovic, B., Mernik, M., Zumer, V.: Self-adapting control parameters in differential evolution: a comparative study on numerical benchmark problems. IEEE Transactions on Evolutionary Computation 10(6), 646–657 (2006)
[18] Das, S., Konar, A., Chakraborty, U.K.: Two improved differential evolution schemes for faster global search. In: ACM-SIGEVO Proceedings of GECCO 2005, Washington, D.C., pp. 991–998 (June 2005)
[19] Storn, R., Price, K.V., Lampinen, J.: Differential Evolution - A Practical Approach to Global Optimization. Springer, Berlin (2005)
[20] Lampinen, J.: A bibliography of differential evolution algorithm. Technical Report. Lappeenranta University of Technology, Department of Information Technology, Laboratory of Information Processing (1999), `http://www.lut.fi/~jlampine/debiblio.htm`
[21] Zaharie, D.: On the explorative power of differential evolution. In: 3rd International Workshop on Symbolic and Numerical Algorithms on Scientific Computing, SYNASC 2001, Timişoara, Romania, October 2-4 (2001)
[22] Zaharie, D.: Critical values for the control parameters of differential evolution algorithms. In: Matouk, R., Oera, P. (eds.) Proceedings of MENDEL 2002, 8th International Mendel Conference on Soft Computing, Brno, Czech Republic, p. 62. Brno University of Technology, Faculty of Mechanical Engineering, Brno, Czech Republic (2002)

[23] Beyer, H.–G.: On the explorative power of ES/EP-like algorithms. Technical Report, University of Dortmund (1998)
[24] Dasgupta, S., Biswas, A., Das, S., Abraham, A.: The population dynamics of differential evolution: a mathematical model. In: IEEE Congress on Evolutionary Computation CEC 2008, IEEE World Congress on Computational Intelligence, WCCI 2008. IEEE Press, USA (2008)
[25] Dasgupta, S., Biswas, A., Das, S., Abraham, A.: Modeling and analysis of the population dynamics of differential evolution algorithm. AI Communications - The European Journal on Artificial Intelligence (2009)
[26] Snyman, J.A.: Practical Mathematical Optimization: An Introduction to Basic Optimization Theory and Classical and New Gradient-Based Algorithms. Springer, Heidelberg (2005)
[27] Ranasinghe, M.: Mixing paradigms: a genetic operator that approximates gradient descent. In: Genetic Algorithms and Genctic Programming at Stanford 2003 (Book of Student Papers from J. R. Koza's Course at Stanford on Genetic Algorithms and Genetic Programming) Stanford University Bookstore (2003)
[28] Hahn, W.: Theory and Application of Lyapunov's Direct Method. Prentice-Hall, Englewood Cliffs (1963)
[29] Haddad, W.M., Chellaboina, V.: Nonlinear Dynamical Systems and Control: A Lyapunov-Based Approach. Princeton University Press, Princeton (2008)
[30] Kirk, W.A., Sims, B.: Handbook of Metric Fixed Point Theory. Kluwer Academic, London (2001)
[31] Anwal, R.P.: Generalized Functions: Theory and Technique, 2nd edn. Birkhâuser, Boston (1998)
[32] Kuo, B.C.: Automatic Control Systems. Prentice-Hall, Englewood Cliffs (1987)
[33] Vesterstrøm, J., Thomson, R.: A comparative study of differential evolution, particle swarm optimization, and evolutionary algorithms on numerical benchmark problems. In: Proc. Sixth Congress on Evolutionary Computation (CEC 2004). IEEE Press, Los Alamitos (2004)
[34] Dasgupta, S., Das, S., Abraham, A., Biswas, A.: Adaptive Computational Chamotaxis in Bacterial Foraging Optimization – An Analysis. IEEE Transactions on Evolutionary Computation (accepted, 2008)
[35] Clerc, M., Kennedy, J.: The particle swarm-explosion, stability, and convergence in a multidimensional complex space. IEEE Transactions on Evolutionary Computation 6(1), 58–73 (2002)
[36] Semenov, M.A., Terkel, D.A.: Analysis of convergence of an evolutionary algorithm with self-adaptation using a stochastic Lyapunov function. Evolutionary Computation, 363–379 (2003)

Chapter 4
Automatic Hard Clustering Using Improved Differential Evolution Algorithm

Abstract. This chapter describes a Differential Evolution (DE) based algorithm for the automatic clustering of large unlabeled datasets. In contrast to most of the existing clustering techniques, the proposed algorithm requires no prior knowledge of the data to be classified. Rather, it determines the optimal number of clusters in the data 'on the run'. Superiority of the new method has been demonstrated by comparing it with two recently developed partitional clustering techniques and one popular hierarchical clustering algorithm. The partitional clustering algorithms are based on Genetic Algorithm (GA) and the Particle Swarm Optimization (PSO) algorithm. An interesting practical application of the proposed method to automatic segmentation of images is also illustrated.

4.1 Introduction

Tremendous research effort has gone in the past few years to evolve the clusters in complex datasets through evolutionary computing techniques. However, little work has been taken up to determine the optimal number of clusters at the same time. Most of the existing clustering techniques, based on evolutionary algorithms, accept the number of clusters k as an input instead of determining the same on the run. Nevertheless, in many practical situations, the appropriate number of groups in a previously unhandled dataset may be unknown or impossible to determine even approximately. For example, while clustering a set of documents arising from the query to a search engine, the number of classes k changes for each set of documents that result from an interaction with the search engine. Also if the dataset is described by high-dimensional feature vectors (which is very often the case), it may be practically impossible to visualize the data for tracking its number of clusters.

The objective of the research work described in this chapter is two-fold. Firstly, it aims at the automatic determination of the optimal number of clusters in any unlabeled dataset. Secondly, it attempts to show that Differential Evolution (DE) [1], with a modification of the chromosome representation scheme, can give very promising results if applied to the automatic clustering problem. DE is easy to implement and requires a negligible amount of parameter tuning to achieve considerably good search results. We changed the algorithm from its classical form to improve its convergence properties. In addition to that, we used a novel representation scheme for the search variables in order to determine the optimal number of clusters. We refer to the new algorithm as the ACDE (Automatic Clustering Differential Evolution) algorithm [2].

S. Das, A. Abraham, and A. Konar: Metaheuristic Clustering, SCI 178, pp. 137–174.
springerlink.com

At this point, we would like to mention that the traditional approach of determining the optimal number of clusters in a dataset is using some specially devised statistical-mathematical function (also known as a clustering validity index) to judge the quality of partitioning for a range of cluster numbers. A good clustering validity index is generally expected to provide global minima/maxima at the exact number of classes in the dataset. Nonetheless, determination of the optimum cluster number using global validity measures is very expensive, since clustering has to be carried out for a variety of possible cluster numbers. In the proposed evolutionary learning framework, a number of trial solutions come up with different cluster numbers as well as cluster center coordinates for the same dataset. Correctness of each possible grouping is evaluated quantitatively with a global validity index (e.g. the CS or DB measure). Then, through a mechanism of mutation and natural selection, eventually the best solutions start dominating the population while the bad ones are eliminated. Ultimately, the evolution of solutions comes to a halt (i.e. converges), when the fittest solution represents a near-optimal partitioning of the dataset with respect to the employed validity index. In this way, the optimal number of classes along with the accurate cluster center coordinates can be found out in an evolutionary search process. A downside to the proposed method is that, its performance depends heavily upon the choice of a suitable clustering validity index. An inefficient validity index may result into many false clusters (due to the over fitting of data) even when the actual number of clusters in the given dataset may be very much tractable. However, with a judicious choice of the validity index, the proposed algorithm can automate the entire process of clustering and yield near optimal partitioning of any previously unhandled dataset in a reasonable amount of time. This is certainly a very desirable feature of a real-life pattern recognition task.

We have compared the ACDE algorithm extensively with two other state-of-the-art automatic clustering techniques [3, 4] based on Genetic Algorithm (GA) [5] and Particle Swarm Optimization (PSO) [6]. In addition, the quality of the final solutions has been compared against a standard agglomerative hierarchical clustering technique. Following performance metrics have been used in the comparative analysis: 1) the accuracy of final clustering results 2) the speed of convergence and 3) the robustness (i.e. ability to produce nearly same results over repeated runs). The test-suite chosen for this study consists of five real life datasets. Finally, an interesting application of the proposed algorithm has been illustrated in context to the automatic segmentation of a few well-known gray scale images.

4.2 The DE-Based Automatic Clustering Algorithm

4.2.1 Vector Representation

In the proposed method, for n data points, each d-dimensional, and for a user-specified maximum number of clusters $K_{\max}$, a chromosome is a vector of real numbers of dimension $K_{\max} + K_{\max} \times d$. The first $K_{\max}$ entries are positive floating-point numbers in [0, 1], each of which controls whether the corresponding cluster is to be activated (i.e. to be really used for classifying the data) or not. The remaining entries are reserved for $K_{\max}$ cluster centers, each d-dimensional. For example, the i-th vector is represented as:

$\vec{X}_i(t) =$

$T_{i,1}$	$T_{i,2}$	.....	$T_{i,K_{\max}}$	$\vec{m}_{i,1}$	$\vec{m}_{i,2}$	....	$\vec{m}_{i,K_{\max}}$

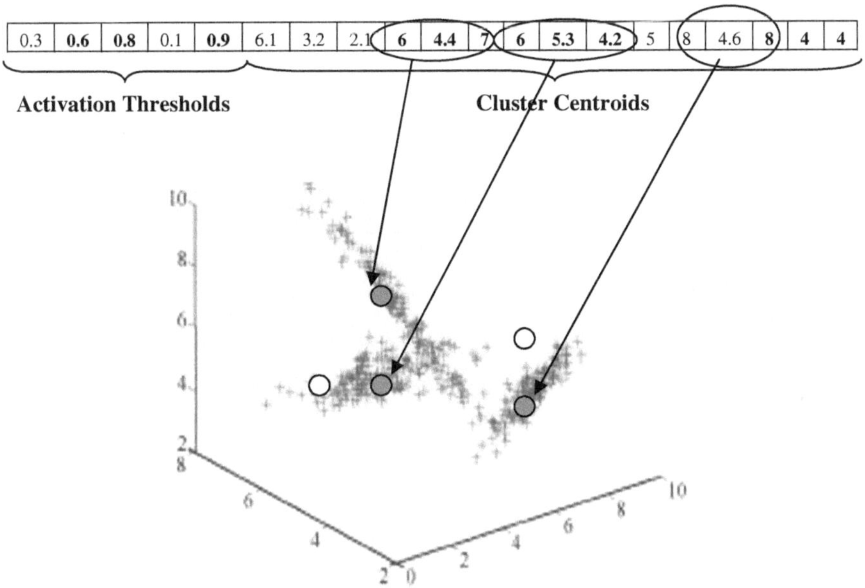

Fig. 4.1 The chromosome encoding scheme in the proposed method. A total of five cluster centers have been encoded for a 3-dimensional dataset. Only the activated cluster centers have been shown as orange circles.

The j-th cluster center in the i-th chromosome is active or selected for partitioning the associated dataset if $T_{i,j} > 0.5$. On the other hand, if $T_{i,j} < 0.5$, the particular j-th cluster is inactive in the i-th vector in DE population. Thus the $T_{i,j}$ s behave like control genes (we call them *activation thresholds*) in the vector governing the selection of the active cluster centers. The rule for selecting the actual number of clusters specified by one vector is:

$$\textbf{IF } T_{i,j} > 0.5 \textbf{ THEN } \text{the } j\text{-th cluster center } \vec{m}_{i,j} \text{ is } \textbf{ACTIVE}$$

$$\textbf{ELSE } \vec{m}_{i,j} \text{ is } \textbf{INACTIVE} \tag{4.1}$$

Example 4.1
As an example, consider one particular vector illustrated in Figure 4.1. There are at most five 3-dimensional cluster centers among which, according to the rule presented in (4.1) the second (6, 4.4, 7), third (5.3, 4.2, 5) and fifth one (8, 4, 4) have been activated for partitioning the dataset. The scheme has been further illustrated in Figure 4.1. The quality of the partition yielded by such a vector can be judged by an appropriate cluster validity index.

When a new trial vector is created using DE-type mutation and crossover, at first the T values are used to select (via (4.1)) the active cluster centroids. If due to mutation, some threshold $T_{i,j}$ in an offspring exceeds 1 or becomes negative, it is forcefully fixed to 1 or zero, respectively. However, if it is found that no flag could be set to one in a vector (all activation thresholds are smaller than 0.5), we randomly select two thresholds and re-initialize them to a random value between 0.5 and 1.0. Thus, the minimum number of possible clusters is 2.

4.2.2 Designing the Fitness Function

One advantage of the ACDE algorithm is that it can use any suitable validity index as its fitness function. After experimenting with a number of validity indices (a brief review of which can be found in section 2.3 of chapter 1), we select the CS measure [7] as the basis of our fitness function, as CS measure deals with clusters of different densities and/or size more efficiently than several other existing validity indices. Before presenting some results from these experiments that establish the superiority of CS measure, we first redefine the CS measure below.

Let the centroid of a cluster be computed by averaging the data vectors belonging to that cluster using the formula,

$$\vec{m}_i = \frac{1}{N_i} \sum_{x_j \in C_i} \vec{Z}_j \tag{4.2}$$

A distance metric between any two data points $\vec{X}_i$ and $\vec{X}_j$ is denoted by $d(\vec{X}_i, \vec{X}_j)$. Then the CS measure can be defined as,

$$CS(k) = \frac{\frac{1}{k}\sum_{i=1}^{k}\left[\frac{1}{|C_i|}\sum_{\vec{X}_i \in C_i} \max_{\vec{X}_q \in C_i}\{d(\vec{Z}_i, \vec{Z}_q)\}\right]}{\frac{1}{k}\sum_{i=1}^{k}\left[\min_{j \in K, j \neq i}\{d(\vec{m}_i, \vec{m}_j)\}\right]}$$

$$= \frac{\sum_{i=1}^{k}\left[\frac{1}{|C_i|}\sum_{\vec{X}_i \in C_i} \max_{\vec{X}_q \in C_i}\{d(\vec{Z}_i, \vec{Z}_q)\}\right]}{\sum_{i=1}^{k}\left[\min_{j \in K, j \neq i}\{d(\vec{m}_i, \vec{m}_j)\}\right]} \tag{4.3}$$

Note that the above measure is a function of the ratio of the sums of within-cluster scatter to between-cluster separation and has the same basic rationale as the DB [8] and the DI [9] measures (given by equations (1.10) and (1.11) of chapter 1). That is, they are to seek clusters that have minimum within-cluster scatter (*i.e.* compact) and maximum between-cluster separation (*i.e.* well-separated). The numerator of (4.3) is basically uses the largest distance between two data points

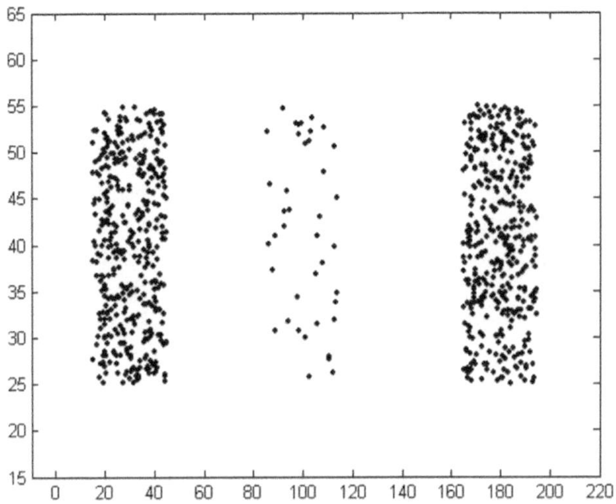

(a) Unlabelled data set

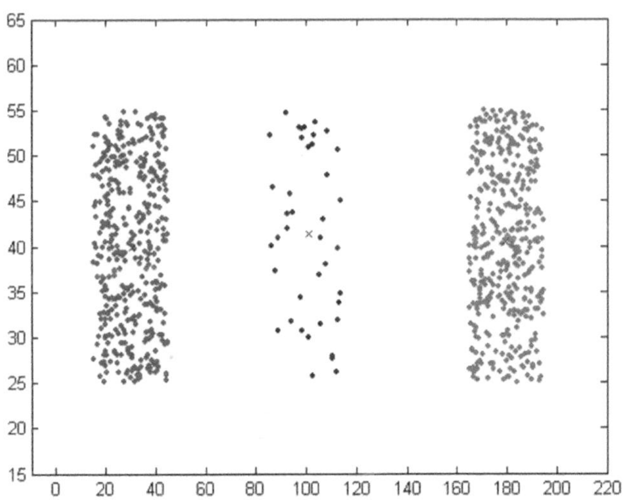

(b) Clustering result with k-means with $k = 2$

Fig. 4.2 Artificial data set of example 4.2

lying in the same cluster to measure the scatter volume. On the other hand the denominator computes the average distance between cluster centers.

The following three examples on three hand-crafted datasets, illustrate the effectiveness of the CS measure in handling clusters of different geometric shapes, densities and sizes over some well-known state-of-the-art validity indices found in literature. For comparison we used the following validity indices: the Dunn's Measure (DI), the Davies-Bouldin's measure (DB) with the parameters $q = t = 2$, the Bezdek's partition coefficient (PC) [10], Bezdek's classification entropy (CE) [11], and the Xie-Beni's index (XB) [12]. Please note that For the DI or PC validity measure, the largest value indicates a valid optimal partition. On the contrary, for the DB, CE, XB, or CS validity measure, the smallest value indicates a valid optimal partition.

The datasets were clustered with either k-means or Gustafson-Kessel (GK) [13] algorithm at each cluster number k for $k = 2$ to $k = 10$. Since both the algorithms are sensitive to initialization, during the clustering procedures we have tried different initializations to cluster the data sets for each cluster number k. Then for each k, the clustering result happened with the highest frequency was chosen to be the clustering result for the cluster number k to be validated by the validity measures. Then all validity measures are computed from the same clustering results.

Example 4.2

The data set shown in Figure 4.2 (a) is used to illustrate that the CS measure can work well for the case of clusters with the similar size but different densities. The data set contains 840 points grouped into 3 rectangular clusters, two of which contain 400 points per cluster while the third one contains only 40 points in it. The performance of each validity measure is given in Table 4.1. In Table 4.1and the others to follow, the bold-faced entries correspond to optimal values of the measures. Note that the CS and DI validity measures find the optimal cluster number k at $k = 3$, but the DB, PC, CE and XB validity measures find the optimal cluster number k at $k = 2$. We find that most of the above validity measures fail to find the correct cluster number in this example.

Table 4.1 Numerical values of the validity measures for example 4.2

k	2	3	4	5	6	7	8	9	10
DI	0.041	**1.637**	0.0156	0.0151	0.0151	0.0151	0.0151	0.0151	0.0151
DB	**0.162**	0.271	0.732	0.615	0.873	0.762	0.826	0.845	0.782
PC	**0.972**	0.984	0.853	0.784	0.679	0.617	0.613	0.591	0.575
CE	**0.056**	0.195	0.358	0.473	0.648	0.766	0.789	0.838	0.919
XB	**0.015**	0.025	0.512	0.358	0.263	0.155	0.295	0.357	0.178
CS	0.527	**0.351**	0.564	0.584	0.888	0.856	0.887	0.928	0.826

Example 4.3

We use the data set shown in Figure 4.3 (a) to demonstrate that the CS measure can also indicate the correct number in the case of clusters with different sizes and

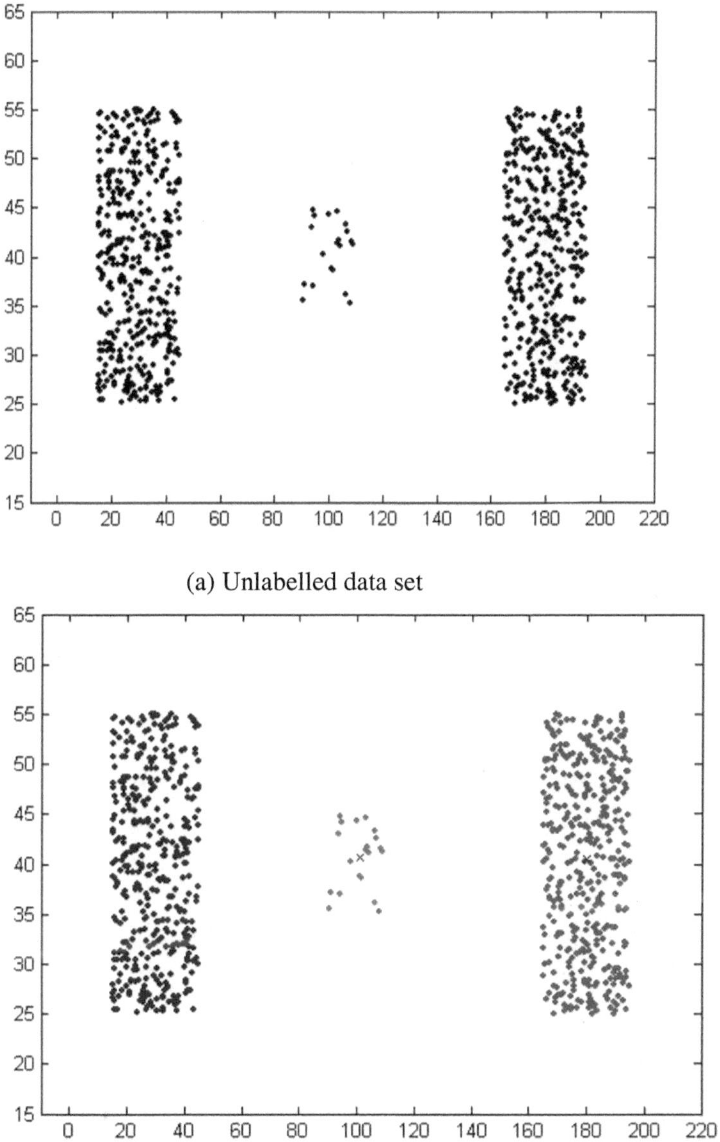

(a) Unlabelled data set

(b) Clustering result with k-means with $k = 2$

Fig. 4.3 Artificial data set of example 4.3

Table 4.2 Numerical values of the validity measures for example 4.3

k	2	3	4	5	6	7	8	9	10
DI	0.026	**2.461**	0.079	0.034	0.037	0.041	0.037	0.034	0.024
DB	**0.145**	0.127	0.685	0.529	0.586	0.673	0.848	0.773	0.735
PC	**0.993**	0.973	0.885	0.812	0.765	0.780	0.648	0.739	0.608
CE	**0.045**	0.097	0.323	0.393	0.467	0.581	0.725	0.658	0.843
XB	**0.013**	0.018	0.439	0.340	0.510	0.561	0.282	1.191	0.216
CS	0.521	**0.222**	0.392	0.381	0.392	0.413	0.678	0.461	0.654

different densities. Table 4.2 shows the values of the six validity measures for $k = 2$ to 10. We can see that only the CS and the DI measures point to the correct choice at $k = 3$. The CS and DI validity measures find the optimal cluster number k at $k = 3$, but the DB, PC, CE and XB validity measures find the optimal cluster number k at $k = 2$. Once again, this example shows that the proposed CS measure can work well for the data set with different densities and sizes.

Example 4.4

We generate a mixture of spherical and ellipsoidal clusters, as shown in Figure 4.4 (a). The total number of data points is 863. The k-means algorithm is applied to cluster these data sets at each cluster number k for $k = 2$ to $k = 10$. The performance of each validity measure is given in Table 4.3. As seen from the table, the DB, S and CS validity measures find the optimal cluster number k is at $k = 4$, but the DI, PC and CE validity measure indicate two clusters as the optimal partition. The clustering result achieved by the GK algorithm at $k = 4$ is shown in Figure 4.4 (b). This example demonstrates that the proposed CS measure can also work well for this case of clusters with different geometrical structures.

The above examples amply demonstrate the effectiveness of CS measure in indicating the optimal cluster numbers for clusters of various densities and/or shape and size. Therefore the fitness function for the i-th parameter vector in ACDE can be written as:

$$f(\vec{X}_i) = \frac{1}{CS_i(k) + eps}, \tag{4.4}$$

where eps is a small bias term equal to 2×10^{-6} and prevents the denominator of the R. H. S of (4.4) from being equal to zero. Thus maximization of the above fitness function corresponds to minimization of the corresponding CS measure and this leads to an optimal partitioning of the dataset under test. Since we compare the clustering performance of ACDE with GCUK, which was designed with a DB-measure based fitness function; we run another set of experiments that uses the following fitness function for all the algorithms:

$$f(\vec{X}_i) = \frac{1}{DB_i(k) + eps}, \tag{4.5}$$

where DB_i is the DB index, evaluated on the partitions yielded by the i-th vector (or the i-th particle for PSO) and eps is the same as before.

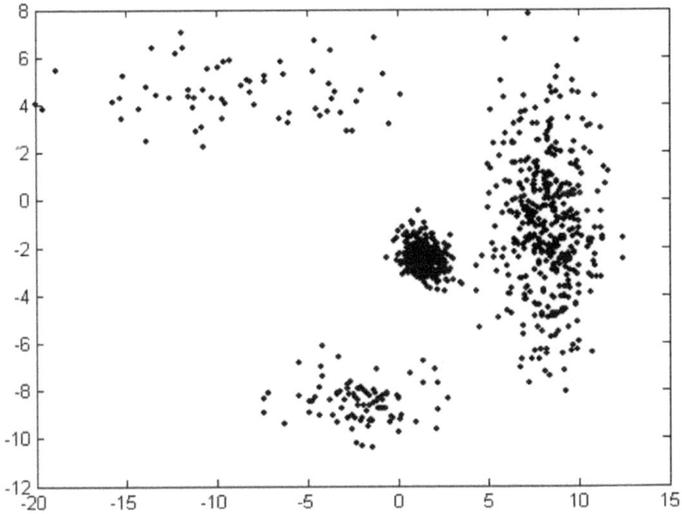

(a) Unlabelled data set

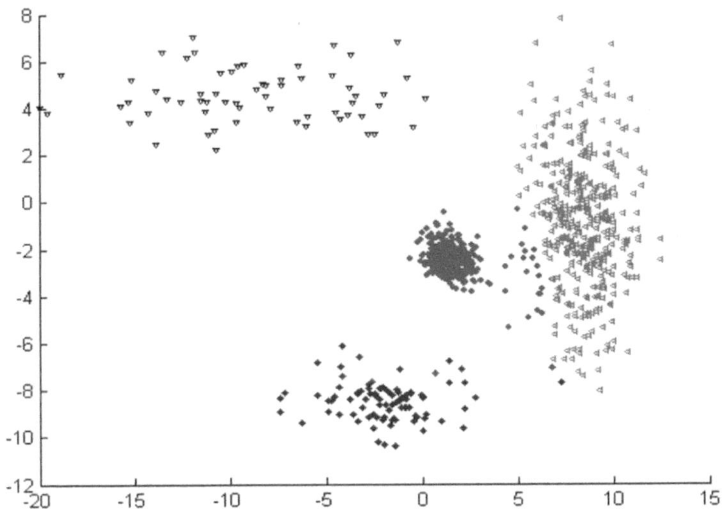

(b) Clustering result with GK with $k = 2$

Fig. 4.4 Artificial data set of example 4.4

Table 4.3 Numerical values of the validity measures for example 4.4

k	2	3	4	5	6	7	8	9	10
DI	**0.241**	0.039	0.091	0.012	0.018	0.035	0.027	0.018	0.024
DB	0.812	**0.624**	0.631	0.772	0.963	0.980	0.812	0.843	0.924
PC	**0.786**	0.714	0.727	0.657	0.602	0.573	0.557	0.527	0.518
CE	**0.371**	0.553	0.572	0.724	0.852	0.942	1.044	1.082	1.121
XB	0.162	0.165	**0.080**	0.176	0.351	0.255	0.250	0.187	0.184
CS	1.058	0.874	**0.771**	0.967	1.132	1.316	1.031	1.20	1.246

4.2.3 Avoiding Erroneous Vectors

There is a possibility that in our scheme, during computation of the CS and/or DB measures, a division by zero may be encountered. This may occur when one of the selected cluster centers in a DE-vector is outside the boundary of distributions of the data set. To avoid this problem we first check to see if any cluster has fewer than two data points in it. If so, the cluster center positions of this special chromosome are re-initialized by an average computation. We put n/k data points for every individual cluster center, such that a data point goes with a center that is nearest to it. The following example demonstrates this process.

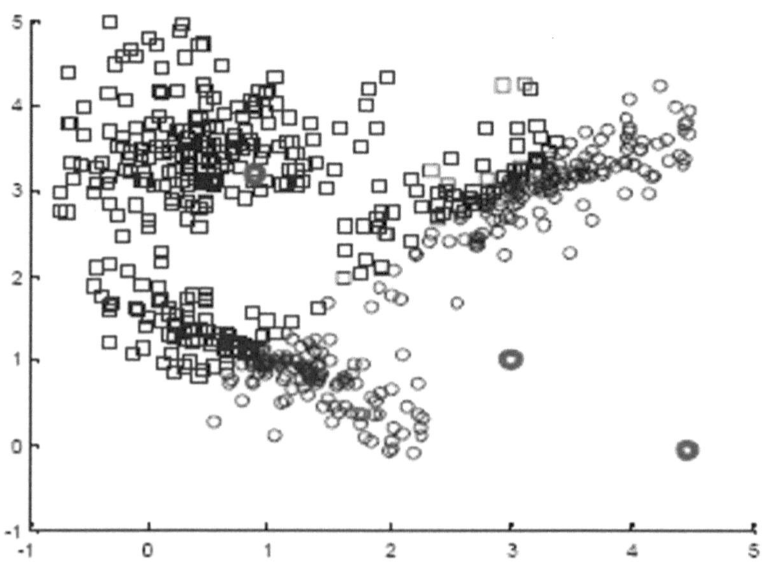

Fig. 4.5 An artificial dataset with trial cluster centroids [(4.46, -0.0521), (3.13, 1.12), (0.8873, 3.1571)]

Example 4.5

For example, Figure 4.5 depict 564 data points in two-dimensional feature space, where the number of the cluster centers is three and it contains mixed spherical and elliptical shapes (Su et al., 1999). If the clustering solution for *i*-th particle is [(4.46, -0.0521), (3.13, 1.12), (0.8873, 3.1571)], it is clear that the location (4.46, -0.0521) will generate an empty cluster which collects zero data point. In a word, no data point will be assigned to the solution (4.46,-0.0521). Based on our updating formula, n/k (i.e. 564/3=188) data points will be assigned to the nearest cluster centers. Repeat this calculation to decide three cluster centers. Therefore, the new replaced cluster centers are [(2.4871, 1.9428), (1.9914, 1.6069), (0.6318, 3.2891)]. According to equation (4.2), the computing result is $\vec{V}_i$ = [(3.1927, 3.0983), (1.1509, 0.9786), (0.3514, 11, 3.3199)].

4.2.4 Modification of the Classical DE

After performing a series of empirical experiments, we propose two parameter tuning strategies for DE in order to improve its convergence behavior over the clustering fitness landscape. In the original DE the difference vector $(\vec{X}_{r_2^i}(t) - \vec{X}_{r_3^i}(t))$ is scaled by a constant factor '*F*'. The usual choice for this control parameter is a number between 0.4 and 1. We propose to vary this scale factor in a random manner in the range (0.5, 1) by using the following relation:

$$F = 0.5 \cdot (1 + rand(0,1)), \tag{4.6}$$

where rand (0, 1) is a uniformly distributed random number within the range [0,1].The mean value of the scale factor is 0.75. This allows for stochastic variations in the amplification of the difference vector and thus helps retain population diversity as the search progresses. In [14] we have already shown that the DERANDSF (DE with Random Scale Factor) can meet or beat the classical DE and also some versions of PSO in a statistically significant manner. In addition to that, here we also decrease the crossover rate *CR* linearly with time from CR_{max} = 1.0 to CR_{min} = 0.5. If *CR* = 1.0, it means that all components of the parent vector are replaced by the difference vector operator according to (2.5). But at the later stages of the optimizing process, if *CR* be decreased, more components of the parent vector are then inherited by the offspring. Such a tuning of *CR* helps to explore the search space exhaustively at the beginning, but adjust the movements of trial solutions finely during the later stages of search, so that they can explore the interior of a relatively small space in which the suspected global optimum lies. The time-variation of *CR* may be expressed in the form of the following equation,

$$CR = CR_{min} + (CR_{max} - CR_{min}) \cdot \left(\frac{MAXIT - iter}{MAXIT} \right), \tag{4.7}$$

where CR_{max} and CR_{min} are the maximum and minimum values of Crossover rate *CR*, *iter* is the current iteration number and MAXIT is the maximum number of allowable iterations.

4.2.5 Pseudo-code of the ACDE Algorithm

The pseudo code for the complete ACDE algorithm can be presented in the following manner.

Step 1: Initialize each search variable vector in DE to contain k number of randomly selected cluster centers and k (randomly chosen) activation thresholds in [0, 1].
Step 2: Find out the active cluster centers in each chromosome with the help of the rule described in (4.1).
Step 3: For $iter$ =1to $MAXITER$ do

i) For each data vector $\vec{Z}_p$, calculate its distance metric $d(\vec{Z}_p, \vec{m}_{i,j})$ from all active cluster centers of the i-th DE-vector $\vec{X}_i$.
ii) Assign $\vec{Z}_p$ to that particular cluster center $\vec{m}_{i,j}$ where
$$d(\vec{Z}_p, \vec{m}_{i,j}) = \min_{\forall b \in \{1,2,\ldots,k\}} \left\{ d(\vec{Z}_p, \vec{m}_{i,b}) \right\}$$
iii) Check if the number of data points belonging to any cluster center $\vec{m}_{i,j}$ is less than 2. If so, update the cluster centers of the chromosome using the concept of average described earlier.
iv) Change the population members according to the DE algorithm with modifications proposed in section 4.2.4. Use the fitness of the vectors to guide the evolution of the population.

Step 4: Report as the final solution the cluster centers and the partition obtained by the globally best vector (one yielding the highest value of the fitness function) at $iter = MAXITER$.

4.3 Experiments and Results for Real Life Datasets

We compare the performance of the ACDE algorithm with two recently developed partitional clustering algorithms and one standard hierarchical agglomerative clustering based on the linkage metric of average link [15]. The former two algorithms are well known as the GCUK (Genetic Clustering with an Unknown number of clusters K) [3] and the DCPSO (Dynamic Clustering PSO) [4]. Moreover, in order to investigate the effects of the changes, made in the classical DE algorithm, we have compared ACDE with an ordinary DE based clustering method which uses the same chromosome representation scheme and fitness function as the ACDE. The classical DE scheme that we have used, is referred in literature as the DE/rand/1/bin [2] where 'bin' stands for binomial crossver method.

4.3.1 The Datasets Used

The following real-life datasets [16, 17] are used in this study. Here n = number of data points, d = number of features and k = number of clusters.

1) Iris plants database (n =150, d = 4, k = 3): This is a well-known database with 4 inputs, 3 classes, and 150 data vectors. The dataset consists of three different species of iris flower: *Iris setosa, Iris virginica* and *Iris versicolour.* For each species, fifty samples with four features each (sepal length, sepal width, petal length and petal width) were collected. The number of objects belonging to each cluster is 50.
2) Glass (n =214, d = 9, k = 6): The data were sampled from six different type of glass: building windows float processed (70 objects), building windows non float processed (76 objects), vehicle windows float processed (17 objects), containers (13 objects), tableware (9 objects), and headlamps (29 objects) with nine features each. These are refractive index, Sodium, Magnesium, Aluminum, Silicon, Potassium, Calcium, Barium and Iron.
3) Wisconsin breast cancer data set (n = 683, d = 9, k = 2): The Wisconsin breast cancer database contains 9 relevant features: clump thickness, cell size uniformity, cell shape uniformity, marginal adhesion, single epithelial cell size, bare nuclei, bland chromatin, normal nucleoli and mitoses. The dataset has two classes. The objective is to classify each data vector into benign (239 objects) or malignant tumors (444 objects).
4) Wine (n = 178, d = 13, k = 3): This is a classification problem with "well behaved" class structures. There are 13 features, 3 classes and 178 data vectors.
5) Vowel Dataset (n = 871, d = 3, k = 6): This dataset consists of 871 Indian Telugu vowel sounds. The dataset has three features F_1, F_2 and F_3 corresponding to the first, second and third vowel frequencies, and six overlapping classes {d (72 objects), a (89 objects), *i* (172 objects), *u* (151 objects), e (207objects), *o* (180 objects)}.

4.3.2 Population Initialization

For the ACDE algorithm, we initialize the activation thresholds (control genes) randomly in [0, 1]. The cluster centroids are also randomly fixed between X_{max} and X_{min}, which denote the maximum and minimum numerical values of any feature of the dataset under test respectively. For example, in case of the grayscale images (discussed in section 4.6) since the intensity value of each pixel serves as a feature, we choose $X_{min} = 0$ and $X_{max} = 255$. To make the comparison fair, the populations for both ACDE and the classical DE (DE/rand/1/bin) based clustering algorithms (for all problems tested) were initialized using the same random seeds. For the GCUK, each string in the population initially encodes the centers of k_i clusters, where $k_i = rand().K_{max}$. Here K_{max} is a soft estimate of the upper bound of the number of clusters. The k_i centers encoded in the chromosome are randomly selected points from the dataset. In case of the DCPSO algorithm, the initial position of the i-th particle $\vec{X}_i(0)$ (for binary PSO) is fixed depending on a user specified probability p_{ini} as follows:

$$x_{i,j}(0) = 0 \quad \text{if } r_{i,j} \leq p_{ini}$$
$$= 1 \quad \text{if } r_{i,j} < p_{ini} \tag{4.7}$$

where $r_{i,j}$ is a uniformly distributed random number in [0, 1]. The initial velocity vector of each particle $\vec{V}_i(0)$ is randomly set in the interval [-5, 5] following [25].

4.3.3 Parameter Setup for the Algorithms Compared

We employ the best possible parameter settings recommended in [3] and [4] for the GCUK and DCPSO algorithms (respectively). For the ACDE algorithm, we choose an optimal set of parameters after experimenting with many possibilities. Table 4.4 summarizes these settings. In Table 4.4, *Pop_size* indicates the size of the population, *dim* implies the dim0ension of each chromosome and P_{ini} is a user specified probability used for initializing the position of a particle in the DCPSO algorithm. For details on this issue, please refer to [4]. Once set, we allow no hand tuning of the parameters to make the comparison fair enough.

Table 4.4 Parameters for the clustering algorithms

GCUK		DCPSO		ACDE		Classical DE	
Parameter	Value	Parameter	Value	Parameter	Value	Parameter	Value
Pop_size	50	Pop_size	100	Pop_size	10**dim*	Pop_size	10**dim*
Cross-over Probability μ_c	0.8	Inertia Weight	0.72	$CR_{\max}$	1.0	CR	0.9
Mutation probability μ_m	0.001	C_1, C_2 P_{ini}	1.494 0.75	$CR_{\min}$	0.5	F	0.8
$K_{\max}$ $K_{\min}$	20 2	$K_{\max}$ $K_{\min}$	20 2	$K_{\max}$ $K_{\min}$	20 2	$K_{\max}$ $K_{\min}$	20 2

4.3.4 Simulation Strategy

In this research, while comparing the ACDE algorithm with other state-of-the-art clustering techniques, we focus on three performance metrics: (a) quality of the solution as determined by the CS and DB measures and the number of misclassified items for each dataset, (b) ability to find the optimal number of clusters, and (c) computational time required to find the solution.

For comparing the speed of the stochastic algorithms like GA, PSO or DE, we require a fair time measurement. The number of iterations or generations cannot be accepted as a time measure since the algorithms perform different amount of works in their inner loops and also they have different population sizes. Hence, we choose *number of fitness function evaluations (FEs)* as a measure of computation time instead of generations or iterations. The advantage of measuring runtime complexity by counting the FEs is that there is a strong correspondence between this measure and the actual processor time as the function complexity increases [18].

Since four of the considered algorithms are stochastic in nature, the results of two successive runs usually do not match for them. Hence, we have taken 40 independent runs (with different seeds of the random number generator) of each algorithm. The results have been stated in terms of the mean values and standard deviations over the 40 runs in each case. As the hierarchical agglomerative algorithm (marked in Table 4.5 as *average-link*), does not use any evolutionary technique, the number of FEs is not relevant to this method. This algorithm is supplied with the correct number of clusters for each problem and we used the Ward updating formula [19] in order to re-compute the cluster distances efficiently.

We used unpaired t-tests to compare the means of the results produced by best and the second best algorithms. Unpaired t-test assumes that the data has been sampled from a normally distributed population. From the concepts of central limit theorem, one may note that as sample sizes increase, the sampling distribution of the mean approaches a normal distribution regardless of the shape of the original population. A sample size around 40 allows the normality assumptions conducive for doing the unpaired t-tests [20]. Please see the appendix for further details about the t-tests.

The four evolutionary clustering algorithms can go with any kind of clustering validity measure serving as their fitness functions. We here run two sets of experiments: one using the CS measure based fitness function shown in (4.4) while the other using the DB measure based fitness function shown in (4.5) with all the four algorithms. For each dataset, the quality of the final solution yielded by four partitional clustering algorithms has been compared with the average-link metric based hierarchical method in terms of the CS and DB measures.

All the algorithms have been developed in Visual C++ platform on a Pentium IV, 2.2 GHz PC, with 512 KB cache and 2 GB of main memory in Windows Server 2003 environment.

4.3.5 Empirical Results

To judge the accuracy of the ACDE, DCPSO, GCUK and the classical DE based clustering algorithms, we let each of them run for a very long time over every benchmark dataset, until the number of FEs exceed 10^6. Then we note the final fitness value, number of clusters found, the inter cluster distance i.e. the mean distance between the centroids of the clusters, (where the objective is to maximize the distance between clusters) and the intra cluster distance i.e. the mean distance between data vectors within a cluster, (where the objective is to minimize the intra-cluster distances). The latter two objectives respectively correspond to crisp, compact clusters that are well separated. In case of the hierarchical algorithm, the CS value has been calculated over the final results obtained after its termination. In columns 3, 4, 5, and 6 of Table 4.5 we report respectively the mean number of classes found, the final CS value, inter and intra cluster distances obtained for each competitor algorithm.

Since the benchmark datasets have their nominal partitions known to the user, we also compute the mean number of misclassified data-points. This is the average number of objects that were assigned to clusters other than according to the nominal classification. Table 4.6 reports the corresponding mean values and standard deviations over the runs obtained in each case of Table 4.5. Table 4.7 shows results of unpaired *t*-tests taken on the basis of the CS measure between the best

Table 4.5 Final solution (Mean and standard deviation over 40 independent runs) after each algorithm was terminated after running for 10^6 function evaluations (FEs) with CS Measure based fitness function

Dataset	Algorithm	Average number of clusters found	CS measure	Mean Intra cluster Distance	Mean Inter cluster Distance
Iris	ACDE	**3.25±0.0382**	**0.6643±0.097**	**3.1164±0.033**	**2.5931±0.027**
	DCPSO	2.23±0.0443	0.7361±0.671	3.6516±1.195	2.2104±0.773
	GCUK	2.35±0.0985	0.7282±2.003	3.5673±2.792	2.5058±1.409
	Classical DE	2.50±0.0473	0.7633±0.039	3.9439±1.874	2.1158±1.089
	Average-link	3.00	0.7863±0.00	3.9808±0.00	2.0817±0.00
Wine	ACDE	**3.25±0.0391**	**0.9249±0.032**	**4.046±0.002**	**3.1483±0.078**
	PSO	3.05±0.0352	1.8721±0.037	4.851±0.184	2.6113±1.637
	GCUK	2.95±0.0112	1.5842±0.328	4.163±1.929	2.8058±1.365
	Classical DE	3.50±0.0143	1.7964±0.802	4.949±1.232	2.6118±1.384
	Average-link	3.00	1.8921±0.00	4.982±0.00	2.5009±0.00
Breast-Cancer	ACDE	**2.00±0.00**	**0.4532±0.034**	**4.2439±0.143**	**3.2577±0.138**
	DCPSO	2.25±0.0632	0.4854±0.009	4.8511±0.373	2.3613±0.021
	GCUK	2.00±0.0083	0.6089±0.016	4.9944±0.904	2.3944±1.744
	Classical DE	2.25±0.0261	0.8984±0.381	4.6944±0.654	2.9635±1.464
	Average-link	2.00	0.9007±0.00	5.0098±0.00	2.2817±0.00
Vowel	ACDE	**5.75±0.0751**	**0.9089±0.051**	**1412.63±0.792**	**2724.85±0.124**
	DCPSO	7.25±0.0183	1.1827±0.431	1482.51±3.973	1923.93±1.154
	GCUK	5.05±0.0075	1.9978±0.966	1495.13±12.334	1944.38±0.747
	Classical DE	7.50±0.0569	1.0844±0.067	1493.72±10.833	2434.45±1.213
	Average-link	6.00	1.7221±0.00	1493.98±0.00	2357.62±0.00
Glass	ACDE	**6.05±0.0148**	**0.3324±0.487**	**563.247±134.2**	**853.62±9.044**
	DCPSO	5.95±0.0093	0.7642±0.073	599.535±10.34	889.32±4.233
	GCUK	5.85±0.0346	1.4743±0.236	594.673±30.62	869.93±1.789
	Classical DE	5.60±0.0754	0.7782±0.643	608.837±20.92	891.82±4.945
	Average-link	6.00	1.0221±0.00	610.033±0.00	895.47±0.00

Table 4.6 Mean Classification Error over nominal partition and standard deviation over 40 independent runs each of which was continued up to 10^6 function evaluations for the first four evolutionary algorithms (using the CS measure)

Dataset	Mean Classification Error				
	ACDE	DCPSO	GCUK	Classical DE	Average-link
Iris	**2.35±0.00**	4.15±0.0	5.00±0.00	3.96±0.00	4.00±0.00
Wine	**36.65±0.0**	99.4±1.09	100.24±1.05	114.50±1.53	134.00±0.00
Breast Cancer	**22.25±0.28**	27.01±1.25	29.00±1.55	29.15±0.50	26.00±0.00
Vowel	**418.75±3.10**	453.58±6.61	476.42±6.92	473.72±4.25	496.00±0.00
Glass	**92.55±0.19**	102.1±0.68	98.21±0.08	105.36±0.54	111.00±0.00

Table 4.7 Results of unpaired t-test between the best and the second best performing algorithm (for each dataset) based on the CS measures of Table 4.5

Dataset	Standard Error	*t*	95% Confidence Interval	Two-tailed *P*	Significance
Iris	0.095	4.0595	-0.142225 to -0.047575	0.0002	**Extremely significant**
Wine	0.074	4.0615	-0.448480 to -0.150120	0.0002	**Extremely significant**
Breast Cancer	0.008	34.835 4	-0.309710 to -0.275690	< 0.0001	**Extremely significant**
Vowel	0.019	6.1344	-0.153616 to -0.077384	< 0.0001	**Extremely Significant**
Glass	3.9214	0.110	-0.654712 to -0.208888	< 0.0001	**Extremely Significant**

two algorithms (standard error of difference of the two means, 95% confidence interval of this difference, the t value, and the two-tailed P value). For all cases in Table 4.7, sample size = 40.

In order to compare the speeds of different algorithms, we selected a threshold value of CS measure for each of the datasets. This cut-off CS value is somewhat larger than the minimum CS value found by each algorithm in Table 4.5. Now we run a clustering algorithm on each dataset and stop as soon as the algorithm achieves the proper number of clusters as well as the CS cut-off value. We then note down the number of fitness function evaluations the algorithm takes to yield the cut-off CS value. A lower number of function evaluations correspond to a faster algorithm. In columns 3, 4, 5, and 6 of Table 4.8 we reported the mean number of function evaluations, CS cut-off value, mean and standard deviations of the final inter and intra cluster distances (on termination of the algorithm) over 40 independent runs for each algorithm In Table 4.9, we reported the misclassification errors (with respect to the nominal classification) for the experiments conducted for Table 4.8. In this table, we exclude the hierarchical average-link algorithm as its time-complexity can not be measured using the number of FEs. It is however noted that, the runtime of a standard hierarchical algorithm scales in a quadratic fashion [15].

Table 4.8 Mean and standard deviations of the number of fitness function evaluations (over 40 independent runs) required by each algorithm to reach a predefined cut-off value of the CS validity index

Dataset	Algorithm	Mean number of function evaluations required	CS Cutoff Value	Mean Intra cluster Distance	Mean Inter cluster Distance
Iris	ACDE	**459888.95±20.50**	0.95	**3.7836±0.509**	**2.0758±0.239**
	DCPSO	679023.85±31.75		3.9637±1.666	2.0093±0.795
	GCUK	707723.70±120.21		3.9992±2.390	1.9243±1.843
	Classical DE	698043.80±93.36		4.5793±2.454	1.9564±1.767
Wine	ACDE	**67384..25**±56.45	1.90	**4.9872±0.148**	**3.1275±0.0357**
	PSO	700473.35±31.42		4.0743±0.093	1.9967±1.828
	GCUK	785333.05±21.75		5.9870±1.349	2.1323±1.334
	Classical DE	675472.95±14.83		4.9927±1.236	2.6842±1.356
Breast-Cancer	ACDE	**292102.50**±29.73	1.10	**4.9744±0.105**	**3.0096±0.246**
	DCPSO	587832.50±7.34		5.6546±0.241	2.1173±0.452
	GCUK	914033.85±24.83		8.0442±0.435	2.0542±1.664
	Classical DE	575484.70±10.26		5.3407±0.652	2.2315±2.885
Vowel	ACDE	**437533.35**±51.73	2.50	**1494.12±0.378**	**2739.85±0.163**
	DCPSO	500493.15±35.47		1575.51±3.786	1923.93±1.154
	GCUK	498354.10±74.60		1593.72±1.789	2633.45±1.213
	Classical DE	667342.80±53.54		1674.13±6.564	1947.38±0.747
Glass	ACDE	**443233.30**±74.65	1.80	**590.572±34.24**	**853.62±0.44**
	DCPSO	566335.80±25.73		619.980±15.98	846.67±0.804
	GCUK	574938.65±82.64		615.88±20.95	857.34±1.465
	Classical DE	542355.95±32.85		608.85±30.62	829.07±1.765

Table 4.9 Mean Classification Error over nominal partition and standard deviation over 40 independent runs, which were stopped as soon as they reached the predefined cut-off CS value

Dataset	Mean Classification Error			
	ACDE	DCPSO	GCUK	Classical DE
Iris	**8.68±0.34**	14.34±0.45	9.34±0.57	9.34±0.00
Wine	**100.32±0.79**	105.23±2.59	139.3±6.34	106.68±1.45
Breast Cancer	**37.50±0.33**	49.67±0.34	43.34±0.69	40.08±1.35
Vowel	**468.00±4.79**	492.34±0.64	491.40±4.25	496.44±1.38
Glass	**98.57±0.77**	102.1±0.68	105.36±0.54	98.21±0.08

Table 4.10 Final solution (Mean and standard deviation over 40 independent runs) when each algorithm was terminated after running for 10^6 function evaluations (FEs) with DB Measure based fitness function

Problem	Algorithm	Avg no. of clusters found	DB measure	Mean Intra cluster Distance	Mean Inter cluster Distance
Iris	ACDE	**3.05±0.0712**	**0.4645±0.022**	**3.1633±0.076**	**2.8387±0.658**
	DCPSO	2.25±0.0593	0.6899±0.008	3.8536±0.122	2.2544±0.039
	GCUK	2.30±0.0738	0.7377±0.065	3.8436±0.076	2.1438±0.022
	Classical DE	2.50±0.0092	0.5822±0.067	3.8876±0.092	2.0358±0.055
	Average-link	3.00	0.8471±0.00	3.9098±0.00	2.2817±0.00
Wine	ACDE	**3.25±0.0931**	**3.0432±0.021**	**4.4212±0.096**	**3.1029±0.047**
	DCPSO	3.05±0.0024	4.3432±0.232	4.8668±0.154	2.6113±1.635
	GCUK	2.95±0.0173	5.3424±0.343	5.1312±1.342	2.7565±2.128
	Classical DE	3.50±0.0143	3.3923±0.092	4.263±1.907	2.8158±1.786
	Average-link	3.00	5.7206±0.00	4.982±0.00	2.5009±0.00
Breast-Cancer	ACDE	2.05±0.0563	0.5203±0.006	4.5463±0.023	3.1002±0.064
	DCPSO	2.50±0.0621	0.5754±0.073	4.9232±0.373	2.2684±0.063
	GCUK	2.50±0.0352	0.6328±0.002	6.5541±0.433	1.8032±0.016
	Classical DE	**2.10±0.0081**	**0.5199±0.007**	**5.2234±0.042**	**2.0236±0.058**
	Average-link	2.00	0.7634±0.00	5.0098±0.00	2.2817±0.00
Vowel	ACDE	**5.75±0.0241**	**0.9224±0.334**	**1449.12±0.834**	**2289.85±0.163**
	DCPSO	7.25±0.0562	1.2821±0.009	1500.57±3.748	1747.76±1.764
	GCUK	5.05±0.0561	2.9482±0.028	1573.23±4.675	2271.89±1.222
	Classical DE	7.50±0.0819	1.4488±0.075	1498.78±2.725	1962.31±0.993
	Average-link	6.00	3.0581±0.00	1493.98±0.00	2357.62±0.00
Glass	ACDE	**6.05±0.0248**	**1.0092±0.083**	**501.757±4.3**	**893.46±3.32**
	DCPSO	5.95±0.0193	1.5152±0.073	514.554±9.5	856.00±8.07
	GCUK	5.85±0.0346	1.8371±0.034	518.903±2.9	852.32±5.43
	Classical DE	5.60±0.0446	1.6673±0.004	514.849±3.4	862.21±2.53
	Average-link	6.00	1.8519±0.00	610.033±0.00	895.47±0.00

Table 4.11 Mean Classification Error over nominal partition and standard deviation over 40 independent runs each of which was continued up to 10^6 function evaluations for the first four evolutionary algorithms (using the DB measure)

Dataset	Mean Classification Error				
	ACDE	DCPSO	GCUK	Classical DE	Average-link
Iris	**2.22±0.00**	2.79±0.55	2.75±0.08	3.14±0.00	4.00±0.00
Wine	**40.15±0.0**	112.5±2.50	118.45±1.77	102.22±1.05	134.00±0.00
Breast Cancer	26.75±0.25	30.23±0.46	26.50±0.80	**25.00±1.09**	26.00±0.00
Vowel	**418.35±7.50**	435.00±3.75	473.46±3.57	475.65±2.67	496.00±0.00
Glass	**8.86±0.42**	14.35±0.26	17.98±0.67	15.69±0.85	111.00±0.00

Tables 4.10, 4.11, 4.12, 4.13, and 4.14 exactly correspond to Tables 4.5, 4.6, 4.7, 4.8, and 4.9 with respect to the experimental results the only difference being that all the experiments conducted for the former group of tables use a DB measure

Table 4.12 Results of unpaired t-test between the best and the second best performing algorithm (for each dataset) based on the DB measures of Table 4.10

Dataset	Standard Error	t	95% Confidence Interval	Two-tailed P	Significance
Iris	0.011	10.5559	-0.139898 to -0.095502	< 0.0001	**Extremely significant**
Wine	0.015	23.3971	-0.378805 to -0.319395	< 0.0001	**Extremely significant**
Breast Cancer	0.001	0.2744	-0.003302 to 0.002502	0.7845	Not significant
Vowel	0.053	6.8087	-0.464875 to -0.254525	< 0.0001	**Extremely Significant**
Glass	0.017	28.9521	-0.540794 to -0.471206	< 0.0001	**Extremely Significant**

Table 4.13 Mean and standard deviations of the number of fitness function evaluations (over 40 independent runs) required by each algorithm to reach a predefined cut-off value of the DB validity index

Dataset	Algorithm	Mean no. of function evaluations required	DB Cutoff Value	Mean Intra cluster Distance	Mean Inter cluster Distance
Iris	ACDE	**504783.45**±12.65	0.80	**3.9928±0.029**	**2.1029±0.842**
	DCPSO	679084.75±16.57		3.7852±1.842	1.7641±0.439
	GCUK	790865.90±10.21		4.4587±3.782	1.9383±1.307
	Classical DE	658796.35±17.28		4.0393±1.542	1.6278±1.681
Wine	ACDE	**464653.35**±5.50	6.00	**4.8292±0.732**	**3.0219±0.069**
	DCPSO	486885.85±2.85		5.1472±0.472	2.1161±1.623
	GCUK	598743.35±8.09		4.9383±1.722	2.9121±0.353
	Classical DE	477869.95±8.12		4.7531±2.043	2.8158±0.389
Breast-Cancer	ACDE	424732.30±8.93	0.90	5.4489±0.342	3.0234±0.683
	DCPSO	467854.60±10.12		5.2885±0.552	2.0124±1.596
	GCUK	678874.90±7.82		6.8832±0.733	2.1637±1.458
	Classical DE	**418765.55±1.23**		**5.8684±0.467**	**1.9235±0.164**
Vowel	ACDE	**435743.05±2.65**	3.00	**1544.92±0.834**	**2081.31±0.679**
	DCPSO	556865.00±4.26		1652.58±2.341	1264.87±3.069
	GCUK	575854.65±1.29		1582.55±7.332	1989.38±7.734
	Classical DE	546859.60±2.05		1608.22±5.866	1604.43±1.674
Glass	ACDE	**506754.00**±12.27	2.00	**132.757±15.8**	**13.46±2.54**
	DCPSO	569787.95±10.83		154.564±39.6	13.56±2.65
	GCUK	687678.75±10.97		155.856±24.7	10.42±4.69
	Classical DE	527585.35±7.50		178.809±30.3	10.21±1.09

Table 4.14 Mean Classification Error over nominal partition and standard deviation over 40 independent runs, which were stopped as soon as they reached the predefined cut-off DB value

Dataset	Mean Classification Error			
	ACDE	DCPSO	GCUK	Classical DE
Iris	**2.22±0.00**	2.79±0.55	2.75±0.08	3.14±0.00
Wine	**40.15±0.0**	112.5±2.50	118.45±1.77	102.22±1.05
Breast Cancer	**26.75±0.25**	30.23±0.46	26.50±0.80	29.00±1.09
Vowel	**418.35±7.50**	435.00±3.75	473.46±3.57	475.65±2.67
Glass	**8.86±0.42**	14.35±0.26	17.98±0.67	15.69±0.85

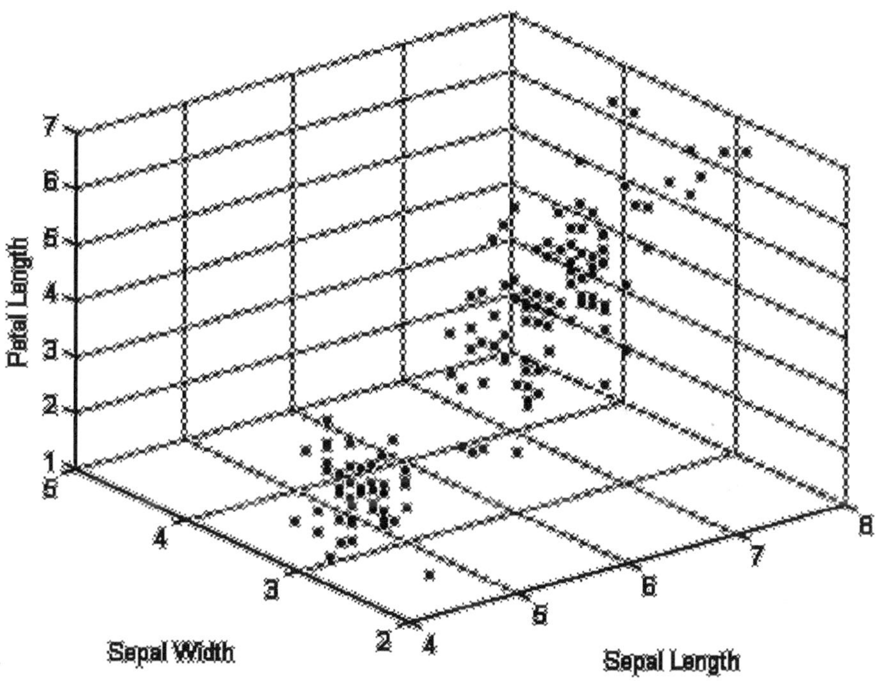

(a)

Fig. 4.6 a) The 3-dimensional plot of the unlabeled Iris data set using first three features. Clustering of Iris data b) by ACDE c) by GCPSO d) by GCUK e) by classical DE f) by Average-link based Hierarchical Clustering Algorithm

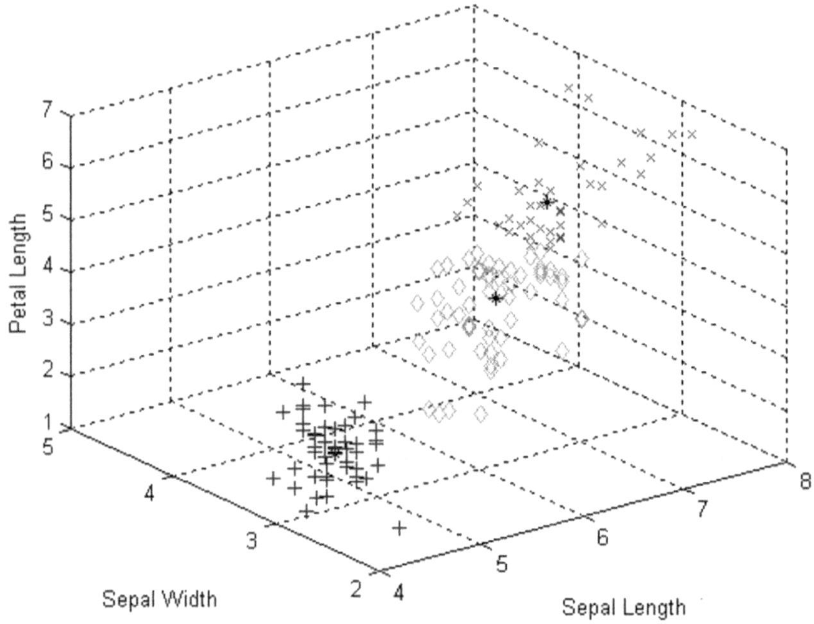

(b)

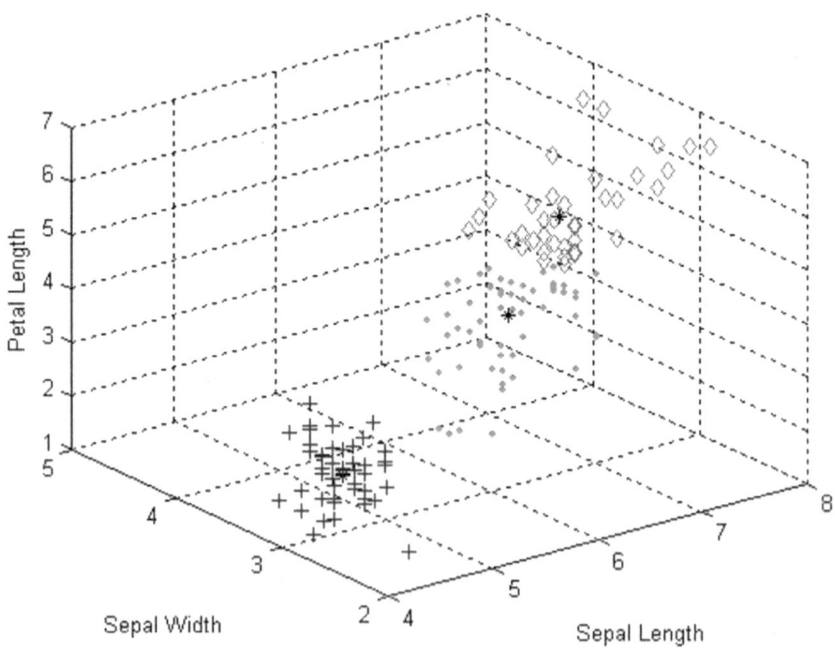

(c)

Fig. 4.6 (*continued*)

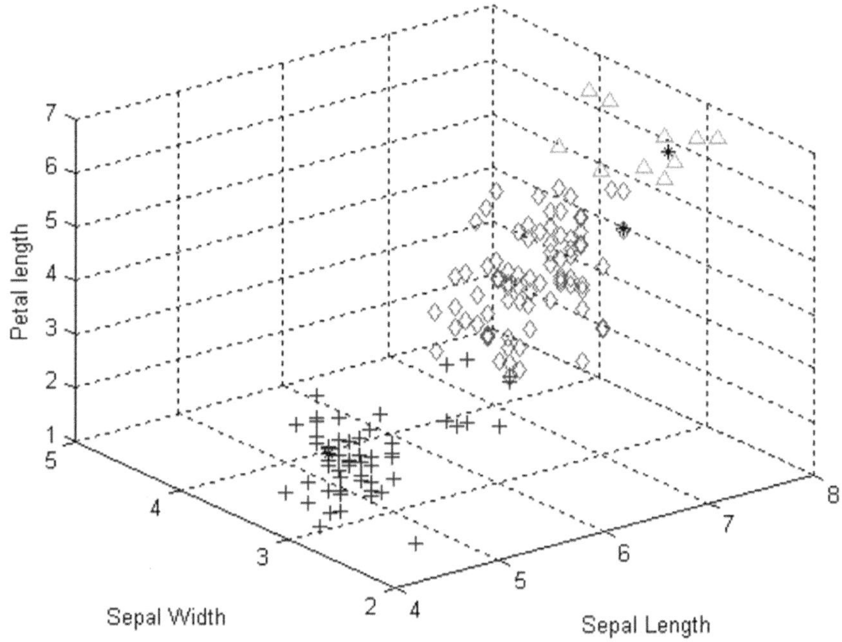

(d)

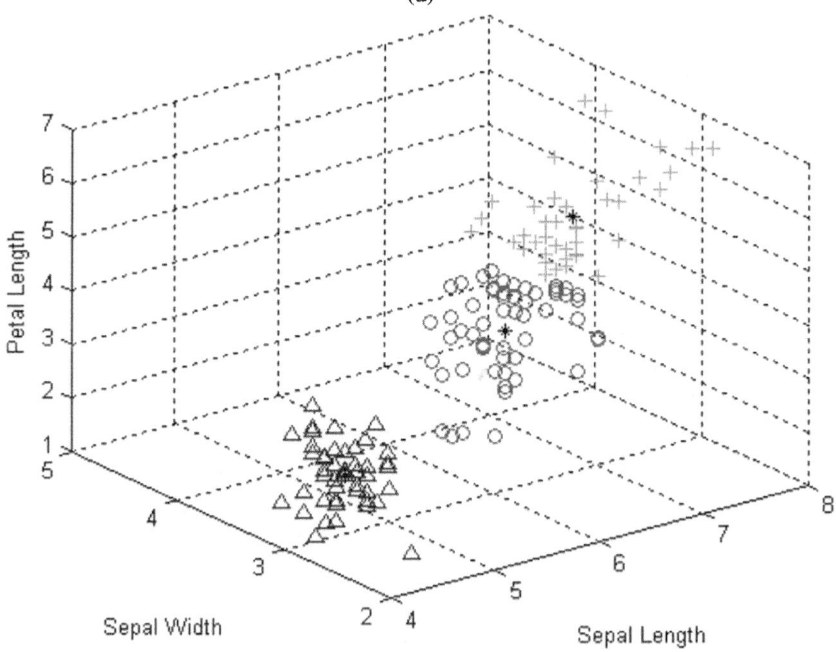

(e)

Fig. 4.6 (*continued*)

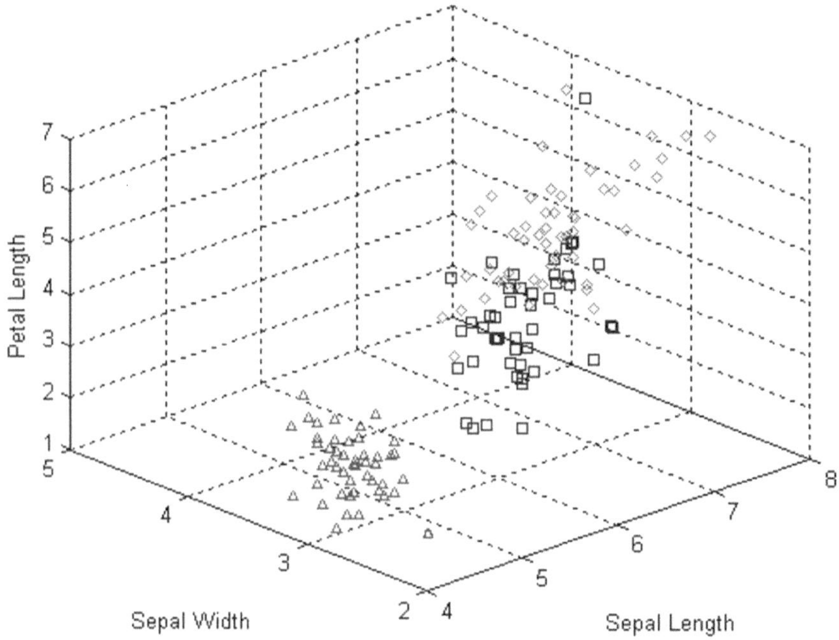

(f)

Fig. 4.6 (*continued*)

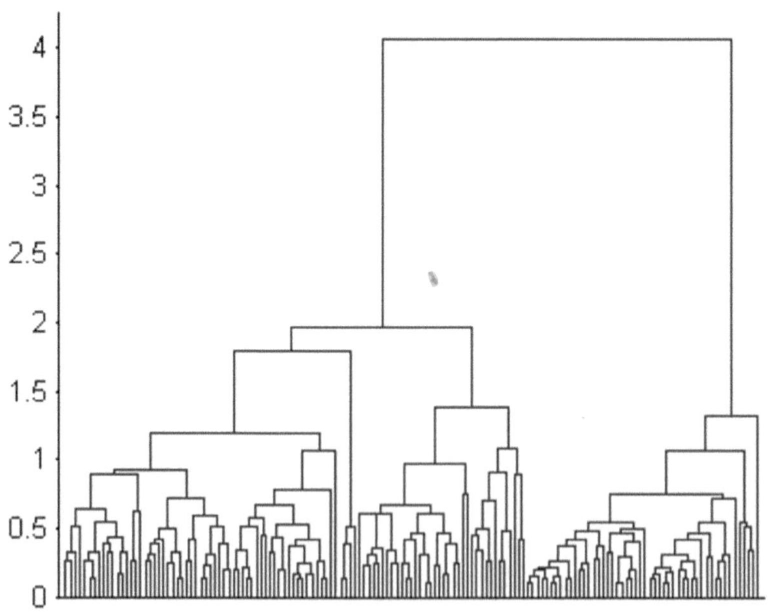

Fig. 4.7 Dendrogram plot for the Iris dataset using the average-link hierarchical algorithm

based fitness function (see (4.5)). In all the tables, the best entries are marked in boldface. In all the tables, the best entries are marked in boldface. Figure 4.6 provides a visual feel of the performance of four clustering methods over the iris dataset. The dataset has been plotted in 3 dimensions using the first three features only. Figure 4.7 shows a dendrogram plot for the Iris dataset using the average-link hierarchical algorithm.

4.3.6 Discussion on the Results (for Real Life Datasets)

A scrutiny of Tables 4.5 and 4.8 reveals the fact that for the iris dataset all the five competitor algorithms terminated with nearly comparable accuracy. The final CS and DB measures were the lowest for the ACDE algorithm. In addition, ACDE was successful in finding the nearly correct number of classes (3 for iris) over repeated runs. However, in Table 4.6, we also find that the GCUK, DCPSO and classical DE yield two clusters on average for the iris dataset. One of the clusters corresponds to the Setosa class while the other corresponds to the combination of Veriscolor and Virginica classes. This happens because the latter two classes are considerably overlapping. There are indices other than the CS or DB measure available in the literature, which yield two clusters for iris dataset [21, 22]. Although the hierarchical algorithm was supplied with the actual number of classes, its performance remained poorer than all the four evolutionary partitional algorithms in terms of the final CS measure, mean intra and inter cluster distance.

Substantial performance differences occur for rest of the more challenging clustering problems with a large number of data-items and clusters as well as overlapping cluster shapes. Tables 4.5 and 4.8 conform to the fact that the ACDE algorithm remains clearly and consistently superior to the other three competitors in terms of the clustering accuracy. For the breast cancer dataset, we observe that both DCPSO and ACDE yield very close final values of the CS index and both find two clusters in almost every run. Entries of Table 4.7 indicate that the ACDE meets or beats its competitors in a statistically significant manner. We also note that the average-link based hierarchical algorithm remained the worst performer over these datasets as well.

In Table 4.10, we find that only in one case (for the breast cancer data) the classical DE based algorithm yields a lower DB measure as compared to ACDE. However, from Table 4.12, we may note that this difference is not statistically significant.

Results from Tables 4.6 and 4.11 reveal that the ACDE yields least number of misclassified items once the clustering is over. In this regard, we would like to mention that despite the convincing performance of all the five algorithms, none of the experiments was without misclassification with respect to the nominal classification, which was what we expected. Interestingly, we found that the final fitness values obtained by our evolutionary clustering algorithms were much better than the fitness of the nominal classification, which shows that the misclassification could not be explained by the optimization performance. Instead, misclassification is the result of the underlying assumptions of the clustering fitness criteria (such as spherical shape of the clusters), outliers in the dataset, errors in collecting data and human errors in the nominal solutions. This is indeed not a negative result. In fact, the differences of a clustering solution based on statistical criteria compared to the nominal classification can reveal interesting data points and anomalies in the dataset. In this way, a clustering algorithm can be used as a very useful tool for data pre-analysis.

From Tables 4.8 and 4.13, we can see that the ACDE was able to reduce both the CS and DB index to the cut-off value within the minimum number of function evaluations for majority of the cases. Both DCPSO and classical DE took lesser computational time than the GCUK algorithm over most of the datasets. One possible reason of this may be the use of less complicated variation operators (like mutation) in PSO and DE as compared to the operators used for GA.

4.4 Application to Image Segmentation

4.4.1 Image Segmentation as a Clustering Problem

Image segmentation may be defined as the process of dividing an image into disjoint homogeneous regions. These homogeneous regions usually contain similar objects of interest or part of them. The extent of homogeneity of the segmented regions can be measured using some image property (e.g. pixel intensity [23]). Segmentation forms a fundamental step towards several complex computer-vision and image analysis applications including digital mammography, remote sensing and land cover study. Segmentation of non-trivial images is one of the most difficult tasks in image processing. Image segmentation can be treated as a clustering problem where the features describing each pixel correspond to a pattern, and each image region (i.e., segment) corresponds to a cluster [23]. Therefore, many clustering algorithms have widely been used to solve the segmentation problem (e.g., k-means [24], Fuzzy C-means [25], ISODATA [26], Snob [27]. Recently the different variants of PSO and DE have emerged as popular optimization techniques for optimal segmentation of gray-scale images, for example see [28 – 31].

4.4.2 Experimental Details and Results

In this section we report the results of applying four evolutionary partitional clustering algorithms (ACDE, DCPSO, GCUK and Classical DE) to the segmentation of five 256×256 grayscale images [2, 31]. The intensity level of each pixel serves as a feature for the clustering process. Hence, although the data points are single dimensional, the number of data-items is as high as 65,536. Finally, the same four algorithms have been applied to classify an image database, which contains 28 small gray scale images of 7 distinct kinds of foliages. The foliage image occurs in the form of a 30×30 digital image. In this case each data-item corresponds to one 30×30 image. Taking the intensity of each pixel as a feature, the dimension of each datapoint becomes 900. We run 2 sets of experiments with 2 fitness functions shown in expressions (4.4) and (4.5). However, to save space we report only the CS measure based results in this subsection. To tackle the high dimensional data points in the last problem mentioned above we use a cosine distance measure described in equation (1.5) following the guidelines in [23]. For the rest of the problems, the Euclidean distance measure is used same as before.

We carried out a thorough experiment with different parameter settings of the clustering algorithms. In Table 4.15, we report an optimal set-up of the parameters that we found best suited for the present image related problems. With these set of parameters, we observed each algorithm to achieve considerably good solutions within an acceptable computational time. Note that the parameter settings do not deviate much for the DCPSO and GCUK algorithms than what is recommended in [3] and [4].

Tables 4.16 and 4.17 summarize the experimental results obtained over five grayscale images in terms of the mean and standard deviations of the number of classes found and the final CS measure reached at by the four adaptive clustering algorithms. Table 4.18 shows results of unpaired *t*-tests taken based on the final

Table 4.15 Parameter setup of the clustering algorithms for the image segmentation problems

GCUK		DCPSO		ACDE		Classical DE	
Parameter	Value	Parameter	Value	Parameter	Value	Parameter	Value
Pop_size	70	Pop_size	40	Pop_size	10*dim	Pop_size	10*dim
Cross-over Probability μ_c	0.85	Inertia Weight	0.75	CR_{max}	1.0	CR	0.95
Mutation probability μ_m	0.005	C_1, C_2	1.494	CR_{min}	0.5	F	0.9
		P_{ini}	0.80				
K_{max} K_{min}	10 2	K_{max} K_{min}	10 2	K_{max} K_{min}	10 2	K_{max} K_{min}	10 2

Table 4.16 Number of classes found over five real life grayscale images and the foliage image database using the CS based fitness function (mean, and standard deviation of the number of classes found over 40 independent runs each run being continued for 10^6 fitness function evaluations)

Image	Optimal no. of Clusters	ACDE	DCPSO	GCUK	Classical DE
Clouds	4	**4.15±0.211**	4.50±0.132	4.75±0.432	3.00±0.00
Robot	3	**4.25±0.428**	2..30±0.012	3.35±0.982	3.00±0.004
Science Magazine	4	**4.05±0.772**	3.25±0.082	6.35±0.093	3.50±0.059
Peppers Image	7	**7.05±0.038**	6.85±0.064	3.90±0.447	8.50±0.067
IRS Mumbai Image	6	**6.10±0.079**	4.65±0.674	7.45±0.043	5.25±0.007
Foliages dataset	7	**7.00±0.00**	10.50±1.132	9.50±0.192	**7.00±0.00**

CS measure of Table 4.18 between the best two algorithms (standard error of difference of the two means, 95% confidence interval of this difference, the t value, and the two-tailed P value). Table 4.19 records the mean number of function evaluations required by each algorithm to reach a predefined cut-off CS value. This table helps in comparing the speeds of different algorithms as applied to image pixel classification.

Figures 4.8 to 4.12 show the five original images and their segmented counterparts obtained using the ACDE, DCPSO, GCUK and the classical DE based clustering algorithms. Figure 4.13 shows the original foliage image database (unclassified). In Table 4.20, we have reported the best classification results achieved with this database using the ACDE algorithm.

Table 4.17 Automatic clustering result over five real life grayscale images and two image datasets using the CS based fitness function (mean, and standard deviation of the final CS measure found over 40 independent runs where each run being continued for 10^6 fitness function evaluations)

Image	CS Measure			
	ACDE	DCPSO	GCUK	Classical DE
Clouds	**0.1317±0.0028**	0.5235±0.0587	0.7806±0.0754	0.1765±0.0332
Robot	**0.19354±0.0028**	0.3478±0.0092	0.9847±0.0846	0.26494±0.00582
Science Magazine	**0.2526±0.0722**	0.4677±0.0493	0.5349±0.0201	0.3509±0.05332
Pepper Image	**0.516±0.0245**	0.9334±0.092	0.7924±0.0160	0.9329±0.0829
IRS Mumbai Image	**0.3892±0.0647**	0.9439±0.0938	0.6992±0.0854	0.3938±0.0693
Foliages dataset	**12.4308±0.3831**	19.8438±0.0921	20.0074±0.3782	14.4965±0.4932

Table 4.18 Results of unpaired t-test between the best and the second best performing algorithm (for each dataset) based on the CS measures of Table 4.17

Image Dataset	Std. Err	t	95% Conf. Intvl	Two-tailed P	Significance
Clouds	0.001	7.1968	-0.0121 to -0.0068	< 0.0001	**Extremely Significant**
Robot	0.002	3.8990	-0.0129 to -0.0040	< 0.0001	**Extremely Significant**
Science Magazine	0.007	34.9267	-0.2665 to -0.2373	< 0.0001	**Extremely Significant**
Pepper Image	0.001	3.0961	-0.0051 to -0.0010	0.0037	**Very Significant**
IRS Mumbai Image	0.003	3.0684	-0.0156 to -0.0032	0.0040	**Very Significant**
Foliages dataset	0.002	3.0584	-0.0109 to -0.0022	0.0041	**Very Significant**

Table 4.19 Mean and standard deviations of the number of fitness function evaluations (over 40 independent runs) required by each algorithm to reach a predefined cut-off value of the CS validity index for the image clustering applications

Image Dataset	Algorithm	CS Cut-off Value	Mean No. of FE required
Clouds	ACDE	10.00	**623210.45**±19.32
	DCPSO		827984.75±16.39
	GCUK		790865.90±18.38
	Classical DE		758796.35±27.57
Robot	ACDE	1.00	**434587.35**±25.59
	PSO		733095.85±20.37
	GCUK		498233.75±12.79
	Classical DE		437369.45±10.38
Science Magazine	ACDE	0.60	384732.35±8.39
	DCPSO		564354.60±12.60
	GCUK		678874.90±12.09
	Classical DE		**448949.55±5.35**
Pepper Image	ACDE	1.00	447874.05±9.34
	DCPSO		548738.55±27.42
	GCUK		475854.65±20.48
	Classical DE		676859.60±17.34
IRS Mumbai Image	ACDE	1.50	**503002.00**±18.48
	DCPSO		906543.95±8.42
	GCUK		554362.75±4.48
	Classical DE		786737.35±3.49
Foliage Image dataset	ACDE	21.00	**903002.00**±52.48
	DCPSO		986543.35±35.29
	GCUK		954974.70±44.84
	Classical DE		976473.40±53.20

4.4.3 Discussion on Image Segmentation Results

From Tables 4.16 – 4.19, as evident, the proposed approach outperforms the state-of-the-art DCPSO and GCUK over a variety of image data sets in a statistically significant manner. Not only does the method find the optimal number of clusters, it also manages to find better clustering of the data points in terms of the two major cluster validity indices used in the literature. From Table 4.20, it is visible that the cluster number of the proposed foliage image patterns is correctly determined by ACDE and the cluster center images can represent common and typical features of each class with respect to different types of foliage.

The remote sensing image of Mumbai (a mega city of India) in Figure 4.12 bears special significance in this context. Usually segmentation of such images

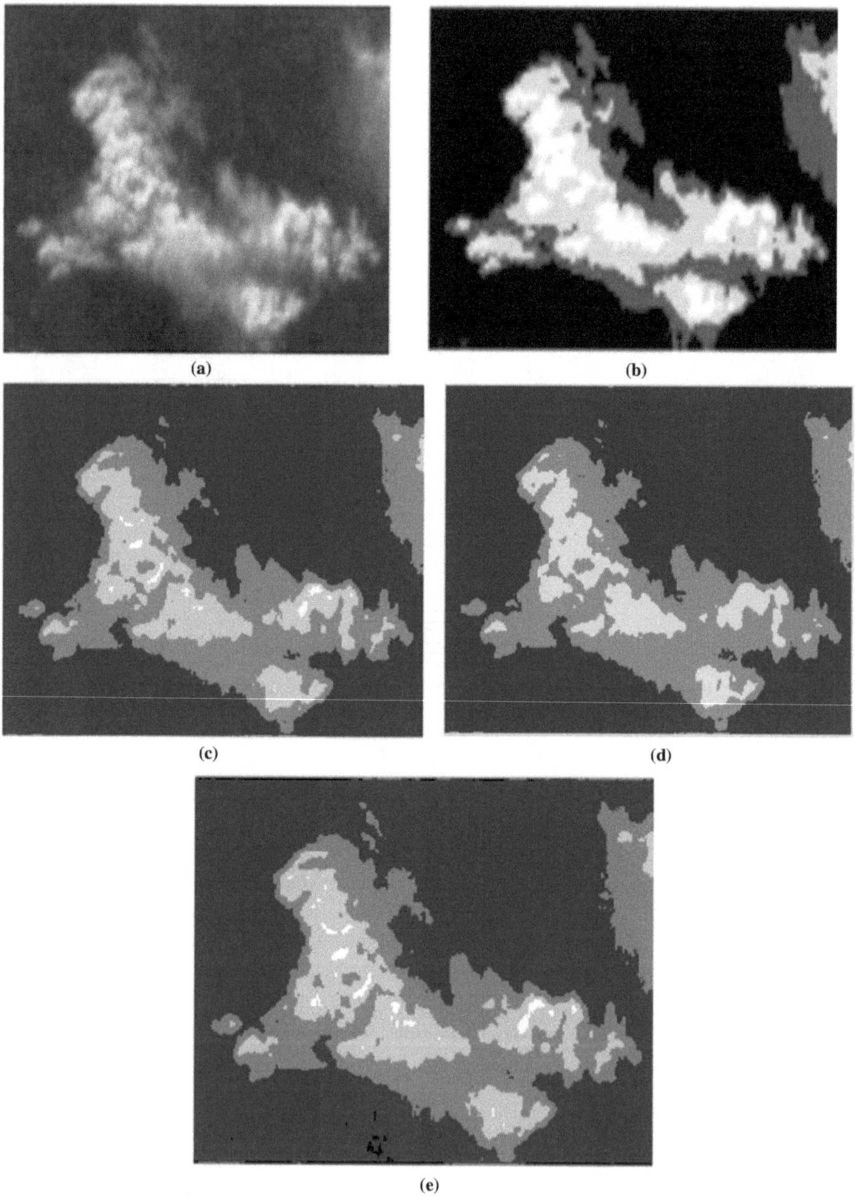

Fig. 4.8 (a) The original Robot image. (b) Segmentation by ACDE (k= 3) (c) Segmentation by DCPSO (k = 2) (d) Segmentation with GCUK (k = 3) (e) Segmentation with Classical DE (k = 3)

(a) (b) (c) (d) (e)

Fig. 4.9 (a) The original Robot image. (b) Segmentation by ACDE (k= 3) (c) Segmentation by DCPSO (k = 2) (d) Segmentation with GCUK (k = 3) (e) Segmentation with Classical DE (k = 3).

helps in the land cover analysis of different areas in a country. The new method yielded six clusters for this image. A close inspection of Figure 4.12 (b) reveals that, most of the land-cover categories have been correctly distinguished in this

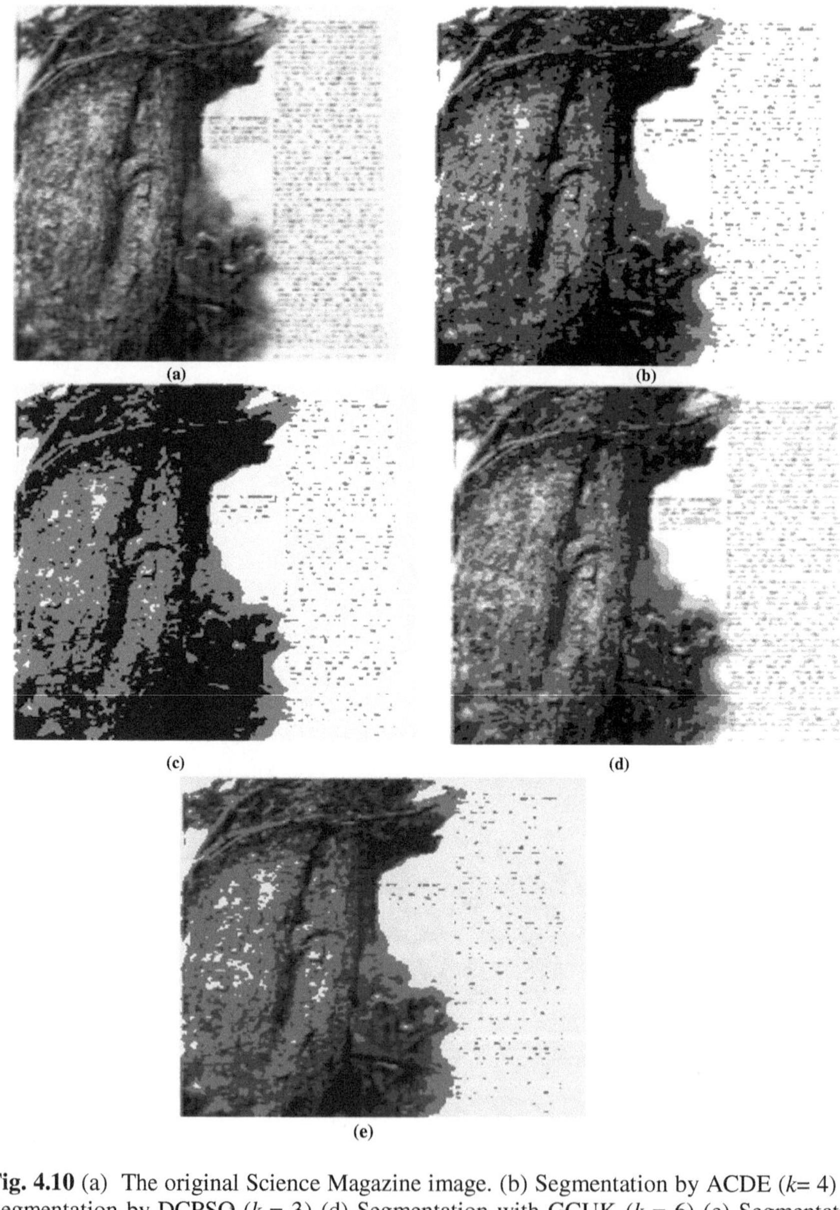

Fig. 4.10 (a) The original Science Magazine image. (b) Segmentation by ACDE (k= 4) (c) Segmentation by DCPSO (k = 3) (d) Segmentation with GCUK (k = 6) (e) Segmentation with Classical DE (k = 3).

image. For example, the *Santa Cruz* airport, dockyard, bridge connecting Mumbai to New Mumbai, and many other road structures have come out distinctly. In addition, the predominance of one category of pixels in the southern part of the image

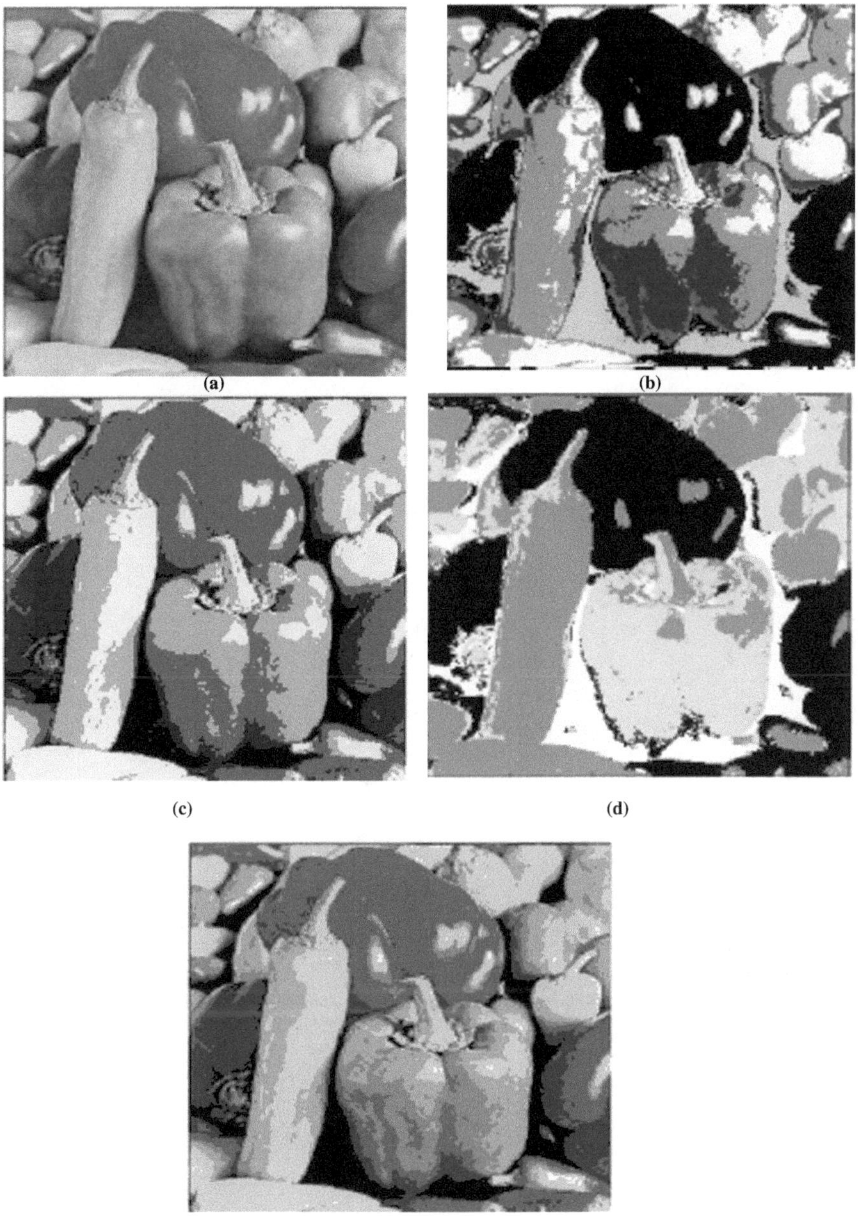

Fig. 4.11 (a) The original Peppers image. (b) Segmentation by ACDE (k= 7) (c) Segmentation by DCPSO (k = 7) (d) Segmentation with GCUK (k = 4) (e) Segmentation with Classical DE (k = 8).

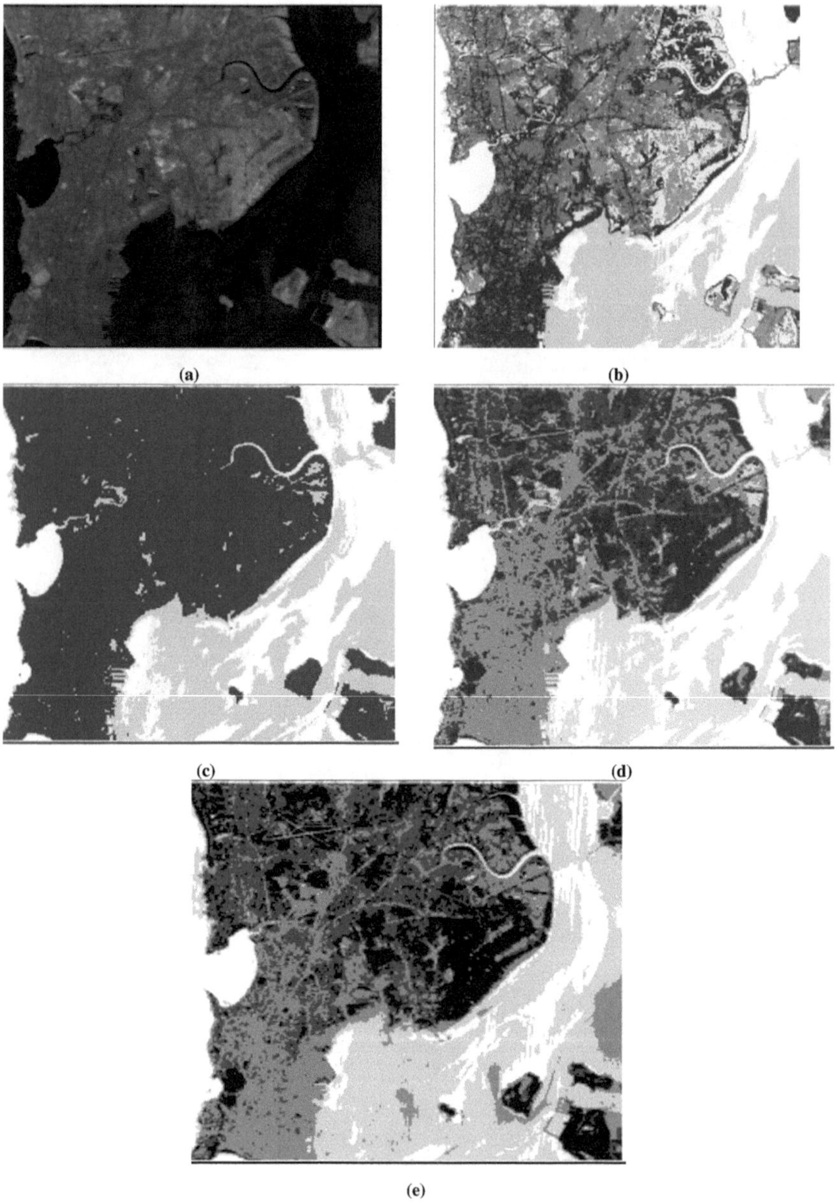

Fig. 4.12 (a) The original IRS image of Mumbai. (b) Segmentation by ACDE (k= 6) (c) Segmentation by DCPSO (k = 4) (d) Segmentation with GCUK (k = 7) (e) Segmentation with Classical DE (k = 5).

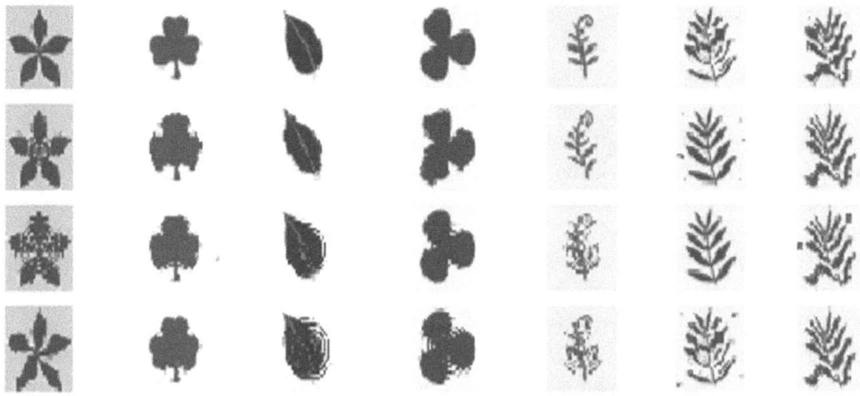

Fig. 4.13 900 dimensional images of seven different kinds of foliages

Table 4.20 Clustering result over the foliage image patterns by the ACDE algorithm

	Class 1	Class 2	Class 3	Class 4	Class 5	Class 6	Class 7
Cluster Centers	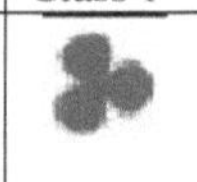				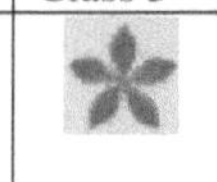		
Members of each class	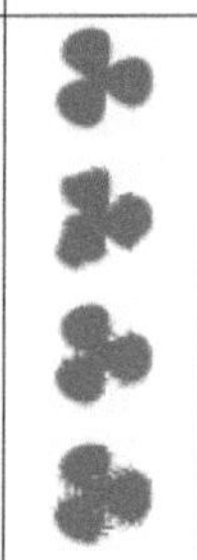		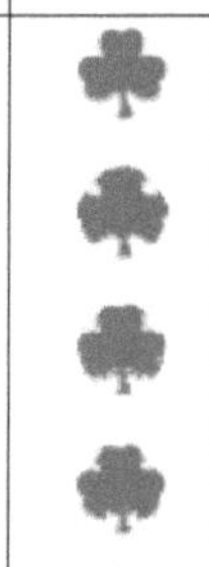		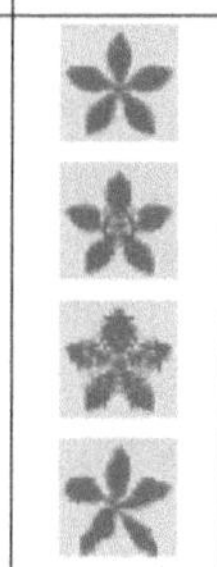	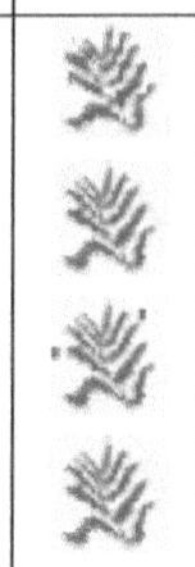	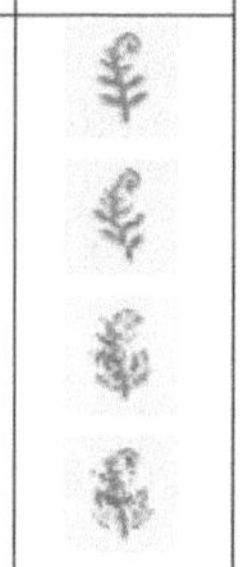

conforms to the ground truth; this part is known to be heavily industrialized, and hence the majority of the pixels in this region should belong to the same class of concrete. The *Arabian Sea* has come out as a combination of pixels of two different classes. The seawater is found to be decomposed into two classes, turbid water 1 and turbid water 2, based on the difference of their reflectance properties.

From the experimental results, we note that the ACDE performs much better than the classical DE based clustering scheme. Since both the algorithms use same chromosome representation scheme and starts with the same initial population, difference in their performance must be due to the difference in their internal operators and parameter values. From this, we may infer that the adaptation schemes suggested for parameters F and CR of DE in equations (4.6) and (4.7) improved the performance of the algorithm considerably at least for the clustering problems covered here.

4.5 Conclusions

This Chapter presented a new, Differential Evolution-based strategy for crisp clustering of real world datasets. An important feature of the proposed technique is that it is able to find the optimal number of clusters automatically (that is, the number of clusters does not have to be known in advance) even for very high dimensional datasets where tracking of the number of clusters may be well nigh impossible. The proposed ACDE algorithm performed very well when compared to other state-of-the-art clustering algorithms in a statistically meaningful way over the considered benchmark datasets. This certainly does not lead us to claim that ACDE may outperform DCPSO or GCUK over every dataset since it is impossible to model all the possible complexities of a real life data with the limited test-suit that we used for testing the algorithms. In addition, the performance of DCPSO and GCUK may also be enhanced with a judicious parameter tuning, which renders itself to further research with these algorithms. However, the only conclusion we can draw at this point is that DE with the suggested modifications can serve as an attractive alternative for dynamic clustering of completely unknown datasets.

To reduce the computational burden further, we feel that it will be more judicious to associate the automatic research of the clusters with the choice of the most relevant features compared to the process used. Often we have a great number of features (especially for a high dimensional dataset like the foliage images) which are not all relevant for a given operation. Hence, future research may focus on integrating the automatic feature-subset selection scheme with the ACDE algorithm. The combined algorithm is expected to automatically project the data to a low dimensional feature subspace, to determine the number of clusters and find out the appropriate cluster centers with the most relevant features at a faster pace.

Appendix: Statistical Tests Used

Optimization strategies used in this thesis like PSO, GA, DE etc. are stochastic in nature. Therefore, it is important to analyze the results from several repeated runs by statistical methods, in order obtain empirical evidence of the capabilities of a given approach. Fundamental for our data analysis is the assumption that all test runs are independent; i.e, one run does not have any influence on subsequent runs. A series of n test-runs yields the observations $x_1, x_2 ... x_n$ from the stochastic variables $X_1, X_2 ... X_n$. All data presented in this thesis is based on 25, 30, 40, or 50 repetitions of test runs (based on the complexities of the problems). For each run we stored relevant information from each generation/time step as the data sample for the analysis - e.g. the fitness value of the best individual from each generation is stored.

To judge the statistical significance of the results, in the present and the following chapters of this Volume, we have used unpaired *t*-tests [20, 32]. The *t*-test is a statistical method to evaluate the significant difference between two algorithms. The t-value will be positive if the first algorithm is better than the second, and it is negative if it is poorer. The t-value is defined as follows:

$$t = \frac{\overline{\alpha}_2 - \overline{\alpha}_1}{\sqrt{\left(\frac{\sigma_2^2}{\xi+1}\right) + \left(\frac{\sigma_1^2}{\xi+1}\right)}} \tag{A.1}$$

where $\overline{\alpha}_1$ and $\overline{\alpha}_2$ are the mean values of the first and second methods, respectively; σ_1 and σ_2 are the standard deviations of the first and second methods, respectively; and ξ is the value of the degrees of freedom. When the t-value is higher than 1.645 ($\xi = 49$), there is a significant difference between the two algorithms with a 95% confidence level. Please note that in this thesis all the t-tests have been performed using the statistical calculator available from the website: http://www.graphpad.com/quickcalcs/ttest1.cfm

References

[1] Price, K., Storn, R., Lampinen, J.: Differential Evolution - A Practical Approach to Global Optimization. Springer, Berlin (2005)

[2] Das, S., Abraham, A., Konar, A.: Automatic clustering using an improved differential evolution algorithm. IEEE Transactions on Systems Man and Cybernetics - Part A 38(1), 218–237 (2008)

[3] Bandyopadhyay, S., Maulik, U.: Genetic clustering for automatic evolution of clusters and application to image classification. Pattern Recognition 35, 1197–1208 (2002)

[4] Omran, M., Salman, A., Engelbrecht, A.: Dynamic clustering using particle swarm optimization with application in unsupervised image classification. In: Fifth World Enformatika Conference (ICCI 2005), Prague, Czech Republic (2005)

[5] Holland, J.H.: Adaptation in Natural and Artificial Systems. University of Michigan Press, Ann Arbor (1975)

[6] Kennedy, J., Eberhart, R.C.: Swarm Intelligence. Morgan Kaufmann, San Francisco (2001)

[7] Chou, C.H., Su, M.C., Lai, E.: A new cluster validity measure and its application to image compression. Pattern Analysis and Applications 7(2), 205–220 (2004)

[8] Davies, D.L., Bouldin, D.W.: A cluster separation measure. IEEE Transactions on Pattern Analysis and Machine Intelligence 1, 224–227 (1979)

[9] Dunn, J.C.: Well separated clusters and optimal fuzzy partitions. Journal of Cybernetics 4, 95–104 (1974)

[10] Bezdek, J.C.: Numerical taxonomy with fuzzy sets. Journal of Math. Biol., 157–171 (1974)

[11] Bezdek, J.C.: Cluster validity with fuzzy sets. Journal of Cybernetics (3), 58–72 (1974)

[12] Xie, X., Beni, G.: Validity measure for fuzzy clustering. IEEE Trans. Pattern Anal. Machine Learning 3, 841–846 (1991)

[13] Gustafson, D., Kessel, W.: Fuzzy clustering with a fuzzy covariance matrix. In: Proc. IEEE CDC, San Diego, CA, USA, pp. 761–766 (1979)
[14] Das, S., Konar, A., Chakraborty, U.K.: Two Improved differential evolution schemes for faster global search. In: ACM-SIGEVO Proceedings of Genetic and Evolutionary Computation Conference (GECCO 2005), Washington DC (2005)
[15] Day, W.H., Edelsbrunner, H.: Efficient algorithms for agglomerative hierarchical clustering methods. Journal of Classification 1, 1–24 (1984)
[16] Blake, C., Keough, E., Merz, C.J.: UCI repository of machine learning database (1998), `http://www.ics.uci.edu/~mlearn/MLrepository.html`
[17] Pal, S.K., Majumder, D.D.: Fuzzy sets and decision-making approaches in vowel and speaker recognition. IEEE Transactions on System, Man and Cybernertics SMC-7, 625–629 (1977)
[18] van den Bargh, F., Engelbrecht, A.P.: A cooperative approach to particle swarm optimization. IEEE Transactions on Evolutionary Computing 8(3) (June 2004)
[19] Olson, C.: Parallel algorithms for hierarchical clustering. Parallel Computing 21(8), 1313–1325 (1995)
[20] Flury, B.: A First Course in Multivariate Statistics, vol. 28. Springer, Heidelberg (1997)
[21] Bezdek, J.C., Pal, N.R.: Some new indexes of cluster validity. IEEE Transactions on Systems, Man, Cybernetics 28, 301–315 (1998)
[22] Kothari, R., Pitts, D.: On finding the number of clusters. Pattern Recognition Letters 20, 405–416 (1999)
[23] Jain, A.K., Murty, M.N., Flynn, P.J.: Data clustering: a review. ACM Computing Surveys 31(3), 264–323 (1999)
[24] Tou, J.T., Gonzalez, R.C.: Pattern Recognition Principles. Addison-Wesley, London (1974)
[25] Trivedi, M.M., Bezdek, J.C.: Low-level segmentation of aerial images with fuzzy clustering. IEEE Trans. on Systems, Man and Cybernetics 16 (1986)
[26] Ball, G., Hall, D.: A clustering technique for summarizing multivariate data. Behavioral Science 12, 153–155 (1967)
[27] Wallace, C.S., Boulton, D.M.: An information measure for classification. Computer Journal 11(2), 185–194 (1968)
[28] Omran, M., Engelbrecht, A., Salman, A.: Particle swarm optimization method for image clustering. International Journal of Pattern Recognition and Artificial Intelligence 19(3), 297–322 (2005)
[29] Das, S., Konar, A., Abraham, A.: Spatial information based image segmentation with a modified particle swarm optimization. In: Sixth International Conference on Intelligent System Design and Applications (ISDA) 2006, Jinan, Shangdong, China. IEEE Computer Society Press, Los Alamitos (2006)
[30] Das, S., Abraham, A., Sarkar, S.K.: A hybrid rough-swarm algorithm for image pixel classification. In: Proceedings of 6th International Conference on Hybrid Intelligent Systems (HIS 2006), AUT Technology Park, Auckland, New Zealand. IEEE Computer Society Press, Los Alamitos (2006)
[31] Das, S., Konar, A.: Automatic Image Pixel Clustering with an Improved Differential Evolution. Applied Soft Computing Journal 9(1), 226–236 (2009)
[32] Press, W.H., Teukolsky, S.l.A., Vetterling, W.T., Flannery, B.P.: Numerical Recipes in C: The Art of Scientific Computing. Cambridge University Press, Cambridge (1992)

Chapter 5
Fuzzy Clustering in the Kernel-Induced Feature Space Using Differential Evolution Algorithm

Abstract. This chapter introduces a scheme for clustering complex and linearly non-separable datasets, without any prior knowledge of the number of naturally occurring groups in the data. The proposed method is based on a modified version of the classical Differential Evolution (DE) algorithm, which uses the neighborhood-based mutation strategy. It also employs a kernel-induced similarity measure instead of the conventional sum-of-squares distance. Use of the kernel function makes it possible to cluster data that is linearly non-separable in the original input space into homogeneous groups in a transformed high-dimensional feature space. The performance of the proposed method has been extensively compared with a few state of the art clustering techniques over a test-suite of several artificial and real life datasets. Based on the computer simulations, some empirical guidelines have been provided for selecting the suitable parameters of the DE algorithm.

5.1 Introduction

The Euclidean distance metric, employed by most of the exisiting partitional clustering algorithms, work well with datasets in which the natural clusters are nearly hyper-spherical and linearly seperable (like the artificial dataset 1 used in this chapter). But it causes severe misclassifications when the dataset is complex, with linearly non-separable patterns (like the synthetic datasets 2, 3, and 4 described in Section 5.5). We would like to mention here that, most evolutionary algorithms could potentially work with an arbitrary distance function and are not limited to the Euclidean distance. In order to illustrate the inefficiency of the classical partitional clustering algorithms using Euclidean distance metric, we test the k-means, fuzzy c-means and the ACDE method developed in the previous chapter on the two-dimensional circle data set. The data set is hand-crafted and contains 400 data points distributed in two linearly non-seperable clusters. Figure 5.1 depicts the results of this clustering algorithm.

S. Das, A. Abraham, and A. Konar: Metaheuristic Clustering, SCI 178, pp. 175–211.
springerlink.com

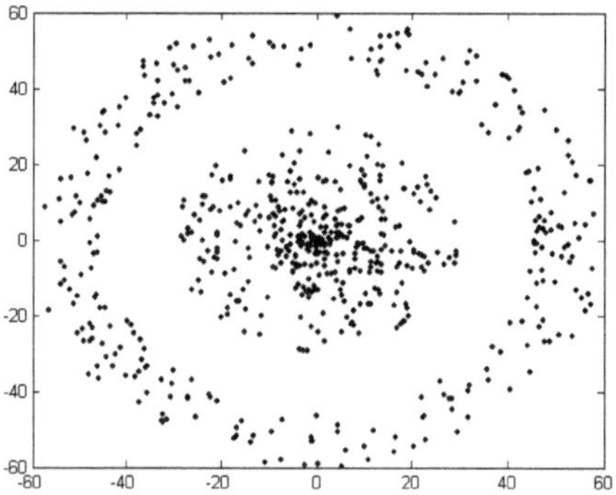

(a) Unlabelled data set

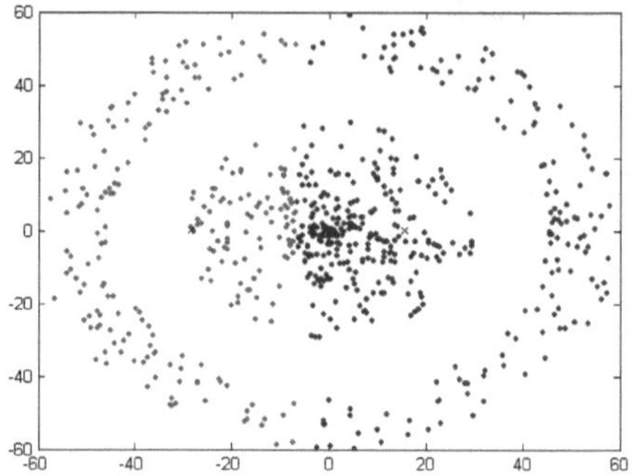

(b) Clustering result with k-means algorithm

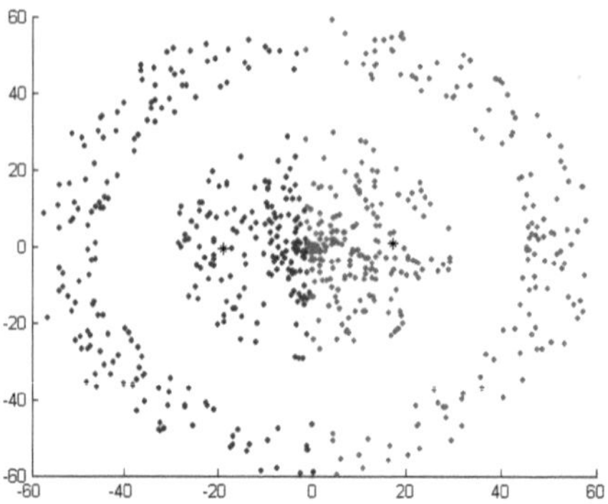

(c) Clustering with fuzzy c-means algorithm

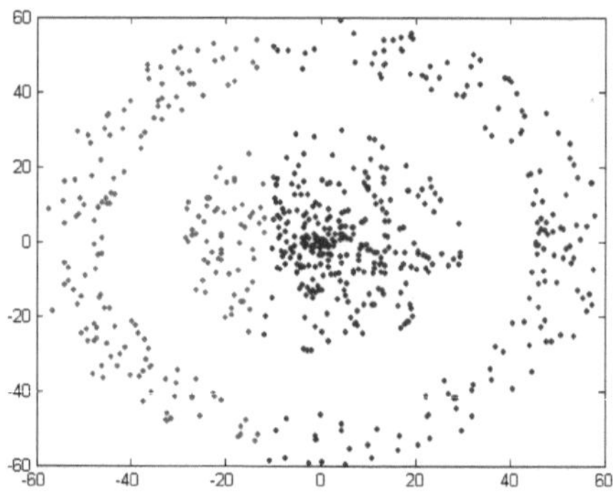

(d) Clustering with ACDE

Fig. 5.1 Results of clustering on the linearly non-separable circle dataset by three partitional clustering algorithms

Figure 5.1 indicates that three partitional clustering algorithms including the two traditional ones (k-means and fuzzy c-means) and the newly developed ACDE end up with comparable results on the linearly nonseparable dataset. The algorithms completely fail to capture the concentric circular grouping in the data as they use the conventional squarred Euclidean distance metric. To circumvent this problem, in this chapter we develop a DE-based clustering technique that first projects the data into a high-dimensional feature space through specially devised kernel functions. A kernel function measures the distance between two data points by implicitly mapping them into a high-dimensional feature space where the data is linearly separable [1]. Not only does it preserve the inherent structure of groups in the input space, but also simplifies the associated structure of the data patterns [1, 2]. Several kernel-based learning methods, including the Support Vector Machine (SVM), have recently been shown to perform remarkably in supervised learning [3 - 6]. The kernelized versions of the k-means and the fuzzy c-means

(FCM) algorithms reported in [5] and [6] respectively, have reportedly outperformed their original counterparts over several test cases.

The main contributions of this Chapter are as follows:

i) We propose a new family of variants of the classical DE algorithm that uses a neighborhood-based mutation strategy [7], taking inspiration from the *l_best* PSO algorithms [8] and apply the same to the kernel induced data clustering problem [9]. The proposed schemes balance between the exploration and exploitation abilities of DE without imposing serious additional burdens in terms of function evaluations. Experiments indicate that the proposed DEGL (DE with global and local neighborhood) algorithm [7] yields more accurate results consuming a lesser amount of computational time than the classical DE in context to the present problem. The schemes have already been discussed in section 2.5.9 of Chapter 2.

ii) We reformulate a recently proposed fuzzy cluster validity index (known as the *Xie-Beni measure orXB*) [10] using the kernelized distance metric. This reformulation eliminates the need to compute the cluster-centroids repeatedly for evaluating *XB* value, due to the implicit mapping via the kernel function. The new *XB* measure forms the objective function to be minimized for optimal clustering.

iii) The new vector representation scheme developed in the last chapter has also been employed here to determine the number of clusters automatically during the run.

We have undertaken extensive performance comparisons in order to establish the effectiveness of the proposed method in detecting clusters from several synthetic as well as real world datasets. Effect of the growth of feature-space dimensionality on the performance of the algorithm was also studied based on the real life datasets. The rest of the Chapter is organised as follows: Section 5.2 describes the kernel distance metric and the reformulation of the *XB* measure. Section 5.3 describes the novel procedure for automatic clustering with DEGL. Experimental results are presented and discussed in Section 5.4. Application of the proposed method to automatic fuzzy image segmentation has been investigated in Section 5.5. Finally conclusions are drawn in Section 5.6.

5.2 The Kernel-Induced Clustering

Given a dataset **Z** in the *d*-dimensional real space $\Re^d$, let us consider a non-linear mapping function from the input space to a high dimensional feature space H:

$$\varphi : \Re^d \rightarrow H \ , \ \vec{Z}_i \rightarrow \varphi(\vec{Z}_i) \tag{5.1}$$

where $\vec{Z}_i = [z_{i,1}, z_{i,2}, \ldots\ldots, z_{i,d}]^T$ and $\varphi(\vec{Z}_i) = [\varphi_1(\vec{Z}_i), \varphi_2(\vec{Z}_i), \ldots\ldots, \varphi_H(\vec{Z}_i)]^T$

The mapping has been schematically illustrated in Figure 5.2.

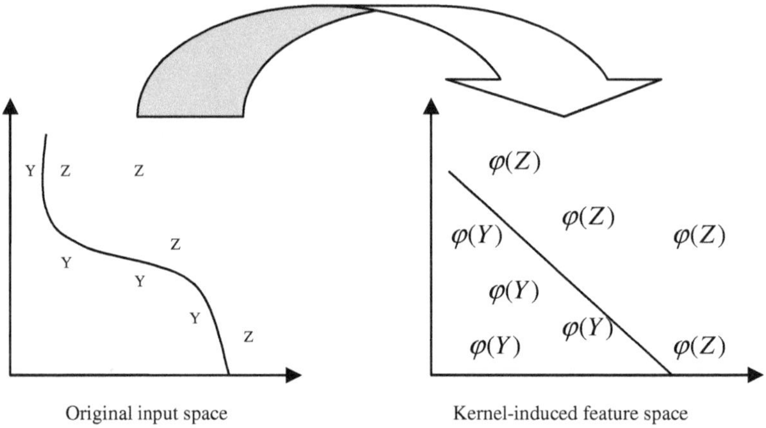

Fig. 5.2 Mapping data points from original input space to a high-dimensional feature

The objective of such a mapping is that the original data which remains linearly non-seperable in the input space can easily be seperated using hyper-planes in higher dimensional feature space. For example consider the concentric circle data depicted in Figure 5.1. If we use a mapping $\varphi : \Re^2 \rightarrow H = \Re^3$, and $[z_{i,1}, z_{i,2}]^T \rightarrow [z_{i,1}, z_{i,2}, z_{i,1}^2 + z_{i,2}^2]^T$, the two circles become linearly seperable in the high dimensional feature space as is evident from Figure 5.3.

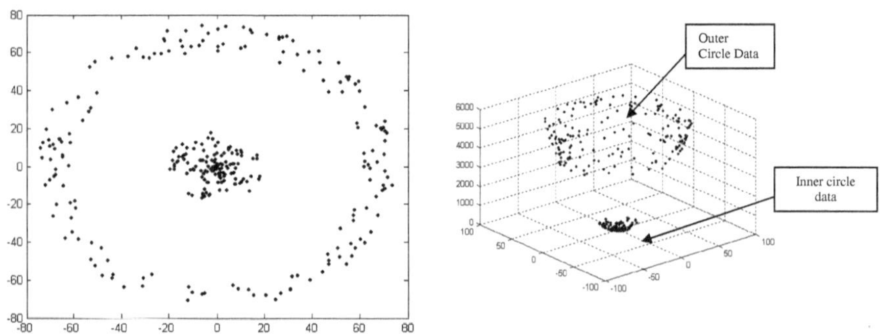

(a) The slightly trimmed circle dataset (b) Three dimensional projection of the same data

Fig. 5.3 Higher dimensional projection of circle dataset using the mapping $[z_{i,1}, z_{i,2}]^T \rightarrow [z_{i,1}, z_{i,2}, z_{i,1}^2 + z_{i,2}^2]^T$

By applying the mapping, a dot product $\vec{Z}_i^T.\vec{Z}_j$ is transformed into $\varphi^T(\vec{Z}_i).\varphi(\vec{Z}_j)$. Now, the central idea in kernel-based learning is that the mapping function φ need not be explicitly specified. The dot product $\varphi^T(\vec{Z}_i).\varphi(\vec{Z}_j)$ in the transformed space can be calculated through the kernel function $K(\vec{Z}_i,\vec{Z}_j)$ in the input space $\Re^d$. Below we define the Mercer kernel [17, 18] for the real vector space formally.

Definition 5.1. Let $\mathbf{Z} = [\vec{Z}_1,\vec{Z}_2,...,\vec{Z}_n]$ be a nonempty set where each $\vec{Z}_i \in \Re^D$. A function K: $\mathbf{Z} \times \mathbf{Z} \rightarrow \Re$ is called a positive definite kernel (or Mercer kernel) if and only if K is symmetric (i.e. $K(\vec{Z}_i,\vec{Z}_j) = K(\vec{Z}_j,\vec{Z}_i)$ and the following equation holds:

$$\sum_{i=1}^{n}\sum_{j=1}^{n} c_i c_j K(\vec{Z}_i,\vec{Z}_j) \geq 0 \qquad \forall n \geq 2, \tag{5.2}$$

where $c_r \in \Re, \ \forall r = 1,....,n$. Each Mercer kernel can be expressed in terms of:

$$K(\vec{Z}_i,\vec{Z}_j) = \varphi^T(\vec{Z}_i).\varphi(\vec{Z}_j),$$

where φ is the mapping function from low dimensional input space to a high dimensional feature space.

Consider the following simple example:

Example 5.1. let $d = 2$ and $H = 3$ and consider the following mapping:

$\varphi : \Re^2 \rightarrow H = \Re^3$, and $[z_{i,1}, z_{i,2}]^T \rightarrow [z_{i,1}^2, \sqrt{2}.z_{i,1}z_{i,2}, z_{i,2}^2]^T$

Now the dot product in feature space H:

$$\begin{aligned}\varphi^T(\vec{Z}_i).\varphi(\vec{Z}_i) &= [z_{i,1}^2,\sqrt{2}.z_{i,1}z_{i,2},z_{i,2}^2].[z_{j,1}^2,\sqrt{2}.z_{j,1}.z_{j,2},z_{j,2}^2]^T. \\ &= [z_{i,1}.z_{j,1} + z_{i,2}.z_{j,2}]^2 \\ &= [\vec{Z}_i^T.\vec{Z}_j]^2 = K(\vec{Z}_i,\vec{Z}_j)\end{aligned}$$

Clearly the simple kernel function K is the square of the dot product of vectors $\vec{Z}_i$ and $\vec{Z}_j$ in $\Re^d$. Hence, the kernelized distance measure between two patterns $\vec{Z}_i$ and $\vec{Z}_j$ is given by:

$$\left\|\varphi(\vec{Z}_i)-\varphi(\vec{Z}_j)\right\|^2 = (\varphi(\vec{Z}_i)-\varphi(\vec{Z}_j))^T(\varphi(\vec{Z}_i)-\varphi(\vec{Z}_j))$$

$$= \varphi^T(\vec{Z}_i).\varphi(\vec{Z}_i)-2.\varphi^T(\vec{Z}_i).\varphi(\vec{Z}_j)+\varphi^T(\vec{Z}_j).\varphi(\vec{Z}_j)$$

$$= K(\vec{Z}_i,\vec{Z}_i)-2.K(\vec{Z}_i,\vec{Z}_j)+K(\vec{Z}_j,\vec{Z}_j) \qquad (5.3)$$

Examples of a few popular Mercer kernels are the following [19]:

linear: $$K(\vec{Z}_i,\vec{Z}_j)=\vec{Z}_i^T\cdot\vec{Z}_j \qquad (5.4)$$

polynomial with degree p: $$K(\vec{Z}_i,\vec{Z}_j)=\left(1+\vec{Z}_i^T\cdot\vec{Z}_j\right)^p,\ p\in \mathrm{N} \qquad (5.5)$$

Gaussian: $$K(\vec{Z}_i,\vec{Z}_j)=\exp\left(-\frac{\left\|\vec{Z}_i-\vec{Z}_j\right\|^2}{2\sigma^2}\right),\ \sigma\in\Re \qquad (5.6)$$

Among the various kernel functions used in literature, in the present context, we have chosen the well-known Gaussian kernel (also referred to as the Radial Basis Function) owing to its better classification accuracy over the linear and polynomial kernels on many test problems [2, 20, 21]. The Gaussian Kernel may also be represented in a slightly different way as:

$$K(\vec{Z}_i,\vec{Z}_j)=\exp\left(\frac{-(\sum_{k=1}^{d}|z_{i,k}-z_{j,k}|^a)^b}{\sigma^2}\right) \qquad (5.7)$$

where $a \ge 0, 1 \le b \le 2$ and $\sigma > 0$ are constants. Clearly, for Gaussian kernel, $K(\vec{Z}_i,\vec{Z}_i)=1$ and thus equation (5.31) reduces to:

$$d^{\phi}(\vec{Z}_i,\vec{Z}_j)=\left\|\varphi(\vec{Z}_i)-\varphi(\vec{Z}_j)\right\|^2=2.(1-K(\vec{Z}_i,\vec{Z}_j)) \qquad (5.8)$$

The kernel trick has recently been used in literature for clustering complex and linearly non separable datasets. . These methods can be broadly divided in three categories, which are based respectively on the:

1) kernelization of the metric [22 - 24],
2) clustering in feature space [25 - 28],
3) description via support vectors [29, 30].

Methods based on kernelization of the metric look for centroids in input space, and the distances between patterns and centroids are computed through kernels:

$$\left\|\varphi(\vec{Z}_k) - \varphi(\vec{m}_i)\right\|^2 = K(\vec{Z}_k, \vec{Z}_k) - 2K(\vec{Z}_k, \vec{m}_i) + K(\vec{m}_i, \vec{m}_i) \qquad (5.9)$$

Clustering in feature space is made by mapping each pattern using the function φ and then computing centroids in feature space. Calling $\vec{m}_i^{\varphi}$ the centroids in feature space, it is possible to compute the distance $\left\|\varphi(\vec{Z}_k) - \varphi(\vec{m}_i)\right\| = \left\|\varphi(\vec{Z}_k) - \vec{m}_i^{\varphi}\right\|$ by means of the kernel trick. The description via support vectors makes use of One Class SVM to find a minimum enclosing sphere in feature space able to enclose almost all data in feature space excluding outliers. The Support Vector Clustering algorithm allows labeling the enclosed patterns in feature space. In the next subsections we will outline these three approaches. A comprehensive foundation of the most important kernel induced clustering algorithms can be found in [31].

5.3 The Kernel-Induced Clustering Technique with DEGL

5.3.1 Kernelization of the Xie-Beni Index

To judge the quality of a partition provided by some clustering algorithm, it is necessary to have a well-defined statistical-mathematical function, called a *cluster validity index.* In chapter 1, we have already reviewed some of the well-known indices available in the literature for fuzzy clustering such as, the partition coefficient, partition entropy, Xie-Beni index, Kwon's index, and the PBMF index. We here briefly describe the Xie-Beni index [10], a kernelized version of which will be used for constructing the fitness function of the proposed method. At this point, we would like to mention that our algorithm, like other evolutionary clustering methods, can be used with any kind of clustering validity index forming its fitness function. However, we here choose the Xie-Beni index owing to its wide popularity among the researchers for correctly finding the number of clusters in many complex datasets [32]. In addition, three main competitor algorithms used here for comparing the performance of the new method also employ the same index to construct their fitness functions. The Xie-Beni index (for fuzzifier q) is given by:

$$XB_q = \frac{\sum_{i=1}^{k}\sum_{j=1}^{n} u_{ij}^{q}\left\|\vec{Z}_j - \vec{m}_i\right\|^2}{n \times \min_{i \neq j}\left\|\vec{m}_i - \vec{m}_j\right\|^2}, \qquad (5.10)$$

where u_{ij} is the fuzzy membership of the j-th data point in i-th cluster and it is given by the usual update rule of the FCM algorithm as:

$$u_{ij} = \frac{\left\| \vec{Z}_j - \vec{m}_i \right\|^{2/(1-q)}}{\sum_{p=1}^{k} \left\| \vec{Z}_j - \vec{m}_p \right\|^{2/(1-q)}}, \tag{5.11}$$

where q is the fuzzifier term. The optimal number of clusters can be obtained by minimizing the above index value. Now using equation (5.8), the kernel version of the membership function shown in (5.11) may be expressed as:

$$u_{ij}^{\phi} = \frac{[1 - K(\vec{Z}_j, \vec{m}_i)]^{1/(1-q)}}{\sum_{p=1}^{k} [1 - K(\vec{Z}_j, \vec{m}_p)]^{1/(1-q)}}. \tag{5.12}$$

Please note that the within cluster sum, which is at the numerator of the Xie-Beni index is also the objective function for the FCM algorithm and is given by:

$$J_q = \sum_{j=1}^{n} \sum_{i=1}^{k} (u_{ij})^q \left\| \vec{X}_j - \vec{m}_i \right\|^2 \tag{5.13}$$

From equations (5.12) and (5.13) the *within cluster sum* function for FCM in kernel space may be written as:

$$J_q^{\phi} = \sum_{j=1}^{n} \sum_{i=1}^{k} \left\{ \frac{[1 - K(\vec{Z}_j, \vec{m}_i)]^{1/(1-q)}}{\sum_{p=1}^{k} [1 - K(\vec{Z}_j, \vec{m}_p)]^{1/(1-q)}} \right\}^q * [2 - 2.K(\vec{Z}_j, \vec{m}_i)] \tag{5.14}$$

$$= 2 * \sum_{j=1}^{n} \sum_{i=1}^{k} \frac{[1 - K(\vec{Z}_j, \vec{m}_i)]^{q/(1-q)}}{\left\{ \sum_{p=1}^{k} [1 - K(\vec{Z}_j, \vec{m}_p)]^{1/(1-q)} \right\}^q} * [1 - K(\vec{Z}_j, \vec{Z}_i)]$$

$$= 2 * \sum_{j=1}^{n} \sum_{i=1}^{k} \frac{[1 - K(\vec{Z}_j, \vec{m}_i)]^{\frac{q}{1-q}+1}}{\left\{ \sum_{p=1}^{k} [1 - K(\vec{Z}_j, \vec{m}_p)]^{1/(1-q)} \right\}^q}$$

$$= 2 * \sum_{j=1}^{n} \frac{\sum_{i=1}^{k} [1 - K(\vec{Z}_j, \vec{m}_i)]^{1/1-q}}{\left\{ \sum_{p=1}^{k} [1 - K(\vec{Z}_j, \vec{m}_p)]^{1/1-q} \right\}^{q}}$$

Substituting i with p,

$$= 2 * \sum_{j=1}^{n} \left\{ \sum_{p=1}^{k} [1 - K(\vec{Z}_j, \vec{m}_p)]^{1/1-q} \right\}^{1-q}. \tag{5.15}$$

We note that J_m^{ϕ} is independent of the membership matrix $\mathbf{U} = [u_{ij}]_{k \times n}$ but depends only on the cluster centroids. Now from (5.14) and (5.10) we may form the kernelized version of the Xie-Beni index as:

$$XB_q^{\phi} = \frac{J_q^{\phi}}{n * \min_{i \neq j} 2 * [1 - K(\vec{m}_i, \vec{m}_j)]}. \tag{5.16}$$

The fitness function for the DEGL algorithm may now be written as:

$$f = \frac{1}{XB_{qi}^{\phi}(c) + eps}, \tag{5.17}$$

where XB_{qi}^{ϕ} is the kernelized Xie-Beni index for partition obtained by the i-th chromosome and eps is a very small constant (we used 0.0002). So maximization of this function means minimization of the corresponding Xie-Beni index.

5.3.2 Summary of the Integrated Clustering Approach

Below we provide a pseudo-code for the complete clustering technique based on DEGL and the kernelized Xie-Beni index.

Step 1: Initialize each search variable vector to contain C_{max} randomly selected cluster centers and C_{max} (randomly chosen) activation thresholds. The thresholds should be initialized between 0 and 1.

Step 2: Find out the active cluster centers in each vector by evaluating the activation thresholds according to (4.1) and set or reset the corresponding flag.

Step 3: For $t = 1$ to t_{max} do

i) For each data vector $\vec{Z}_p$, Calculate its kernel induced distance metric $d^{\phi}(\vec{Z}_p, \vec{m}_{i,j})$ from all active cluster centers $\vec{m}_{i,j}$ of a parameter vector $\vec{X}_i$ $(i = 1,2,\ldots.NP)$.

ii) Assign $\vec{Z}_p$ to cluster center $\vec{m}_{i,j}$ such that

$$d^{\phi}(\vec{Z}_p, \vec{m}_{i,j}) = \min_{\forall b \in \{1,2,\ldots,k\}} \left\{ d^{\phi}(\vec{Z}_p, \vec{m}_{i,b}) \right\}$$

iii) Check if the number of data points belonging to any cluster center is less than 2. If so, update the cluster centers of the DE vector using the concept of average described in subsection 4.2.3.

iv) Now perform mutation on each population member $\vec{X}_i(t)$ of DE using equations (2.36)-(2.42) to form the corresponding donor vectors $\vec{V}_i(t)$. Note that in this work the DEGL/SAW scheme is used. Exchange components of the donor with the target vector $\vec{X}_i(t)$ according to (2.5) to form the trial vectors $\vec{U}_i(t)$.

v) Locate the active cluster centers of the trial vectors thus formed, by applying rule (4.1) and set or reset the associated flags correspondingly.

vi) Repeat steps (i), (ii) and (iii) for each trial vector.

vii) Evaluate fitness of both the target and trial vectors according to equation (5.17). Use only the active cluster centers in both the vectors (that is the ones with the flag set to 1). Replace the target vector $\vec{X}_i(t)$ with the trial vector $\vec{U}_i(t)$ only if the latter yields a higher value of the fitness function.

Step 4: Report as the final solution the cluster centers and the partition obtained by the globally best DE vector (the one yielding the highest value of the fitness function) at time $t = t_{\max}$.

5.4 Experimental Results

5.4.1 General Comparison with Other Clustering Algorithms

To test the effectiveness of the proposed method, we compare its performance with six other clustering algorithms using a test-bed of five artificial and three real world datasets. Among the considered algorithms, there are three recently developed automatic clustering algorithms well-known as the FVGA (Fuzzy Clustering with Variable length GA) [33], the DCPSO (Dynamic Clustering PSO) [34] and the MEPSO (Multi-Elitist PSO) algorithms [35]. The rest of the algorithms are the kernel fuzzy c-means algorithm (KFCM) [5] and a kernelized fuzzy ant clustering algorithm (KFAC) [36, 37]. Only KFCM was provided with the correct number of classes as it is non-automatic in nature. The DEGL and MEPSO use same parameter encoding scheme and the same fitness function. The particle representation scheme and fitness function of DCPSO and FVGA follow their respective literatures. These algorithms do not use the kernel trick for clustering data.

We used datasets with a wide variety in the number and shape of clusters, number of datapoints and the count of features of each datapoint. The real life datasets used here are the Glass, the Wisconsin breast cancer, the image segmentation, the Japanese vowel and the automobile [38]. The synthetic datasets included here, comes with linearly non-separable clusters of different shapes (like elliptical, concentric circular dish and shell, rectangular etc). Brief details of the datasets have been provided in Table 5.1. Scatterplot of the synthetic datasets have also been shown in Figure 5.4.

The clustering results were judged using Huang's accuracy measure [39]:

$$r = \frac{\sum_{i=1}^{k} n_i}{n}, \tag{5.18}$$

where n_i is the number of data occurring in both the i-th cluster and its corresponding true cluster, and n is the total number of data points in the data set. According to this measure, a higher value of r indicates a better clustering result, with perfect clustering yielding a value of $r = 1$.

We used $\sigma = 1.1$ for all the artificial datasets, $\sigma = 0.9$ for breast cancer dataset and $\sigma = 2.0$ for the rest of the real life datasets for the RBF kernel following [29].In these experiments, KFCM was run 100 times with the initial centroids randomly selected from the data set. A termination criterion of $\varepsilon = 0.001$was set

Table 5.1 Description of the Datasets

Dateset	Number of Datapoints (n)	Number of clusters (k)	Data-dimension (d)
Synthetic_1	500	2	2
Synthetic_2	52	2	2
Synthetic_3	400	4	3
Synthetic_4	250	5	2
Synthetic_5	600	2	2
Glass	214	6	9
Wine	178	3	13
Breast Cancer	683	2	9
Image Segmentation	2310	7	19
Japanese Vowel	640	9	12

Table 5.2 Parameter Settings for the contestant algorithms

DCPSO		FVGA		DEGL		KFAC		MEPSO	
Parameter	Value	Parameter	Value	Parameter	Value	Parameter	Value	Parameter	Value
Pop_size	100	Pop_size	20	Pop_size	10*dim	Number of ants	30	Pop_ size	40
Inertia Weight	0.72	Crossover Probability P_c	0.6	F	0.8	Memory per ant	5	Inertia Weight	0.794
C_1, C_2	1.494	Mutation Probability P_m	0.05	CR	0.9	Iterations per epoch	50	C_1, C_2	0.35→2.4 2.4→0.35
P_{ini}	0.75	K_{max}	20	K_{max}	20	Epochs	1000	K_{max}	20
K_{max}	20	K_{min}	2	K_{min}	2	P_{rest}	0.75	K_{min}	2
K_{min}	2					$P_{continue}$	0.20		
				Neighbor-hood Size	10% of the population size.	$P_{continueCurrent}$	0.001		
						D_{max}	0.001		
						D_{min}	0.01		

Table 5.3 Mean and standard deviation of the clustering accuracy (%) achieved by each clustering algorithm over 40 independent runs (Each run continued up to 50, 000 FEs for FVGA, DCPSO, Kernel_MEPSO and DEGL)

Datasets	Algorithms					
	KFCM	KFAC	FVGA	DC-PSO	Kernel_ MEPSO	DEGL
Synthetic_1	82.63 (0.017)	87.28	54.98 (0.88)	57.84 (0.065)	97.56 (0.281)	**99.89 (0.002)**
Synthetic_2	74.34 (0.017)	75.73	65.82 (0.146)	59.91 (0.042)	79.25 (0.042)	**80.92 (0.0051)**
Synthetic_3	90.36 (0.41)	94.03	97.75 (0.632)	97.94 (0.093)	97.38 (0.131)	**99.31 (0.001)**
Synthetic_4	67.65 (0.104)	80.25	74.30 (0.239)	75.83 (0.033)	87.85 (0.638)	**96.03 (0.471)**
Synthetic_5	82.26 (0.197)	84.33	54.45 (0.348)	52.55 (0.209)	89.46 (0.472)	**99.75 (0.001)**
Glass	71.63 (0.052)	73.92	76.27 (0.327)	79.45 (0.221)	90.57 (0.832)	**95.09 (0.630)**
Wine	76.43 (0.245)	59.36	80.64 (0.621)	85.81 (0.362)	90.64 (0.277)	**93.82 (0.053)**
Breast Cancer	69.84 (0.344)	70.54	73.22 (0.437)	78.19 (0.336)	92.58 (0.342)	**98.35 (0.211)**
Image Segmentation	56.83 (0.641)	70.93	78.84 (0.336)	81.93 (1.933)	85.32 (0.483)	**90.72 (0.982)**
Japanese Vowel	50.89 (0.772)	61.83	70.23 (1.882)	82.57 (0.993)	81.32 (2.428)	**84.93 (2.292)**
Average	72.28	75.16	74.48	76.49	87.58	**93.65**

for this algorithm. For all the competitive algorithms, we have selected their best parameter settings as reported in the corresponding literatures. The control parameters for DEGL were chosen after performing a series of trial experiments. The same set of parameters were used for all the test problems and for all the algorithms. These parameter settings have been reported in Table 5.2.

Table 5.3 compares the algorithms on the quality of the optimum solution as judged by the Huang's measure. The mean and the standard deviation (within parentheses) for 40 independent runs (with different seeds for the random number generator) of each of the six algorithms are presented in Table 5.3. Missing values of standard deviation in this table indicate a zero standard deviation. The best solution in each case has been shown in bold. Table 5.4 shows results of unpaired t-tests between the better of the new algorithm (DEGL) and the best of the other five in each case (standard error of difference of the two means, 95% confidence interval of this difference, the t value, and the two-tailed P value). Tables 5.5 and 5.6 present the mean and standard deviation of the number of classes found by the three automatic clustering algorithms. In Figure 5.4 we present the clustering results on the synthetic datasets by the new DEGL algorithm.

Table 5.4 Results of unpaired t-tests on the data of Table 5.3

Datasets	Std. Err	t	95% Conf. Intvl	Two-tailed P	Significance
Synthetic_1	0.005	976.36	(-5.01, -4.98)	< 0.0001	**Extremely significant**
Synthetic_2	0.001	9094.7	(-7.19, -7.18)	< 0.0001	**Extremely significant**
Synthetic_3	0.015	129.88	(-1.94, -1.88)	< 0.0001	**Extremely significant**
Synthetic_4	0.057	132.61	(-7.70, -7.48)	<0.0001	**Extremely Significant**
Synthetic_5	0.075	137.88	(10.14, 10.44)	<0.0001	**Extremely Significant**
Glass	0.105	120.17	(-12.77 , -12.35)	<0.0001	**Extremely Significant**
Wine	0.057	134.52	(-7.81, -7.58)	< 0.0001	**Extremely significant**
Breast Cancer	0.063	130.07	(8.04, 8.28)	< 0.0001	**Extremely significant**
Image Segmentation	0.173	19.6495	(3.055, 3.744)	< 0.0001	**Extremely significant**
Japanese Vowel	0.395	5.9780	(-3.147, -1.574)	< 0.0001	**Extremely significant**

Table 5.5 Mean and standard deviation (in parenthesis) of the number of clusters found over the synthetic datasets for four automatic clustering algorithms over 40 independent runs

Algorithms	Synthetic _1	Synthetic _2	Synthetic _3	Synthetic _4	Synthetic _5
FVGA	2.50 (0.021)	3.05 (0.118)	4.15 (0.039)	9.85 (0.241)	4.25 (0.921)
DCPSO	2.45 (0.121)	2.80 (0.036)	4.25 (0.051)	9.05 (0.726)	6.05 (0.223)
Kernel_MEPSO	2.50 (0.026)	2.65 (0.126)	4.10 (0.062)	9.35 (0.335)	2.25 (0.361)
DEGL	**2.10 (0.033)**	**2.15 (0.102)**	**4.00 (0.00)**	**10.05 (0.021)**	**2.05 (0.001)**

For comparing the speed of the stochastic algorithms like GA, PSO or DE, we choose *number of fitness function evaluations (FEs)* as a measure of computation time instead of generations or iterations. From the data provided in table 5.3, we choose a threshold value of the classification accuracy for each dataset. This threshold value is somewhat larger than the minimum accuracy attained by each automatic clustering algorithm. Now we run an algorithm on each dataset and stop as soon as it achieves the proper number of clusters as well as the threshold accuracy. We then note down the number of fitness function evaluations the algorithm takes. A lower number of FEs corresponds to a faster algorithm. The speed comparison results are provided in Table 5.7. The KFCM method is not included in this table, as it is non-automatic and does not employ evolutionary operators as in FVGA and PSO based methods.

Table 5.6 Mean and standard deviation (in parenthesis) of the number of clusters found over the synthetic datasets for four automatic clustering algorithms over 40 independent runs

Algorithms	Glass	Wine	Breast Cancer	Image Segmentation	Japanese Vowel
GCUK	5.85 (0.035)	4.05 (0.021)	2.25 (0.063)	7.05 (0.008)	9.50 (0.218)
DCPSO	5.60 (0.009)	3.75 (0.827)	2.25 (0.026)	7.50 (0.057)	10.25 (1.002)
Ordinary PSO	5.75 (0.075)	**3.00 (0.00)**	**2.00 (0.00)**	7.20 (0.025)	9.25 (0.822)
Kernel_MEPSO	**6.05 (0.015)**	**3.00 (0.00)**	**2.00 (0.00)**	**7.00 (0.00)**	**9.05 (0.021)**

Table 5.7 Mean and standard deviations of the number of fitness-function evaluations (over 40 successful runs) required by each algorithm to reach a predefined cut-off value of the classification accuracy

Dateset	Threshold accuracy (in %)	FVGA	DCPSO	Kernel_ MEPSO	DEGL
Synthetic_1	50.00	48000.05 (21.43)	42451.15 (11.57)	43812.30 (2.60)	**37029.65 (17.48)**
Synthetic_2	55.00	41932.10 (12.66)	45460.25 (20.97)	40438.05 (18.52)	**36271.05 (10.41)**
Synthetic_3	85.00	40000.35 (4.52)	35621.05 (12.82)	37281.05 (7.91)	**32035.55 (4.87)**
Synthetic_4	65.00	46473.25 (7.38)	43827.65 (2.75)	42222.50 (2.33)	**36029.05 (6.38)**
Synthetic_5	50.00	43083.35 (5.30)	39392.05 (7.20)	42322.05 (2.33)	**35267.45 (9.11)**
Glass	65.00	47625.35 (6.75)	40382.15 (7.27)	38292.25 (10.32)	**37627.05 (12.36)**
Wine	55.00	44543.70 (44.89)	43425.00 (18.93)	3999.65 (45.90)	**35221.15 (67.92)**
Breast Cancer	65.00	40390.00 (11.45)	37262.65 (13.64)	35872.05 (8.32)	**32837.65 (4.26)**
Image Segmentation	55.00	39029.05 (62.09)	40023.25 (43.92)	35024.35 (101.26)	**34923.22 (24.28)**
Japanese Vowel	40.00	40293.75 (23.33)	28291.60 (121.72)	29014.85 (21.67)	**24023.95 (20.62)**

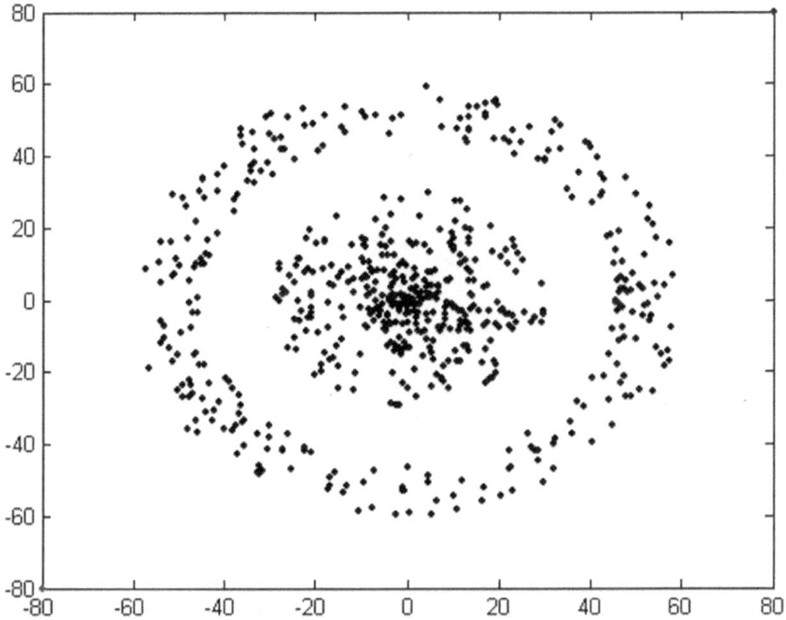

(a) Unlabeled Synthetic_1

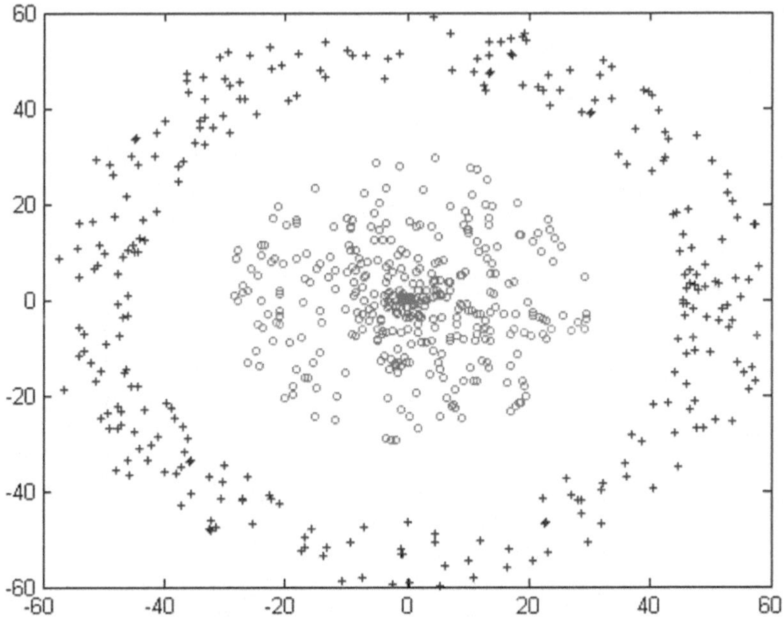

(a) Synthetic_1 Clustered with DEGL

Fig. 5.4 Two- and three-dimensional synthetic datasets clustered with DEGL

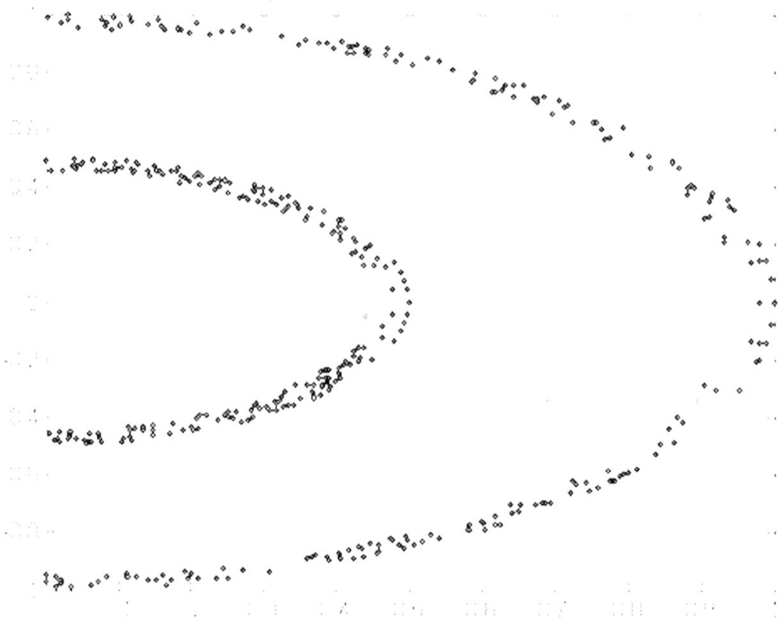

(b) Unlabeled Synthetic_2

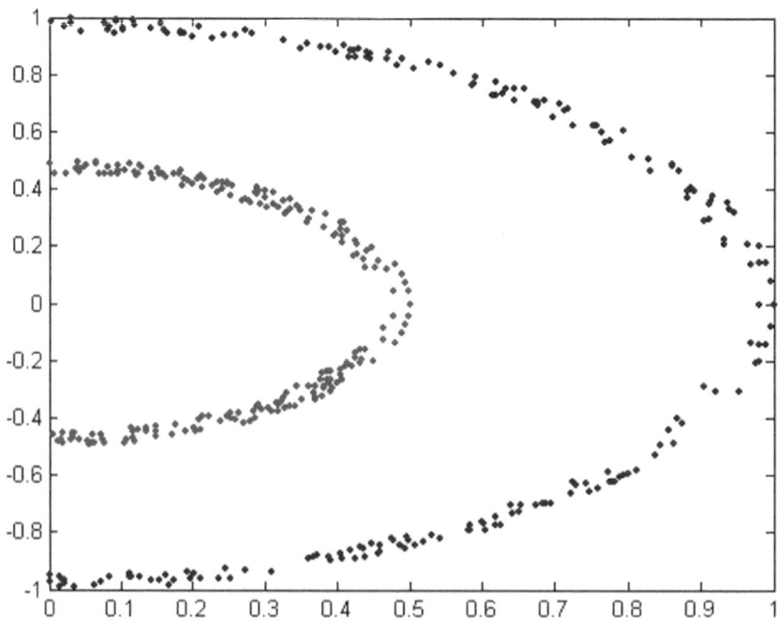

(b) Synthetic_2 Clustered with DEGL

Fig. 5.4 (*continued*)

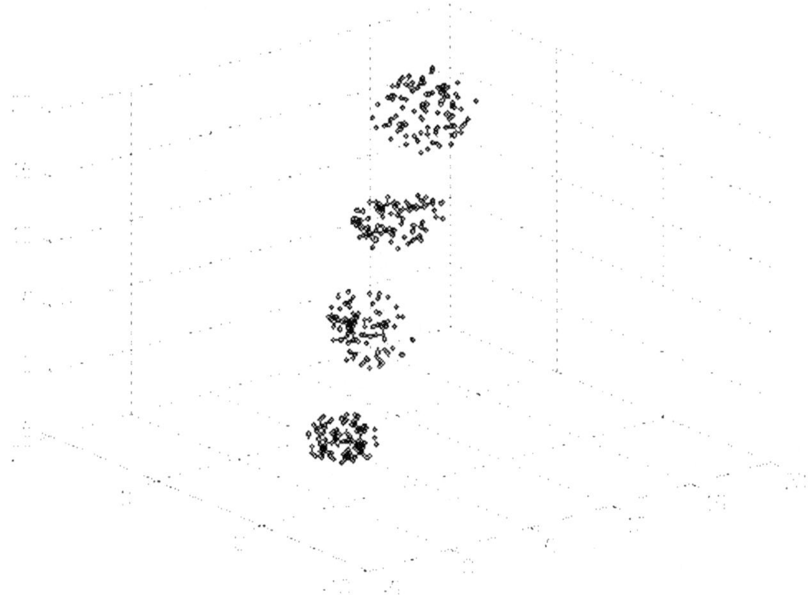

(c) Unlabeled Synthetic_3

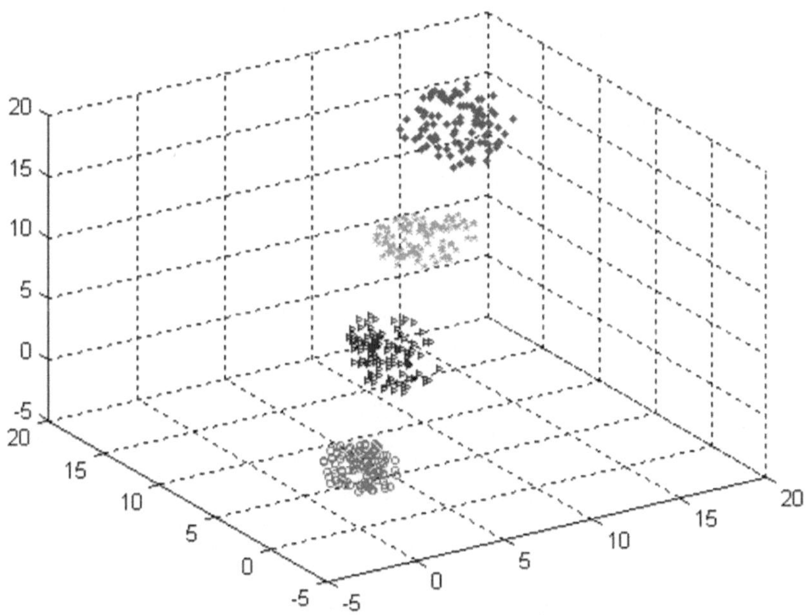

(c) Synthetic_3 Clustered with DEGL

Fig. 5.4 (*continued*)

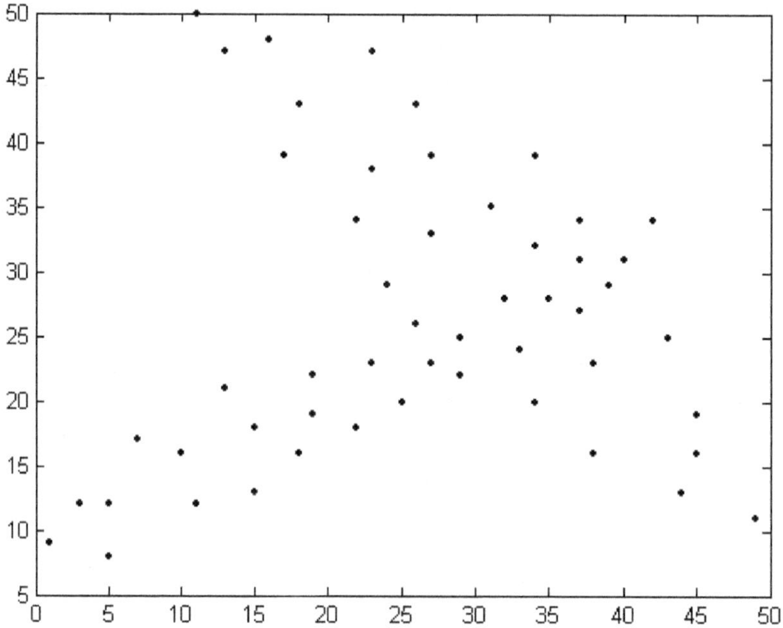

(d) Unlabeled Synthetic_4

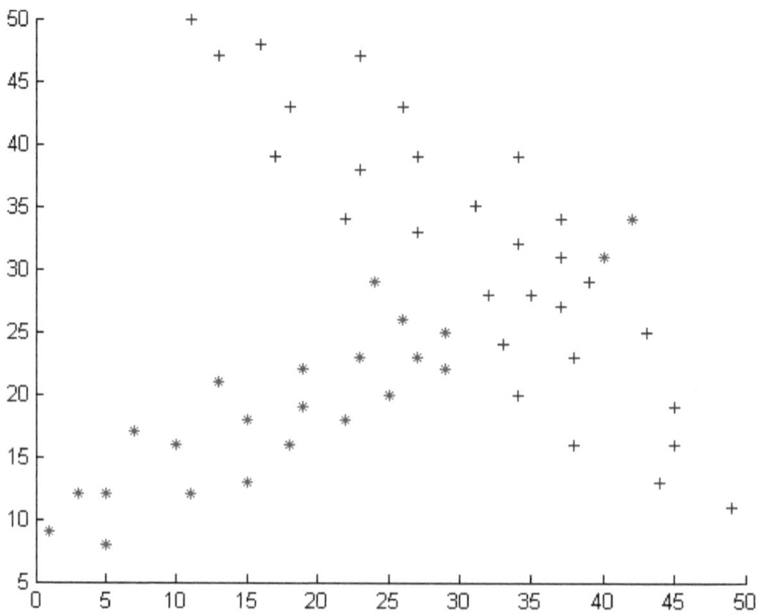

(d) Synthetic_4 Clustered with DEGL

Fig. 5.4 (*continued*)

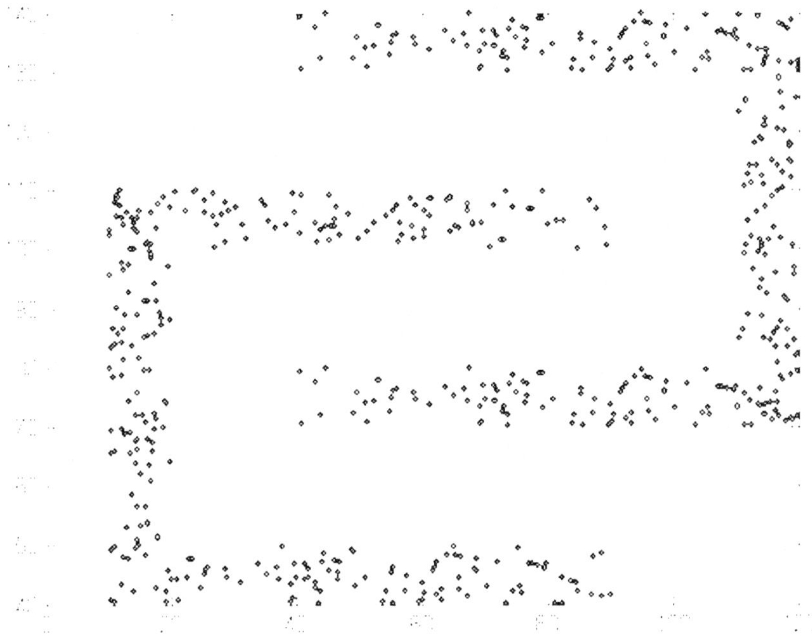

(e) Unlabeled Synthetic_5

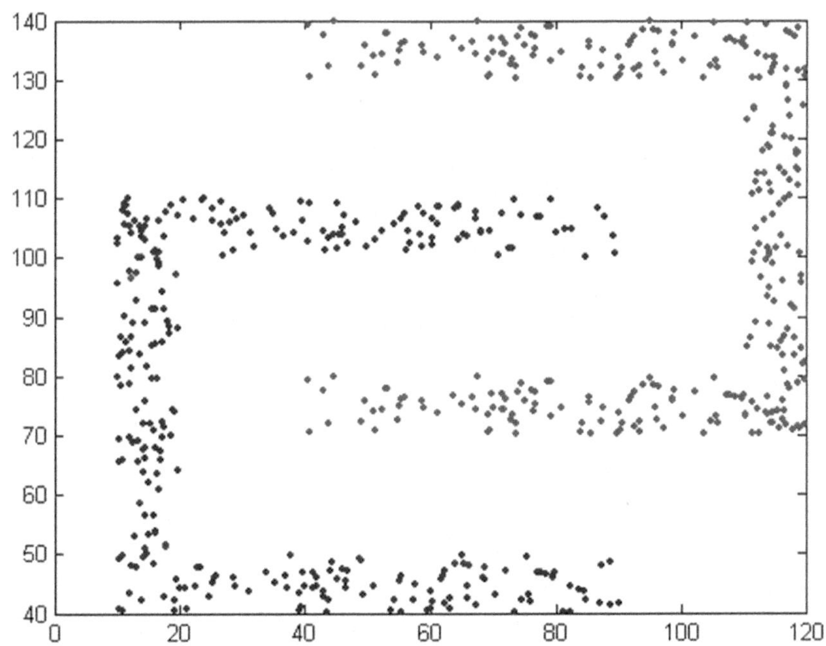

(e) Synthetic_5 Clustered with DEGL

Fig. 5.4 (*continued*)

A close scrutiny of Tables 5.3 to 5.7 reveals that the kernel based DEGL algorithm performed markedly better as compared to the other competitive clustering algorithms, both in terms of accuracy and convergence speed. We note that in general, the kernel based clustering methods outperform the DCPSO or FVGA algorithms (which do not use the kernelized fitness function) especially on linearly non-separable artificial datasets like synthetic_1, synthetic_2 and synthetic_5. The PSO based methods (especially MEPSO) on average took lesser computational time than the FVGA algorithm over most of the datasets. One possible reason of this may be the use of less complicated variation operators (like mutation) in PSO as compared to the operators used for GA. But the DEGL took least number of FEs to reach a threshold fitness value in all test cases, due to the better trade-off between exploration and exploitation abilities achieved by its neighborhood-based mutation operators.

5.4.2 Scalability of the DEGL-Based Clustering Algorithm

From Subsection 5.4.1, it is evident that the DEGL algorithm outperformed all other considered algorithms over the 2 and 3-dimensional synthetic datasets as well as over the real-life datasets with dimensions ranging up to 19. We further investigate the effect of growth of the dimensionality of the feature space on the clustering performance of DEGL. For this purpose, we used a few highly complex

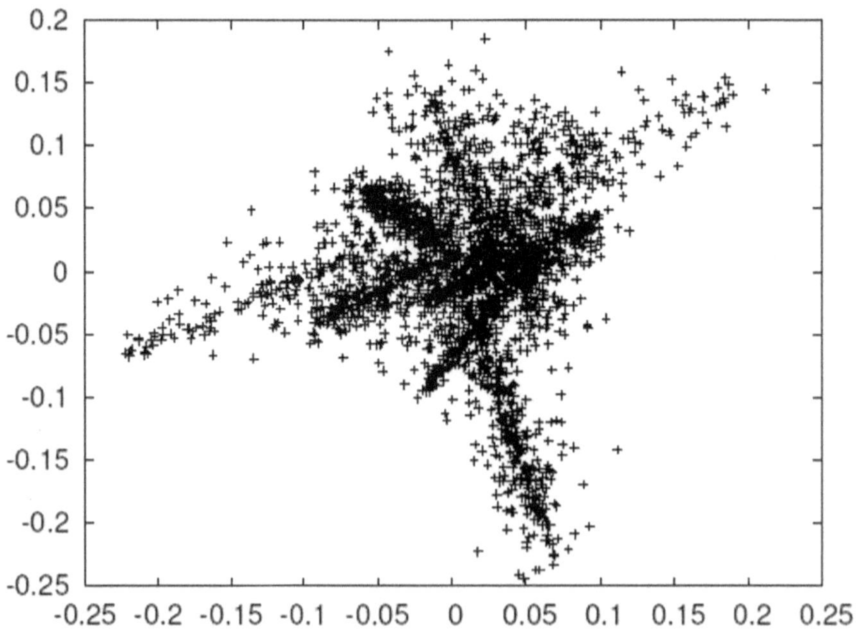

Fig. 5.5 100-dimensional data set containing 10 clusters (projected to two dimensions)

synthetic datasets of dimensions 20, 40, 60, 80, and 100 obtained from the Gaussian and ellipsoidal cluster generators made available through the website www.dbk.ch.umist.ac.uk/handl/generators/. The detailed statistical description of the generators can be found in [41]. Figure 5.5 shows a 100 dimensional dataset with 10 elongated and arbitrarily oriented clusters used in our experiment, when projected in the first two dimensions.

Table 5.8 Mean and standard deviation of the number of classes found and clustering accuracy (%) achieved by each Clustering algorithm over 25 independent runs (Each run continued up to 500, 000 FEs for DEGL, FVGA, MOCK and Kernel_MEPSO)

Dateset	FVGA		Kernel_MEPSO		BIRCH	MOCK		DEGL	
	No. of classes found	% Accuracy	No. of classes found	% Accuracy	% Accuracy	No. of classes found	% Accuracy	No. of classes found	% Accuracy
20d_10c	10.20	67.83 (2.8371)	10.50	75.88 (2.4918)	79.86	10.12	92.84 (3.2913)	**10.00**	**93.29 (1.8382)**
40d_10c	8.40	58.34 (5.9382)	10.08	91.84 (3.9932)	75.73	10.20	93.88 (4.9291)	**10.13**	**92.9 (3.5728)**
60d_10c	6.24	54.28 (4.0933)	8.36	53.25 (7.3855)	82.65	**10.44**	**87.12 (3.289)**	9.20	82.02 (1.9928)
80d_10c	5.88	43.23 (4.8256)	6.50	46.35 (5.30)	76.05	**10.04**	**86.93 (5.9961)**	8.40	67.82 (4.6617)
100d_10c	6.04	40.08 (1.4528)	5.60	42.35 (6.7587)	81.15	**10.28**	**84.56 (2.0928)**	7.44	58.93 (2.0931)

In these experiments, we compare the DEGL with Kernel_MEPSO, FVGA, and two scalable clustering algorithms including BIRCH [42] and MOCK (Multi Objective Clustering with automatic K determination) [43]. We exclude the subspace clustering algorithms like CLIQUE [44] from this comparison, as they, unlike the algorithms being dealt here, seek for meaningful clusters in a lower dimensional *subspace* (a subset of the feature space dimensions).

Since our primary objective is to study the scalability issues of the DEGL, we kept the number of clusters as well as the number of data points same for all the synthetic datasets generated for this purpose. Among the algorithms compared, only BIRCH was provided with the true number of classes in each case. Rest of the competitors could determine the number of clusters on the run. In Table 5.8, we report the mean number of classes found and the mean classification accuracy over 25 independent runs of all the competitor algorithms except BIRCH which was run once for each problem as it is deterministic and non-automatic. Figure 5.6 shows how the CPU time consumed by different stochastic algorithms varies with the data dimensionality.

For BIRCH we used the default parameter values provided in [58]. The CPU time was measured on a Pentium IV, 2.2 GHz PC, with 512 KB cache and 2 GB of main memory in Windows Server 2003 environment. The results suggest that

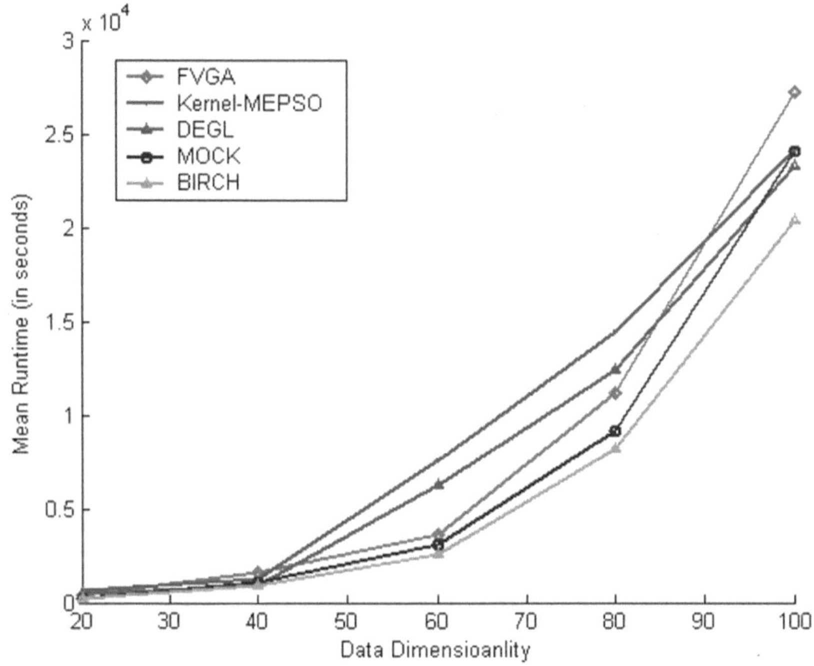

Fig. 5.6 The variation of CPU time with growth of data dimensionality

for data dimensionality lesser than 40, the ability of DEGL to separate elliptical and overlapped clusters is best among all the methods compared. But as the dimension exceeds 60, the accuracy of DEGL becomes comparable to BIRCH and worse than the MOCK which employs a multi-objective evolutionary algorithm. The version of MOCK used here employs novel schemes for initialization and mutation that enable a more efficient exploration of the search space. It also modifies the null data model that is used as a basis for selecting the most significant solution from the Pareto front. We also note that although the final accuracy of DEGL remains better than FVGA or MEPSO, above 60 dimensions, the former over fits the data, finding a larger number of clusters (than the true number of classes) by projecting data points to an even higher dimensional space. Figure 5.6 indicates that the runtime of DEGL to converge to a threshold accuracy value above 40 dimensions, becomes worse than BIRCH and above 60 dimensions the speed of the algorithm is hardly acceptable in comparison to that of both MOCK and BIRCH.

The deterioration of the performance of DEGL and kernel_MEPSO is in accordance with [44] which indicates that; higher dimensionality confounds the process of kernel based learning especially in presence of unbalanced classes. The curse of dimensionality may be overcome in kernel-based clustering by utilizing the subspace models. An efficient feature selection technique may also be incorporated in the framework of the Kernel-induced DEGL clustering for this purpose.

5.5 Application to Image Pixel Clustering

A clustering or partitioning of unlabeled feature vectors obtained from an image provides a segmentation of the image into unlabeled regions. The resultant regions can then be labeled by hand or by some automated process. Therefore many clustering algorithms have widely been applied to solve the multi-class segmentation problem (e.g., k-means, FCM, ISODATA, and Snob, appropriate citations for which can be found in chapter 1). FCM seems to be the most popular candidate in the field of fuzzy clustering algorithms. Many researchers have attempted modifications of the classical FCM and applications to image segmentation in the past few years [45 - 50]. FCM-type algorithms, however, assume a prior knowledge of the number of class c, while in many practical situations, the appropriate number of classes is unknown or impossible to determine even approximately. Another disadvantage of the FCM-type algorithms is that they often make use of a Euclidean distance measure, which is fine for most noise-free data. However, difficulties arise when data is corrupted by noise or outliers. In order to deal with these kinds of data sets, kernel methods have widely been studied. The fundamental idea of kernel based techniques is to transform the original low-dimensional input space into a higher dimensional feature space through some (nonlinear) mapping, therefore some complex linear problems in the original space can possibly be solved in the transformed high-dimensional space. In this section we investigate the performance of the kernel-induced DEGL clustering technique on a few complex gray-scale and multi-spectral images. Intensity of each pixel has been treated as a single feature for clustering the pixels into homogeneous groups.

We compare the performance of the proposed DEGL algorithm with a kernel FCM with Spatial Constraints (KFCM_S). In [51] Chen *et al.* presented a kernel-based non-automatic fuzzy clustering algorithm that exploits the spatial contextual information in image data. The algorithm is realized by modifying the objective function in the conventional FCM algorithm using a kernel-induced distance metric and a spatial penalty term that takes into account the influence of the neighboring pixels on the centre pixel. The list of considered algorithms also includes FVGA (which was used for image segmentation in [33]), KFAC, and kernel-MEPSO as well.

5.5.1 Parametric Setup for the Contestant Algorithms

All the considered algorithms use their best possible set of parameters as described in the relevant literatures. Same parameter set was employed for all the test problems, without allowing any hand tuning. The three kernel based algorithms employ the Gaussian kernel with $a = 2$, $b = 1$and $\sigma = 150$ (in equation (5.15)) for all images following [51].

5.5.2 The Test-Suite for Comparison

The considered algorithms were tested on five gray-scale images of varying complexity. The test images are marked as *test_image_1* to *5*. *Test_image_5* is a sample of high resolution Quick-bird satellite image [52, 53]. The resolution is 2.4 metres and its size is 401×401pixels. It represents a scene mixing urban and rural areas. The image is multi-spectral and comes in four spectral bands: blue, green, red and near infrared. The rest of the images are mono-spectral and comes in 256×256 pixels. Rest of the test images are real life images of varying complexities and have been taken from the Berkley segmentation dataset [54].

5.5.3 Quantitative Validation of Clustering Results

Since test images are too heterogeneous to allow for visual validation, some statistical tools have to be defined to evaluate the segmentation results on a quantitative basis. However, a ground truth representation of the image is necessary to compute the statistical scores. Fortunately, the manually segmented and hand-labeled ground truth images are available for all the five test images used here. In case of *test_image_5*, the ground-truth image is hand-labeled to represent the following classes: wooded area, farming type 1, farming type 2, open land, poor vegetation area, and urban area. As segmentation only amounts to defining homogenous regions, it is not possible to compare it directly with the manual classification. Each region in a segmented image has to be manually labeled, thus creating a completely classified image, which could then be compared with the ground truth image. Below we describe the statistical measures employed here for the quantitative evolution of the final clustering results using the notations inspired from [55]. Let n be the total number of pixels, and n_{ij} be the number of pixels classified into class i in the experimental classification and into class j in the real classification. Let $n_{i+} = \sum_j n_{ij}$ be the number of pixels classified into cluster i in the experiment, and $n_{j+} = \sum_j n_{ij}$ be the number of samples classified into class j in the ground truth image.

***1) Overall Accuracy* [55]:** This is the most widely used statistical score function for the validation. The overall accuracy is the percentage of **correctly classified** pixels in the image. It is given by:

$$p_o = \frac{\sum_i n_{ii}}{n} \tag{5.19}$$

***2) Kappa Index* [55]:** This measure is very similar to the overall accuracy, but it introduces a chance agreement. A zero value would indicate that the classification agrees with the reference as bad as an arbitrary classification. The Kappa index is computed by:

$$Ka = \frac{n.\sum_i n_{ii} - \sum_i n_{i+}n_{+i}}{n^2 - \sum_i n_{i+}n_{+i}} \tag{5.20}$$

***3) Adjusted Rand Index* [56]:** This index comes as a generalization of the Rand Index [57]. It introduces a statistically induced normalization in order to yield values close to 0 for random partitions. Using a representation based on contingency tables, the Adjusted Rand Index is given by:

$$R = \frac{\sum_{i,j}\binom{n_{ij}}{2} - [\sum_i \binom{n_{i+}}{2}.\sum_j \binom{n_{j+}}{2}]/\binom{n}{2}}{\frac{1}{2}[\sum_i \binom{n_{i+}}{2} + \sum_j \binom{n_{j+}}{2}] - [\sum_i \binom{n_{i+}}{2}.\sum_j \binom{n_{j+}}{2}]/\binom{n}{2}} \tag{5.21}$$

***4) Dice Similarity index* [58]:** It is a measure of agreement between the regions of each class. Let $n\{R_c\}$ be the number of pixels belonging to class c in the reference data, $n\{C_c\}$ the number of pixels belonging to the same class c in the experimental classification. We can also define $n\{R_c \cap C_c\}$ as the number of pixels which belong to the intersection of R_c and C_c. The dice similarity index is then given by

$$S_c = 2.[\frac{n\{R_c \cap C_c\}}{n\{R_c\} + n\{C_c\}}] \tag{5.22}$$

5.5.4 The Simulation Strategy

All the algorithms have been developed from in Visual C++ on a Pentium IV, 1.2 GHz PC, with 512 KB cache and 2 GB of main memory with Windows Server 2003 environment. For each image data, each run continues until the number of function evaluations (FEs) reaches 100,000 for DE/current-to-best/1/bin, DEGL and FVGA. KFAC was run for 1000 epochs over each problem following [34]. Thirty independent runs (with different seeds for the random number generator) have been taken for all the algorithms. The results have been stated in terms of the mean best-of-run values and standard deviations over these 30 runs in each case.Comparisons are made in terms of three performance metrics: (a) quality of the solution as determined by the statistical score functions (b) ability to find the optimal number of clusters, and (c) time required to find the solution. Since the KFCM_S algorithm cannot handle an unknown number of clusters, it was fed with the correct number of classes for each image (by *correct* number of classes, we mean the number of classes shown in the ground truth image). For all cases of KFCM_S, the algorithm is executed until the difference between the cluster centers in two consecutive iterations falls below 0.00001.

5.5.5 Experimental Results

Figures 5.7 to 5.11 show the five original unlabeled images, the corresponding ground truth images and the DEGL-based segmentation results.

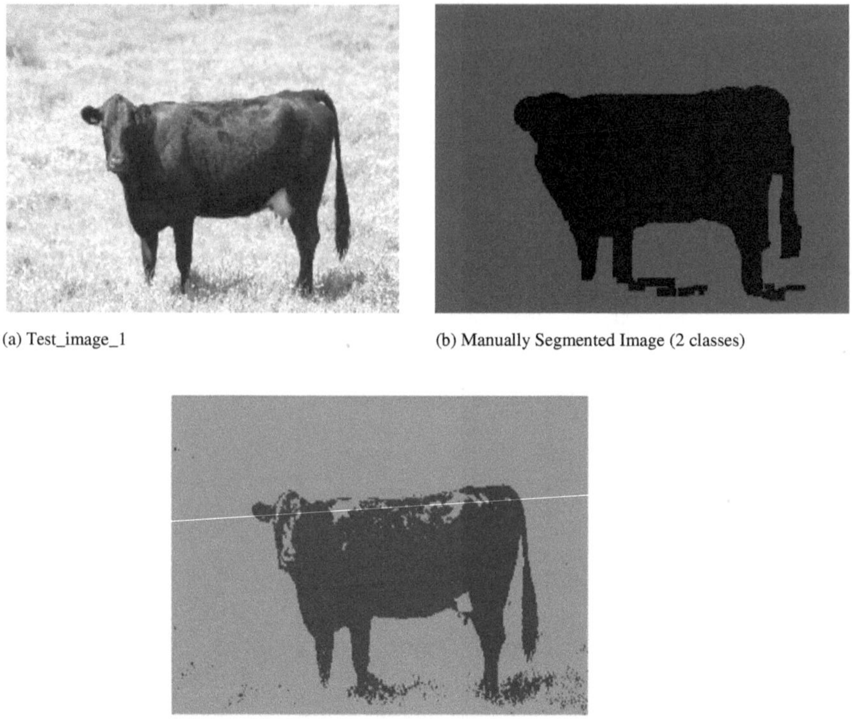

(a) Test_image_1

(b) Manually Segmented Image (2 classes)

(c) Segmentation with DEGL (2 classes)

Fig. 5.7 Segmentation result for Test Image 1

Table 5.9 contains the mean and standard deviations of the number of classes obtained by the four automatic clustering algorithms. In Table 5.10, we report the mean value (and standard deviations) of the Xie Beni index, the overall accuracy (in %), the kappa index (in %) and the adjusted rand index, all of which were calculated over the 'best-of-run' solutions in each case. We used unpaired t-tests to compare the means of the results produced by best and the second best algorithms. Table 5.11 shows results of unpaired t-tests between the best algorithm and the second best in each case (in terms of the final Xie-Beni index value) in each case (standard error of difference of the two means, 95% confidence interval of this difference, the t value, and the two-tailed P value). For all cases in Table 5.11, sample size = 30 and degrees of freedom = 58. Table 5.12

(a) Test Image 2

(b) Manually Segmented Image (2 classes)

(c) Segmentation with DEGL (2 classes)

Fig. 5.8 Segmentation result for Test Image 2

reports the dice similarity index value, which was calculated for each class in the ground truth image and the image segmented by each of the contestant algorithms. In order to save space, we show dice index values only for the remote sensing test_image_5, where the ground truth consists of six manually labeled classes. Table 5.13 reports the mean time taken by each algorithm to terminate on the image data. Since, for all the images, the minimum and maximum number of clusters are 2 and 10 respectively, the KFCM_S algorithm is executed nine times (once for each given number of clusters) by supplying to it the number of clusters as 2 to 10, and the total running time is reported in the last column of Table 5.14. For all cases of KFCM_S, the algorithm is executed until the difference between the cluster centers in two consecutive iterations falls below 0.0001.

Finally a box plot of the best-of-the-run overall accuracy over 30 independent runs of the five contestant algorithms has been shown in Figure 5.12. To save space, box plot has only been provided for the multi-spectral quick-bird image (test_image_5) as this may be considered as the most difficult image for segmentation in the present test-suite and the box plots for other images too show a similar trend.

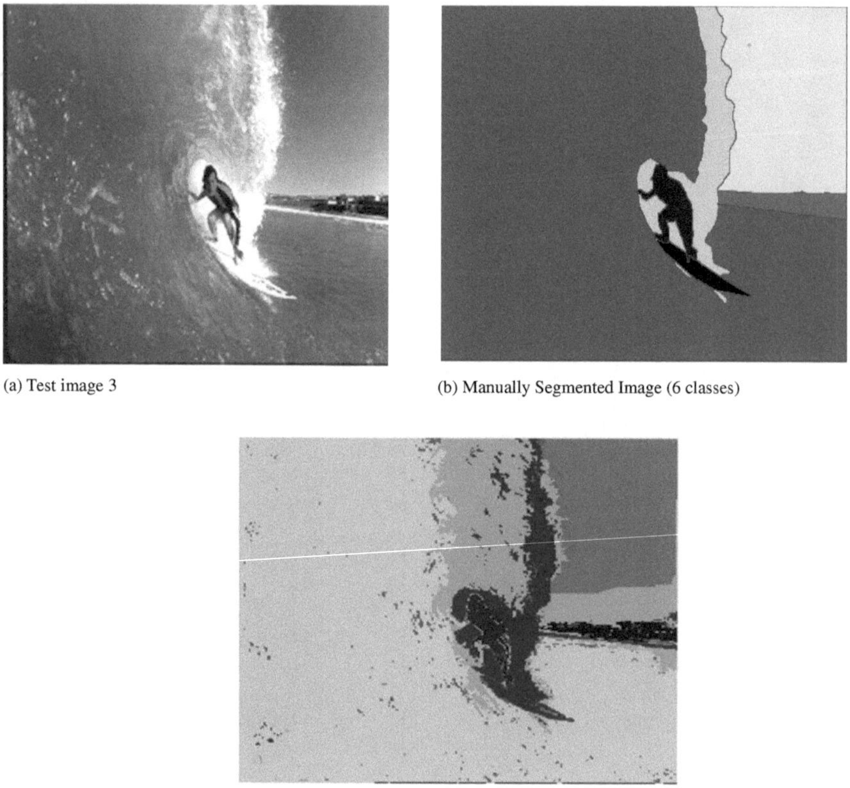

(a) Test image 3

(b) Manually Segmented Image (6 classes)

(c) Segmentation with DEGL (6 classes)

Fig. 5.9 Segmentation result for Test Image 3

5.5.6 *Discussion on the Results*

In this research, the proposed DEGL has been compared with three automatic image segmentation algorithms, based on GA, PSO and the Ant Colony Systems (ACS) and a kernelized version of the classical FCM algorithm. To make the performance evaluation/comparison meaningful and effective, all the considered test images are real-life images and for each one of them, a manually segmented counterpart (the ground truth image) exists. A segmentation result obtained

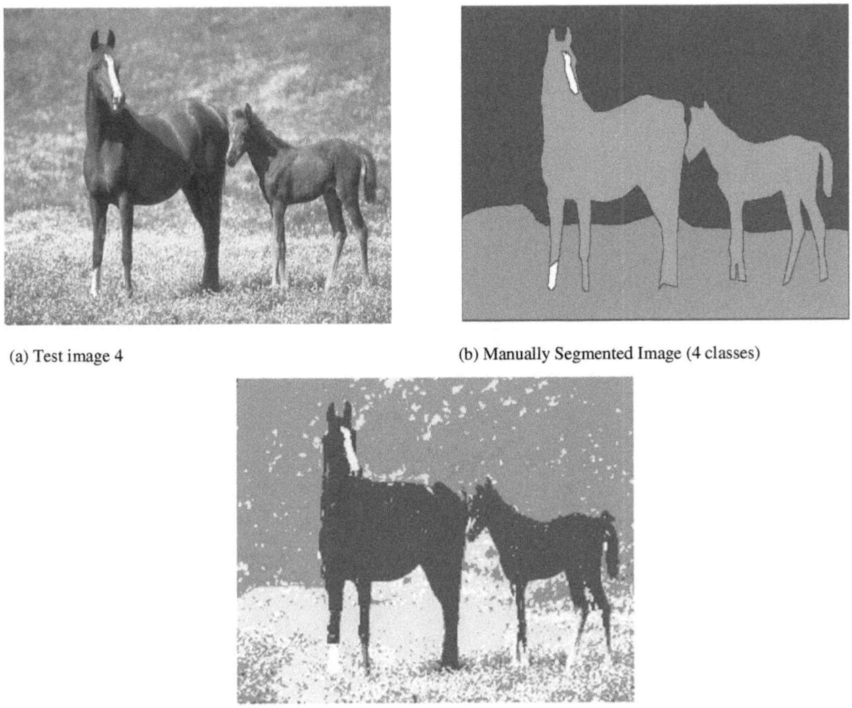

(a) Test image 4

(b) Manually Segmented Image (4 classes)

(c) Segmentation with DEGL (4 classes)

Fig. 5.10 Segmentation result for Test Image 4

through the grouping of the pixels, can be considered as *good* if it closely matches the ground truth image [59]. Here we adopted three well-known statistical measures (the overall accuracy, kappa index and the adjusted rand index) for evaluating the similarity between the reference image and the final segmentation results in each case. We also employed the dice similarity index in order to measure the correspondence between each segmented region in the output image from an algorithm and the ground truth image. Note that the quantitative evaluation does not guarantee the absolute accuracy of the results since the ideal classifier does not exist in practice, nevertheless, it shows the error provided by the segmentation and that will not be corrected with the classification/labeling step. In other words, it indicates the accuracy of the segmentation to keep the frontiers between the ground objects.

From Tables 5.9, 5.10, and 5.11, it is evident that our approach outperforms the state-of-the-art FVGA, KFAC, KFCM_S and DE/current-to-best/1/bin over all the five test images in a statistically significant manner. Not only does the method find the optimal number of clusters, but also it manages to find segmentations closest to the ground truth as evident from the higher values of the overall accuracy, kappa index and the adjusted rand index in Table 5.10. It is interesting

(a) Test image 5

(b) Reference Image with 6 class labels histogram equalization

(c) Segmentation with DEGL (6 classes)

Fig. 5.11 Segmentation result for Test Image 5

Table 5.9 Mean number of classes found (and standard deviation) by the four automatic clustering algorithms over five real life gray-scale images

Image	Optimal No. of Clusters	Mean and Std Dev of the number of classes estimated by the competitor algorithms			
		KFAC	FVGA	kernel-MEPSO	DEGL
test_image_1	2	3.67±0.134	2.33±0.541	3.03±0.172	**2.16±0.346**
test_image_2	2	3.13±0.212	3.43±0.563	2.65±0.334	**2.03±0.428**
test_image_3	6	8.26±0.883	8.15±0.447	7.03±0.982	**6.17±0.772**
test_image_4	4	5.33±0.459	4.93±0.762	4.33±0.504	**4.03±0.868**
test_image_5	6	7.86±1.423	8.67±2.704	7.33±1.538	**6.23±0.913**

to see from Table 5.10 that, a lower value of the Xie-Beni index, which means the clusters are compact and well separated, also corresponds to a well segmented image, matching closely with the ground truth. Although it happens for all the five test images covered here, a better value of the clustering validity index may *not* always correspond to a good correspondence between the ground truth and the output image, especially when the objective of the segmentation is to extract certain objects from the background [60].

Table 5.10 Mean and standard deviation (in parentheses) of four statistical validity measures calculated on the final clustering results produced by five contestant algorithms over 30 independent runs

Image	Clustering validity measures	Mean and Std Dev of the validity indices over the final clustering results of 30 independent runs				
		KFCM_S	KFAC	FVGA	kernel-MEPSO	DEGL
Test_image_1	Xie-Beni index	0.6807 (0.0513)	0.6242 (0.1948)	0.7372 (0.0932)	0.5928 (0.1022)	**0.5131 (0.0218)**
	Overall Accuracy (%)	64.09 (0.0043)	73.92 (0.0026)	57.09 (0.0732)	79.31 (0.0883)	**88.43 (0.0002)**
	Kappa Index (%)	53.83 (0.0231)	66.38 (0.0159)	42.83 (0.0438)	69.54 (0.1003)	**82.61 (0.0017)**
	Adjusted Rand Index	0.7525 (0.0293)	0.8213 (0.0298)	0.7015 (0.0492)	0.9107 (0.1103)	**0.9247 (0.0782)**
Test_image_2	Xie-Beni index	0.1902 (0.0062)	0.1619 (0.0183)	0.1584 (0.0039)	0.1063 (0.0082)	**0.1321 (0.0104)**
	Overall Accuracy (%)	61.95 (0.1146)	79.01 (1.231)	74.32 (0.5382)	82.09 (0.5341)	**87.18 (0.0317)**
	Kappa Index (%)	49.39 (0.3921)	66.22 (0.1096)	64.37 (0.0934)	76.45 (0.0842)	**81.12 (0.0492)**
	Adjusted Rand Index	0.6621 (0.0036)	0.7358 (0.0382)	0.7126 (0.5361)	0.8837 (0.1051)	**0.9028 (0.0482)**
Test_image_3	Xie-Beni index	0.10119 (0.0001)	0.0967 (0.0043)	0.0937 (0.0058)	0.0917 (0.0037)	**0.08078 (0.0019)**
	Overall Accuracy (%)	58.51 (0.0038)	74.60 (0.0971)	71.82 (0.1193)	79.62 (0.1193)	**83.72 (0.0137)**
	Kappa Index (%)	41.13 (0.0884)	62.38 (0.0292)	60.81 (0.0827)	71.74 (0.0362)	**89.57 (0.0912)**
	Adjusted Rand Index	0.6821 (0.0928)	0.7503 (0.0023)	0.7118 (0.1928)	0.8612 (0.1002)	**0.9029 (0.0842)**
Test_image_4	Xie-Beni index	0.3236 (0.0043)	0.2637 (0.0191)	0.2771 (0.0461)	0.2109 (0.0038)	**0.1943 (0.0102)**
	Overall Accuracy (%)	68.52 (0.0032)	88.47 (0.0339)	84.09 (0.1083)	90.32 (0.0837)	**92.22 (0.0022)**
	Kappa Index (%)	56.91 (0.0281)	71.53 (0.0039)	65.93 (0.0332)	83.03 (0.0391)	**86.18 (0.0112)**
	Adjusted Rand Index	0.7352 (0.0624)	0.7738 (0.0928)	0.7509 (0.0021)	0.8298 (0.0239)	**0.8925 (0.0500)**
Test_image_5	Xie-Beni index	0.2032 (0.0029)	0.1943 (0.0391)	0.1821 (0.0033)	0.1493 (0.0049)	**0.1192 (0.0086)**
	Overall Accuracy (%)	68.52 (0.0211)	74.08 (0.0034)	63.85 (0.039)	83.55 (0.0409)	**93.64 (0.0083)**
	Kappa Index (%)	54.33 (0.0051)	62.58 (0.0109)	50.98 (0.0821)	76.93 (0.0075)	**86.83 (0.0131)**
	Adjusted Rand Index	0.4932 (0.0095)	0.5638 (0.0048)	0.4394 (0.0051)	0.8051 (0.0118)	**0.9132 (0.0506)**

Table 5.11 Unpaired *t*-test Results on the data of table 3 (for Xie-Beni Index values)

Image	Std. Err	*t*	95% Conf. Intvl	Two-tailed *P*	Significance
test_image _1	0.019	4.1774	0.04158 to 0.11789	< 0.0001	**Extremely Significant**
test_image _2	0.013	5.0453	-0.0922 to -0.0394	< 0.0001	**Extremely Significant**
test_image _3	0.002	17.965	-0.0452 to -0.0360	< 0.0001	**Extremely Significant**
test_image _4	0.005	6.4431	-0.0419 to -0.0219	< 0.0001	**Extremely Significant**
test_image _5	0.002	16.6564	0.0264 to 0.0337	< 0.0001	**Extremely Significant**

Table 5.12 Mean Dice similarity index computed on the final clustering results of 30 independent runs only for the multi-spectral quick bird image (Test_image_5)

Image	Algorithms	Mean Dice similarity index computed on the final clustering results of 30 independent runs					
		wooded	farm 1	farm 2	op. land	poor veg.	urban
Test_ image _5	KFCM_S	70.67	95.48	92.56	94.09	86.38	45.95
	KFAC	78.05	95.10	95.42	94.43	87.60	45.97
	FVGA	65.71	85.34	79.14	74.85	80.02	44.05
	DE/current-to-best/1/bin	86.09	96.23	95.97	95.85	90.31	57.55
	DEGL	83.66	97.01	96.21	96.29	91.52	67.96

Table 5.13 Comparison among the mean execution time (in seconds) and standard deviation taken by the competitor algorithms

Image	KFAC	FVGA	DE/current-to-best/1/bin	DEGL	FCM fed with correct number of clusters	FCM – total time for 2 to 10 clusters
test_image _1	291.24 (22.27)	517.25 (12.16)	206.05 (23.77)	**232.35 (24.37)**	6.35 (0.021)	52.24 (0.018)
test_image _2	287.65 (16.31)	634.65 (32.29)	254.15 (23.16)	**224.15 (14.36)**	12.35 (0.001)	102.50 (0.018)
test_image _3	567.85 (23.81)	867.85 (24.17)	509.20 (14.21)	**459.20 (33.61)**	15.75 (0.002)	120.45 (0.002)
test_image _4	**375.50 (54.49)**	667.50 (37.49)	384.55 (44.36)	392.58 (40.33)	18.35 (0.001)	144.20 (0.03)
test_image _5	806.53 (52.16)	1226.54 (140.16)	854.03 (92. 06)	**793.33 (34.56)**	24.60 (0.001)	197.25 (0.21)

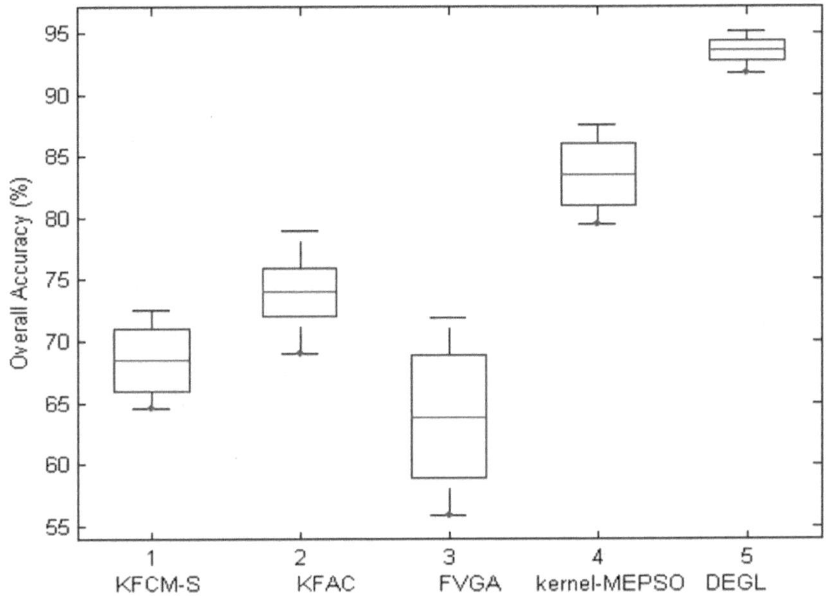

Fig. 5.12 A box-plot showing performance of the algorithms w.r.t. the overall accuracy on the test_image_5 (Best of the run results over 30 independent runs)

The remote sensing quick-bird image in Figure 5.11 bears special significance in the present context. Usually segmentation of such images helps in the land cover analysis of different areas in a country. The new method yielded six clusters on average for this image, and this corresponds exactly to the number of classes indicated in the reference image. From Table 5.12, one may see that each class of the DEGL segmented image shows most similarity to each corresponding class in the reference image. Except for the FVGA algorithm, we can see that farming, poor vegetation and open land areas have a very good dice similarity index. Urban and wooded regions are less accurate and this is probably due to the high heterogeneity of these regions. From Tables 5.9 and 5.11, we note that in general, the kernel based clustering methods outperform the FVGA algorithm (which does not make use of a kernelized fitness function). The performance of the FVGA method may be improved with the use of a kernel based Xie-Beni or some other clustering validity index. This renders itself to further research with FVGA for automatic pixel clustering task.

Table 5.13 indicates that our algorithm has a worse run-time than KFCM_S, but clearly this is an unfair comparison as the KFCM_S has to be fed with the correct number of clusters, whereas the proposed method finds it itself. It can be further observed that both DEGL and kernel_MEPSO algorithms consume lesser computational time than the FVGA for all the five images and the KFAC for four images (excluding the test_image_4). One possible reason of this may be the use of less complicated variation operators (like differential mutation) in DE and PSO as compared to the operators used for GA or fuzzy ant systems.

5.6 Conclusions

This Chapter has presented a new, modified DE-based strategy for fuzzy clustering of complex data. An important feature of the proposed technique is that it is able to find the optimal number of clusters automatically (that is, the number of clusters does not have to be known in advance) for complex and linearly non-separable datasets. The proposed kernel-induced DEGL algorithm has been shown to outperform some of other state-of-the-art clustering algorithms in a statistically meaningful way over several benchmark datasets considered for experiments. This certainly does not lead us to claim that it may outperform algorithms like kernel_MEPSO, FVGA, and KFAC over any dataset, since it is impossible to model all the possible complexities of a real life data with the limited test-suit that we used for testing the algorithms. In addition, the performance of FVGA and DCPSO may also be enhanced by modifying there fitness functions with a kernel induced distance metric. This renders itself to further research with these algorithms.

To investigate the effect of the growth of data dimensionality, we compared the performance of the DEGL with one scalable clustering algorithm BIRCH and a multi objective optimization based clustering algorithm MOCK which uses special operators to handle the curse of dimensionality issue. The test data used for these experiments included five synthetic datasets of dimensions ranging from 20 to 100. The number of data points and that of the true classes was kept same in all the datasets. It was observed that, the final classification accuracy and the mean number of classes found by the DEGL deteriorate considerably when the feature space dimensionality exceeds 40. Future research should focus on improving the performance of the algorithm over high dimensional datasets by incorporating some feature selection mechanism in it. Automatic clustering in lower dimensional subspaces with DEGL may also be a worthy topic of further research.

References

[1] Muller, K.R., Mika, S., Ratsch, G., Tsuda, K., Scholkopf, B.: An introduction to kernel-based learning algorithms. IEEE Transactions on Neural Networks 12(2), 181–202 (2001)

[2] Girolami, M.: Mercer kernel-based clustering in feature space. IEEE Transactions on Neural Networks 13(3), 780–784 (2002)

[3] Scholkopf, B., Smola, A.J.: Learning with Kernels. MIT Press, Cambridge (2002)

[4] Vapnik, V.N.: Statistical Learning Theory. Wiley, New York (1998)

[5] Zhang, D.Q., Chen, S.C.: Clustering incomplete data using kernel-based fuzzy c-means algorithm. Neural Process Letters 18, 155–162 (2003)

[6] Zhang, R., Rudnick, A.I.: A large scale clustering scheme for kernel k-means. In: The Sixteenth International Conference on Pattern Recognition, pp. 289–292 (2002)

[7] Das, S., Konar, A., Abraham, A., Chakraborty, U.K.: Differential Evolution Using a Neighborhood based Mutation Operator. In: IEEE Transactions on Evolutionary Computation. IEEE Press, USA (2008)
[8] Parsopoulos, K.E., Vrahatis, M.N.: UPSO: a unified particle swarm optimization scheme. In: Proceedings of the Int. Conf. Comput. Meth. Sci. Eng (ICCMSE 2004). Lecture Series on Computer and Computational Sciences, vol. 1, pp. 868–873. VSP International Science Publishers, the Netherlands (2004)
[9] Das, S., Sil, S., Chakraborty, U.K.: Kernel induced pixel clustering with differential evolution. In: IEEE Congress on Evolutionary Computation CEC 2008, IEEE World Congress on Computational Intelligence, WCCI 2008. IEEE Press, USA (2008)
[10] Xie, X., Beni, G.: Validity measure for fuzzy clustering. IEEE Transactions on Pattern Analysis and Machine Learning 3, 841–846 (1991)
[11] Mezura-Montes, E., Velázquez-Reyes, J., Coello, C.A.C.: A comparative study of differential evolution variants for global optimization. In: Genetic and Evolutionary Computation Conference (GECCO 2006), pp. 485–492 (2006)
[12] Storn, R., Price, K.V., Lampinen, J.: Differential Evolution - A Practical Approach to Global Optimization. Springer, Berlin (2005)
[13] Zielinski, K., Peters, D., Laur, R.: Run time analysis regarding stopping criteria for differential evolution and particle swarm optimization. In: Proc. of the 1st International Conference on Experiments/Process/SystemModelling/Simulation/Optimization, Athens, Greece (2005)
[14] Aho, A.V., Hopcroft, J.E., Ullman, J.D.: Data Structures and Algorithms. Addison-Wesley, Reading (1983)
[15] Cormen, T.H., Leiserson, C.E., Rivest, R.L.: Introduction to Algorithms, 1st edn. MIT Press and McGraw-Hill (1990)
[16] Collet, P., Louchet, J., Lutton, E.: Issues on the optimization of evolutionary algorithms code. In: Proc. of the 2002 Congress on Evolutionary Computation (CEC 2002), Honolulu, HI, USA (2002)
[17] Aronszajn, N.: Theory of reproducing kernels. Transactions of the American Mathematical Society 68(3), 337–404 (1950)
[18] Saitoh, S.: Theory of Reproducing Kernels and its Applications. Longman Scientific & Technical, Harlow, England (1988)
[19] Vapnik, V.N.: The Nature of Statistical Learning Theory. Springer, New York (1995)
[20] Pirooznia, M., Deng, Y.: SVM Classifier – a comprehensive java interface for support vector machine classification of microarray data. In: Proc. of Symposium of Computations in Bioinformatics and Bioscience (SCBB 2006), Hangzhou, China (2006)
[21] Hertz, T., Bar, A., Daphna Weinshall, H.: Learning a Kernel Function for Classification with Small Training Samples. In: Proceedings of the 23rd International Conference on Machine Learning, Pittsburgh, PA (2006)
[22] Wu, Z.D., Xie, W.X., Yu, J.P.: Fuzzy c-means clustering algorithm based on kernel method. In: ICCIMA (2003)
[23] Zhang, D.Q., Chen, S.C.: Kernel based fuzzy and possibilistic c-means clustering. In: Proceedings of the International Conference Artificial Neural Network, pp. 122–125, Turkey (2003)
[24] Zhang, D.Q., Chen, S.C.: A novel kernelized fuzzy c-means algorithm with application in medical image segmentation. Artificial Intelligence in Medicine 32(1), 37–50 (2004)

[25] Graepel, T., Obermayer, K.: Fuzzy topographic kernel clustering. In: Proceedings of the 5th GI Workshop Fuzzy Neuro Systems 1998, pp. 90–97 (1998)
[26] Macdonald, D., Fyfe, C.: The kernel self-organising map. In: Fourth International Conference on Knowledge-Based Intelligent Engineering Systems and Allied Technologies, 2000, vol. 1, pp. 317–320 (2000)
[27] Qinand, A.K., Suganthan, P.N.: Kernel neural gas algorithms with application to cluster analysis. In: ICPR, pp. 617–620 (2004)
[28] Zhang, D.Q., Chen, S.C.: Fuzzy clustering using kernel method. In: The 2002 International Conference on Control and Automation, pp. 162–163 (2002)
[29] Camastra, F., Verri, A.: A novel kernel method for clustering. IEEE Transactions on Pattern Analysis and Machine Intelligence 27(5), 801–804 (2005)
[30] Hur, A.B., Horn, D., Siegelmann, H.T., Vapnik, V.N.: Support vector clustering. Journal of Machine Learning Research 2, 125–137 (2001)
[31] Filippone, M., Camastra, F., Masulli, F., Rovetta, S.: A survey of kernel and spectral methods for clustering. Pattern Recognition 41(1), 176–190 (2008)
[32] Pal, N.R., Bezdek, J.C.: On cluster validity for the fuzzy c means model. IEEE Transactions on Fuzzy Systems 3(3), 370–379 (1995)
[33] Maulik, U., Bandyopadhyay, S.: Fuzzy Partitioning Using Real Coded Variable Length Genetic Algorithm for Pixel Classification. IEEE Transactions on Geosciences and Remote Sensing 41(5), 1075–1081 (2003)
[34] Omran, M.G., Engelbrecht, A.P., Salman, A.: Dynamic Clustering using Particle Swarm Optimization with Application in Unsupervised Image Classification. In: Proceedings of World Academy of Science, Engineering and Technology, vol. 9 (November 2005)
[35] Das, S., Abraham, A., Konar, A.: Automatic Kernel Clustering with Multi-Elitist Particle Swarm Optimization Algorithm. Pattern Recognition Letters, Elsevier Science 29, 688–699 (2008)
[36] Dorigo, M., Maniezzo, V., Colorni, A.: Ant System: Optimization by a colony of cooperating agents. IEEE Transactions on Systems, Man, and Cybernetics-Part B 26(1), 29–41 (1996)
[37] Gu, Y., Hall, L.O.: Kernel Based Fuzzy Ant Clustering with Partition validity. In: IEEE International Conference on Fuzzy Systems. IEEE Press, Los Alamitos (2006)
[38] Blake, C., Keough, E., Merz, C.J.: UCI repository of machine learning database (1998), `http://www.ics.uci.edu/~mlearn/MLrepository.html`
[39] Huang, Z., Ng, M.K.: A fuzzy k-modes algorithm for clustering categorical data. IEEE Transactions on Fuzzy Systems 7(4), 446–452 (1999)
[40] Handl, J., Knowles, J.: Multiobjective clustering around medoids. In: Proceedings of the Congress on Evolutionary Computation (CEC 2005), vol. 1, pp. 632–639 (2005)
[41] Zhang, T., Ramakrishnan, R., Livny, M.: BIRCH: An efficient data clustering method for very large databases. In: Proc. of the ACM SIGMOD Conference on Management of Data, Montreal, Canada (1996)
[42] Handl, J., Knowles, J.: An evolutionary approach to multiobjective clustering. IEEE Transactions on Evolutionary Computation 11(1), 56–76 (2007)
[43] Agrawal, R., Gehrke, J., Gunopulos, D., Raghavan, P.: Automatic subspace clustering of high dimensional data for data mining applications. In: Proc. ACM-SIGMOD Int. Conf. Management of Data, Seattle, Washington (1998)

[44] Evangelista, P.F., Embrecht, M.J., Szymanski, B.K.: Taming the Curse of Dimensionality in Kernels and Novelty Detection. In: Advances in Soft Computing: Applied Soft Computing Technologies: The Challenge of Complexity, pp. 425–438. Springer, Heidelberg (2006)

[45] Hall, L.O., Özyurt, I.B., Bezdek, J.C.: Clustering with a genetically optimized approach. IEEE Transactions on Evolutionary Computing 3(2), 103–112 (1999)

[46] Gath, I., Geva, A.: Unsupervised optimal fuzzy clustering. IEEE Transactions on PAMI 11, 773–781 (1989)

[47] Bensaid, A.M., Hall, L.O., Bezdek, J.C., Clarke, L.P.: Partially supervised clustering for image segmentation. Pattern Recognition 29, 859–871 (1996)

[48] Clark, M.C., Hall, L.O., Goldgof, D.B., Clarke, L.P., Velthuizen, R.P., Silbiger, M.S.: MRI segmentation using fuzzy clustering techniques. IEEE Eng. Med. Biol. 13, 730–742 (1994)

[49] Ahmed, M.N., Yamany, S.M., Mohamed, N., Farag, A.A., Moriarty, T.: A modified fuzzy c-means algorithm for bias field estimation and segmentation of MRI data. IEEE Transactions on Medical Imaging 21, 193–199 (2002)

[50] Wang, X., Wang, Y., Wang, L.: Improving fuzzy c-means clustering based on feature-weight learning. Pattern Recognition Letters 25, 1123–1132 (2004)

[51] Chen, S., Zhang, D.: Robust image segmentation using FCM with spatial constraints based on new kernel-induced distance metric. IEEE Transactions on System, Man and Cybernetics-Part B 34(4), 1907–1916 (2004)

[52] Gigandet, X., Cuadra, M.B., Pointet, A., Cammoun, L., Caloz, R., Thiran, J.: Region-based satellite image classification: method and validation ICIP (3), 832–835 (2005)

[53] `http://www.satimagingcorp.com/gallery-quickbird.html`

[54] Martin, D., Fowlkes, C., Tal, D., Malik, J.: A Database of Human Segmented Natural Images and its Application to Evaluating Segmentation Algorithms and Measuring Ecological Statistics. In: Proc. 8th International Conference on Computer Vision (2) 416—423 (2001)

[55] `http://www.eecs.berkeley.edu/Research/Projects/CS/vision/grouping/segbench`

[56] Congalton, R.G., Green, K.: Assessing the Accuracy of Remotely Sensed Data: Principles and Practices, Lewis, ch. 5, pp. 45–50 (1999)

[57] Hubert, L., Arabie, P.: Comparing partitions. Journal of Classification, 193–218 (1985)

[58] Rand, W.M.: Objective criteria for the evaluation of clustering methods. Journal of the American Statistical Association 66, 846–850 (1971)

[59] Dice, L.R.: Measures of the amount of ecologic association between species. Ecology. 26, 297–302 (1945)

[60] Unnikrishnan, R., Pantofaru, C., Hebert, M.: Toward Objective Evaluation of Image Segmentation Algorithms. IEEE Transactions on Pattern Analysis and Machine Intelligence (PAMI) 29(6), 929–944 (2007)

[61] Pal, N.R., Pal, S.K.: A Review on Image Segmentation Techniques. Pattern Recognition 26, 1277–1294 (1993)

Chapter 6
Clustering Using Multi-objective Differential Evolution Algorithms

Abstract. This chapter introduces the task of fuzzy clustering in a Multi-objective Optimization (MO) framework. It compares the performances of four recently developed multi-objective variants of Differential Evolution (DE) algorithm over the fuzzy clustering problem, where two conflicting fuzzy validity indices are simultaneously optimized. The resultant Pareto optimal set of solutions from each algorithm consists of a number of non-dominated solutions, from which the user can choose the most promising ones according to the problem specifications. A real-coded representation of the search variables, accommodating variable number of cluster centers, is used for DE. The performances of four DE variants have also been contrasted to that of two most well-known schemes of MO clustering namely the Non Dominated Sorting Genetic Algorithm (NSGA II) and Multi-Objective Clustering with an unknown number of Clusters K (MOCK). Experimental results over four artificial and four real life datasets of varying range of complexities indicates that DE holds immense promise as a candidate algorithm for devising MO clustering schemes .

6.1 Introduction

Optimization-based automatic clustering algorithms greatly rely on a cluster validity function (optimization criterion) the optima of which appear as proxies for the unknown "correct classification" in a previously unhandled dataset. Different formulations of the clustering problem vary in the optimization criterion used. Most existing clustering methods, however, attempt, explicitly or otherwise, to optimize just one such clustering criterion modeled by a single cluster validity index. This often results into considerable discrepancies observable between the solutions produced by different algorithms on the same data. A single cluster validity measure is hardly able to judge the correctness of clustering for a wide variety of real life datasets. A wrong choice of the validity measure may lead to poor clustering results. Thus, the single-objective clustering method may prove futile (as judged by means of expert's knowledge) in a context where the criterion employed is inappropriate. In situations where the best solution corresponds to a tradeoff between different conflicting objectives the common sense advocates for a multi-objective framework for clustering. In the case of iterative optimization

S. Das, A. Abraham, and A. Konar: Metaheuristic Clustering, SCI 178, pp. 213–238.
springerlink.com

algorithms, it is possible that a single-objective approach would visit such tradeoff solutions during a run, but it would not recognize them as good and would almost surely discard them. The following example illustrates the need for modeling the clustering task as a Multi-objective Optimization (MOO) problem.

Consider a situation in which a firm would like to identify consumer segments that have different levels of proneness to deals. The firm conjecture is that there may be customers prone to deals in general and some other customers prone to specific type of deals. Suppose that the firm collects survey data (e.g., Likert-scale measurements) regarding the proneness of customers to deals, such as coupons, buy-one-get-one-free, free gifts and rebates. The firm has data on demographics, such as age, household-income, education, and gender. Also, additional data are available on customers' behaviors pertaining deal responsiveness, such as coupon redemption frequency and quantity of sale items purchased. With these data, the firm could apply standard (single-criterion) clustering techniques to find clusters that are homogenous with respect to the customers' demographics. Homogeneity is given by the selection of the objective function to be optimized when solving the associated clustering problem. As we discuss later, there are a number of criteria that could be used to obtain homogenous clusters. The firm, instead, may choose to cluster customers in order to maximize the explained variation in the actual customers' responses to deals. It is possible, however, that neither of these clustering schemes turn out to be useful to the firm that might be looking for clusters that are both homogenous according to some criterion and at the same time explain the variation related to one or more dependent variables.

Although there has been a plethora of papers reporting several single-objective evolutionary clustering techniques, very few research works have so far been undertaken towards the application of Evolutionary Multi-objective Optimization Algorithms (EMOA) for pattern clustering.

A state-of-the-art literature survey indicates that DE has already proved itself as a promising candidate in the field of Evolutionary Multi-objective Optimization (EMO) [3 – 5]. Unlike NSGA II or PESA II, however, the multi-objective variants of DE have not been applied to the general data clustering problems till date, to the best of our knowledge. This chapter primarily aims at comparing the performances of four most representative multi-objective DE algorithms on the multi-objective fuzzy clustering problem [6]. The multi-objective DE-variants considered here are namely the Pareto DE (PDE) [7], the Multi-objective DE (MODE) [8], DE for Multi-objective Optimization (DEMO) [9], and Non-Dominated Sorting DE (NSDE) [10]. Since DE, by nature, is a real-coded population-based optimization algorithm, we resort to centroid-based representation scheme for the search variables. Note that in contrast to single objective optimization that yields a single best solution, in MOO, a number of, often conflicting, objective functions are optimized simultaneously and thus an MOO algorithm, in general, ends up with a number of Pareto optimal solutions. None of these Pareto-optimal solutions can be improved any further on any objective without degrading it on another. For assessing the performance of the DE-variants, we consider the Xie-Beni index [11] and a penalized form of the FCM measure (J_m) [12] as the objective functions. Note that any other and any number of objective functions could be used in the proposed MOO clustering framework. The best clustering solutions have been selected from the Pareto-optimal set using the gap-statistics [13] method for all DE-variants. The performance of the multi-objective DE-variants have also been contrasted with two best-known EMOA-based clustering methods till date. The first one of these is MOCK by Handl and Knowles [14] while the second one is based on NSGA II and was used by Bandyopadhyay *et al.* for pixel clustering in remote sensing satellite image data [15]. Although we experimented with a large variety of datasets, here we report the results for eight representative datasets including the microarray Yeast sporulation data [16].

The rest of the Chapter is organized in the following way. Section 6.2 reviews the multi-objective extensions of DE and discusses the four contestant MO variants of DE in sufficient details. The fitness functions, scheme of search variable representation and selection of best solutions from the Pareto front have been discussed in section 6.3. Experimental setup and the results have been presented and discussed in section 6.4. Finally the chapter is concluded in section 6.5 with a discussion on the future research directions.

6.2 Multi-objective Optimization Using Differential Evolution Algorithm

In this section we outline four most frequently used variants of DE, specially devised for tackling the MOO problems.

6.2.1 The Pareto Differential Evolution (PDE)

One of the recent approaches of evolutionary multi-objective optimization is Pareto differential evolution algorithm [7].This algorithm was designed for EMO problems with continuous variables and achieved very competitive results compared to other evolution algorithms in EMO literature. However, there is no obvious way to select best crossover and mutation rates apart from running the algorithm with different rates. It handles only one (main) population. Reproduction is undertaken only among non-dominated solutions, and offspring are placed into the population if they dominate the main parent. A distance metric relationship is used to maintain diversity.

The PDE algorithm is similar to the DE algorithm with some modifications:

- The initial population is initialized according to a Gaussian distribution N (0.5, 0.15).
- The step-length parameter F is generated from a Gaussian distribution N (0, 1).
- Reproduction is undertaken only among non-dominated solutions in each generation.

- The boundary constraints are preserved either by reversing the sign if the variable is less than 0 or keeping subtracting 1 if it is greater than 1 until the variable is within its boundaries.
- Offspring are placed into the population if they dominate the main parent.

The algorithm works as follows. An initial population is generated at random from a Gaussian distribution with mean 0.5 and standard deviation 0.15. All dominated solutions are removed from the population. The remaining non-dominated solutions are retained for reproduction. If the number of non-dominated solutions exceeds some threshold, a distance metric relation (will be described in the next paragraph) is used to remove those parents who are very close to each others [17, 18]. Three parents are selected at random. A child is generated from the three parents and is placed into the population if it dominates the first selected parent; otherwise a new selection process takes place.

This process continues until the population is completed .A maximum number of non-dominated solutions in each generation was set to 50. If this maximum is exceeded, the following nearest neighbor distance function is adopted:

$$D(\vec{X}) = \frac{(\min\|\vec{X} - \vec{X}_i\| + \min\|\vec{X} - \vec{X}_j\|)}{2}, \tag{6.1}$$

where $\vec{X} \neq \vec{X}_i \neq \vec{X}_j$. This means the nearest neighbor distance is the average Euclidean distance between the closest two points. The non-dominated solution with the smallest neighbor distance is removed from the population until the total number of non-dominated solutions is retained to 50.

6.2.2 The Multi-Objective Differential Evolution (MODE)

MODE was proposed by Xue *et al.* in [8]. This algorithm uses a variant of the original DE, in which the best individual is adopted to create the offspring. A Pareto-based approach is introduced to implement the selection of the best individual. If a solution is dominated, a set of non-dominated individuals can be identified and the "best" turns out to be any individual (randomly picked) from this set. Also, the authors adopt ($\mu + \lambda$) selection, Pareto ranking and crowding distance in order to produce and maintain well-distributed solutions. Xue *et al.* used MODE to solve five high dimensionality unconstrained problems with 250,000 evaluations and the results are compared only to those obtained by SPEA [19].

6.2.3 Differential Evolution for Multi-objective Optimization (DEMO)

DEMO was proposed by Robic and Filipic [9]. This algorithm combines the advantages of DE with the mechanisms of Pareto-based ranking and crowding distance sorting. DEMO only maintains one population and it is extended when newly created candidates take part immediately in the creation of the subsequent candidates. This enables a fast convergence towards the true Pareto front, while the use of non-dominated sorting and crowding distance (derived from the NSGA-II [1]) of the extended population promotes the uniform spread of solutions.

DEMO is implemented in three variants (DEMO/parent, DEMO/closest/dec and DEMO/closest/obj) [9]. Below we provide a pseudo-code for MODE/parent:

1. Evaluate the initial population $\boldsymbol{P}$ of random individuals.
2. While stopping criterion not met, do:
   2.1. For each individual $\vec{X}_i$ (i = 1 . . . *NP*) from $\boldsymbol{P}$ repeat:
      (a) Create candidate $\vec{U}_i$ from parent $\vec{X}_i$.
      (b) Evaluate the candidate.
      (c) If the candidate dominates the parent, the candidate replaces the parent. If the parent dominates the candidate, the candidate is discarded. Otherwise, the candidate is added in the population.
   2.2. If the population has more than population size *NP* individuals, truncate it.
   2.3. Randomly enumerate the individuals in $\boldsymbol{P}$.

In DEMO the candidate replaces the parent if it dominates it. If the parent dominates the candidate, the candidate is discarded. Otherwise (when the candidate and the parent is non-dominated with regard to each other), the candidate is added to the population. This step is repeated until *NP* number of candidates is created. After that, we get a population of the size between *NP* and 2.*NP*. If the population has enlarged, we have to truncate it to prepare it for the next step of the algorithm.

The truncation consists of sorting the individuals with non-dominated sorting and then evaluating the individuals of the same front with the crowding distance metric. The truncation procedure keeps in the population only the best *NP* individuals (with regard to these two metrics). The described truncation is derived from NSGA-II. DEMO incorporates two crucial mechanisms. The immediate replacement of the parent individual with the candidate that dominates it is the core of DEMO. The newly created candidates that enter the population (either by replacement or by addition) instantly take part in the creation of the following candidates. This emphasizes elitism within reproduction, which helps achieving the first goal of multi objective optimization – convergence to the true Pareto front. The second mechanism is the use of non-dominated sorting and crowding distance metric in truncation of the extended population. Besides preserving elitism, this mechanism stimulates the uniform spread of solutions. This is needed to achieve the second goal – finding as diverse non-dominated solutions as possible. DEMO's selection scheme thus efficiently pursues both goals of multi objective optimization.

The other two variants were inspired by the concept of Crowding DE is recently introduced by Thomsen [20]. When optimizing functions with many optima, we would sometimes like not only to find one optimal point, but also discover and maintain multiple optima in a single algorithm run. For this purpose, Crowding DE can be used. Crowding DE is basically conventional DE with one important difference. Usually, the candidate is compared to its parent. In Crowding DE, the candidate is compared to the most similar individual in the population. The applied similarity measure is the Euclidean distance between the two solutions.

The second, DEMO/closest/dec, works in the same way as DEMO/parent, with the exception that the candidate solution is compared to the most similar individual in decision space. If it dominates it, the candidate replaces this individual; otherwise it is treated in the same way as in DEMO/parent. The applied similarity measure is the Euclidean distance between the two solutions in decision space. In the third variant, EMO/closest/obj, the candidate is compared to the most similar individual in objective space. DEMO/closest/dec and DEMO/ closest/obj need more time for one step of the procedure than DEMO/ parent. This is because at every step they have to search for the most similar individual in the decision and objective space, respectively.

DEMO is compared in five high-dimensionality unconstrained problems outperforming in some problems to the NSGA-II [1], PDEA [21], PAES [22], SPEA [19] and MODE [8].

6.2.4 Non-dominated Sorting DE (NSDE)

This approach was proposed by Iorio and Li [10], which is a simple modification of the NSGA-II [1]. The only difference between this approach and the NSGA-II is in the method for generating new individuals. The NSGA-II uses a real-coded crossover and mutation operator, but in the NSDE, these operators were replaced with the operators of Differential Evolution. In the NSGA-II the better individuals remain unknown until all candidates are sorted and assigned a crowding distance and non-domination level. Parents are selected for mating using a tournament selection operator which uses the rank and crowding distance of individuals. Within the NSGA-II framework, the DE variants are used to generate N offspring from the selected parents. The offspring individuals are then evaluated on the objective functions. Following this, they are combined with the parent generation. The combined population is then sorted into dominance ranks, as was mentioned previously. Each individual also has a crowding distance associated with it. New candidates are generated using the DE/current-to-rand/1 strategy. NSDE is used to solve rotated problems with a certain degree of rotation on each plane. The results of the NSDE outperformed those produced by the NSGA-II.

6.3 The Multi-objective Clustering Scheme

6.3.1 Search-Variable Representation

Since we aim at simultaneously finding the number of clusters and the optimum partitions in the dataset, to accommodate a variety of choices for the number of clusters, we keep the vector representation scheme in DE same as the one described in Section 4.2.1.

6.3.2 Selecting the Objective Functions

The performance of a multi-objective clustering algorithm critically depends upon the clustering objectives it tries to optimize simultaneously. A judicious choice of these objective functions can lead to remarkable performance, whereas arbitrary or unintelligent selection of objectives may lead to poor performances. The selection of objectives should be such that they can balance each other critically and are contradictory in nature. Conflict among the objective functions is often beneficial since it guides to globally optimal solutions. It also ensures that no single clustering objective is optimized leaving other probable significant objectives unnoticed.

In this work we choose the Xie-Beni index XB_q and a penalized form of the FCM objective function J_q as the two objectives. For convenience we present the two functions here once again. The FCM measure J_q may be defined as:

$$J_q = \sum_{j=1}^{n} \sum_{i=1}^{k} u_{ij}^q \cdot d^2(\vec{Z}_j, \vec{m}_i), \quad 1 \le q \le \infty \tag{6.1}$$

where q is the fuzzy exponent, d indicates a distance measure between the j-th pattern vector and i-th cluster centroid, and u_{ij} denotes the membership of j-th pattern in the i-th cluster. It is well known that J_m becomes a minimum when $n = k$. Naturally a minimization of J_q without specifying the number of clusters tends to increase the cluster number towards a maximum. In order to balance this tendency, we propose to use a slightly modified version of J_q, which is penalized (i.e. increased) to some extent for a large number of clusters. The modified FCM index is presented below:

$$J_q^{'} = \sqrt{k+1} \left(\sum_{j=1}^{n} \sum_{i=1}^{k} u_{ij}^q \cdot d^2(\vec{Z}_j, \vec{m}_i) \right), \tag{6.2}$$

where k is the number of clusters.

The XB index is defined as a function of the ratio of the total variation σ (which in fact is similar to J_q) to the minimum separation sep of the clusters. Here σ and sep may be written as:

$$\sigma = \sum_{i=1}^{k} \sum_{p=1}^{n} u_{ip}^2 \cdot d(\vec{m}_i, \vec{Z}_p) \tag{6.3}$$

$$\text{and} \quad sep(Z) = \min_{i \ne j} \{ d^2(\vec{m}_i, \vec{m}_j) \} \tag{6.4}$$

The XB index is then written as:

$$XB_q = \frac{\sigma}{n \times sep(Z)} = \frac{\sum_{i=1}^{k}\sum_{p=1}^{n} u_{ip}^{q} \cdot d^2(\vec{m}_i, \vec{Z}_p)}{n \times \min_{i \neq j}\{d^2(\vec{Z}_i, \vec{Z}_j)\}} \tag{6.5}$$

Note that when the partitioning is compact and the individual clusters are well separated, value of σ should be low while *sep* should be high, thereby yielding lower values of XB_q index. The objective therefore is to minimize the XB_q index.

For computing the measures described in equations (6.2) and (6.5), the centers encoded in a DE vector are first extracted. Let the set of centers be denoted by $M = \{\vec{m}_1, \vec{m}_2, ..., \vec{m}_k\}$. The membership value of the *j*-th pattern in *i*-th cluster $u_{ij}, i = 1,2,....k$ and $j = 1,2,....,n$ are computed as:

$$u_{ij} = \frac{1}{\sum_{p=1}^{k}\left(\frac{d(\vec{m}_i, \vec{Z}_j)}{d(\vec{m}_p, \vec{Z}_j)}\right)^{\frac{2}{q-1}}} \tag{6.6}$$

Note that while computing the u_{ij} s, using equation (6.6), if $d(\vec{m}_p, \vec{Z}_j)$ is equal to zero for some *p*, then u_{ij} is set to zero for all $i = 1,2,....k$, $i \neq j$, while u_{pj} is set equal to one. Subsequently the centers encoded in a vector are updated using the following equation:

$$\vec{m}_p = \frac{\sum_{j=1}^{n}(u_{pj})^q \cdot \vec{Z}_j}{\sum_{j=1}^{n}(u_{pj})^q} \tag{6.7}$$

and the cluster membership values are recomputed.

Note that the XB_q index is a combination of global (numerator) and particular (denominator) situations. The numerator is similar to J_m but the denominator has a factor that gives the separation between to minimum distant clusters. Hence this factor only considers the worst case, i.e. which two clusters are closest to each other and forgets about the other partitions. Here, greater value of the denominator (lower value of whole index) signifies a better partitioning. Thus it is evident that J_m and XB_q indices should be simultaneously minimized in order to get good solutions. The two terms at the numerator and the denominator of XB_q may not attain their best values for the same partitioning when the data has complex and overlapping clusters, such as remote sensing image data.

6.3.3 Selecting the Best Solutions from Pareto-front

Multi-objective clustering does not return a single solution, but a set of clustering solutions. These individual groupings correspond to different tradeoffs between the two objectives and, in our case, also consist of different numbers of clusters. Several researchers have already investigated the identification of promising solutions from Pareto front approximations recently [23 - 25]. These works have primarily dealt with the reduction of the size of the approximation set in absence of additional expert's knowledge. For choosing the most interesting solutions from the Pareto front, we apply Tibshirani *et al.*'s Gap statistic [13], a statistical method to determine the number of clusters in a data set. The Gap statistic is based on the expectation that the most suitable number of clusters shows in a significant "knee" when plotting the performance of a clustering algorithm (in terms of a selected internal evaluation measure) as a function of the number of clusters.

In order to compute the Gap statistic, the clustering problem is solved for a range of different values of k and for each k, the resulting data-grouping $C = \{C_1, C_2, ..., C_k\}$ is evaluated by means of the intra-cluster variance, which is given by:

$$V(C) = \sum_{C_i \in C} \sum_{\vec{Z}_j \in C_i} d^2(\vec{Z}_j, \vec{m}_i) \tag{6.8}$$

The intra-cluster variance is usually affected by the number of clusters. Thus a plot $Var(k)$ showing the evolution of $V(C)$ as a function of the input parameter k exhibits a decreasing trend that is solely caused by the finer partitioning and not by the actual capturing of structure within the data. The Gap statistic overcomes this effect through a normalization of the performance curve. B reference curves $R_b(k)$ (with $b \in \{1, ..., B\}$) are computed, which are the performance curves obtained with the same clustering algorithm for uniform random reference distributions. Using these, the normalized performance curve ("Gap curve") for $Var(k)$ is given as:

$$Gap(k) = \frac{1}{B} \sum_{b=1}^{B} \log[R_b(k)] - \log[Var(k)]. \tag{6.9}$$

The best suited number of clusters is then determined as the smallest k, such that

$$Gap(k) \geq Gap(k+1) - s_{k+1}, \tag{6.10}$$

where $$s_k = sd_k \sqrt{1 + \frac{1}{B}},$$

and sd_k is the standard deviation within the set of the B log-scaled reference points $R_b(k)$.

In our case, we employed a Poisson's distribution in eigen-space for the null model (as suggested in [13]). We compute the performance curves for a range of different numbers of clusters, and, for each k, we generate B = 10 reference distributions.

6.3.4 Evaluating the Clustering Quality

In this research, the final clustering quality is evaluated using two external measures. Specifically we choose the Adjusted Rand Index [26] (which is a generalization of the Rand index [27]) and the Sihouette index [60]. The Adjusted Rand index has already been described in equation (5.29) of chapter 5. Silhouette width reflects the compactness and separation of the clusters. Given a set of data points $Z = \{\vec{Z}_1, ..., \vec{Z}_n\}$ and a given clustering solution $C = \{C_1, C_2, ..., C_k\}$, the silhouette width $s(\vec{Z}_j)$ for each data $\vec{Z}_j$ belonging to cluster C_i indicates a measure of the confidence of belongingness, and it is defined as:

$$s(\vec{Z}_j) = \frac{b(\vec{Z}_j) - a(\vec{Z}_j)}{\max(a(\vec{Z}_j), b(\vec{Z}_j))}. \tag{6.11}$$

Here $a(\vec{Z}_j)$ denotes the average distance of data point $\vec{Z}_j$ from the other data points of the cluster to which the data point $\vec{Z}_j$ is assigned (i. e. cluster C_i). On the other hand, $b(\vec{Z}_j)$ represents the minimum of the average distances of data point $\vec{Z}_j$ from the data points belonging to clusters C_r, $r = 1,2,..., k$ and $r \neq i$. The value of $s(\vec{Z}_j)$ lies between -1 and +1. Large values of $s(\vec{Z}_j)$ (near to 1) indicates that the data point $\vec{Z}_j$ is well clustered. Value of $s(\vec{Z}_j)$ around 0 means that the data point lies between two clusters and a negative value of $s(\vec{Z}_j)$ indicates that the data point $\vec{Z}_j$ is probably placed in a wrong cluster. Overall silhouete index $s(C)$ of a clustering solution $C = \{C_1, C_2, ..., C_k\}$ is defined as the mean silhouette width over all the data points:

$$s(C) = \frac{1}{n} \sum_{j=1}^{n} s(\vec{Z}_j) \tag{6.12}$$

Greater values of $s(C)$ (near to 1) reflect that most of the data points are correctly clustered and this in turn indicates a better clustering solution. Silhouette index can be evaluated for any distance measure.

6.4 Experiments and Results

6.4.1 Datasets Used

The experimental results showing the effectiveness of multi-objective DE based clustering has been provided for six artificial and four real life datasets. The artificial datasets are named as Dataset_1 to Dataset_6, where the first five datasets are in two dimensions and the last one is three dimensional with number of clusters varying from 3 to 10. Table 6.1 presents the number of objects, dimensionality and the number of clusters for each data. The real-life datasets are iris, wine, breast-cancer and the yeast sporulation data. While iris, wine and Wisconsin's breast cancer datasets have already been described in section 4.2.1 of Chapter 4, we briefly discuss the microarray dataset of yeast.

We consider the microarray data on the transcriptional program of sporulation in budding yeast, the collection and analysis of which have been described in [16]. The sporulation dataset is available publicly from the website: *http://cmgm.* stanford.edu/pbrown/sporulation. This dataset consists of 6118 genes measured across 7 time points (0, 0.5, 2, 5, 7, 9 and 11.5 h) during the sporulation process of budding yeast. The data are then log-transformed. Among the 6118 genes, the genes, whose expression levels did not change significantly during the harvesting, have been ignored from further analysis. This is determined with a threshold level of 1.6 for the root mean squares of the log2-transformed ratios. The resulting set consists of 474 genes. Please note that for the yeast sporulation dataset, we have used the Pearson correlation coefficient based distance measure [29], described in equation (1.6) of Chapter 1, as it has been shown to be more effective for clustering microarray datasets [29].

6.4.2 Parameters for the Algorithms

All the multi-objective DE variants have been used with 40 parameter vectors in each generation and each run of each algorithm was continued for 100 generations. After conducting a series of experiments, we find that choosing the value of scale factor F randomly between 0.5 and 1 while keeping the value of Cr at 0.9 gives reasonably good results for all the four DE variants considered here. In order to make the comparison fair enough, we ensure that in each run over each problem, all the multi-objective De variants may start from the same initial population. In this way, any performance difference between these algorithms is actually attributed to the internal operators of these algorithms and not to the initial population which may accidentally hit a very good solution for some algorithm.

The parameters for the multi-objective GA (NSGA II) based clustering are fixed as follows: number of generations = 100, population size = 50, crossover probability = 0.8, mutation probability $= \dfrac{1}{Chromosome\_length}$. Please note that

Table 6.1 Details of the datasets used

Dataset	Number of points	Number of clusters	Number of Characteristics
Dataset_1	900	9	2
Dataset _2	76	3	2
Dataset _3	400	4	3
Dataset _4	300	6	2
Dataset _5	500	10	2
Dataset_ 6	810	3	2
Iris	150	3	4
Wine	178	3	13
Breast-Cancer	683	2	9
Yeast Sporulation	474	7	7

the four DE variants and the NSGA II use the same parameter representation scheme. Clustering with MOCK was performed with the source codes available from http://dbkgroup.org/handl/mock/.

6.4.3 *Presentation of Results*

The mean Silhouette index values of the best-of-run solutions provided by six contestant algorithms over the 10 datasets have been provided in Table 6.2. The Table also shows the mean number of classes determined by each algorithm. All the results presented in this Table are averages over 30 independent runs of each algorithm. The best entries have been marked in boldface in each row. The best

Table 6.2 Average Silhouette index and number of clusters found and standard deviations (in parentheses) by six contestant algorithms over 30 independent runs on ten datasets

Dataset	Algorithms											
	MODE		DEMO		PDE		NSDE		NSGA2		MOCK	
	k	Silhouette Index	k	Silhouette Index	k	Silhouette Index	k	Silhouette Index	k	Silhouette Index	k	Silhouette Index
Dataset_1	**9.12 (1.46)**	**0.670726 (0.0321)**	9.43 (0.843)	0.670523 (0.08263)	9.83 (0.35)	0.656594 (0.05372)	8.81 (2.837)	0.661661 (0.00483)	9.37 (1.72)	0.669317 (0.0892)	8.52 (2.81)	0.66342 (0.0736)
Dataset_2	3.36 (0.65)	0.678734 (0.03716)	3.74 (0.363)	0.670530 (0.05381)	3.15 (0.726)	0.663812 (0.00366)	**3.07 (0.052)**	**0.683382 (0.00082)**	3.16 (0.072)	0.674393 (0.00927)	3.33 (1.03)	0.658921 (0.004731)
Dataset_3	4.14 (0.36)	0.771363 (0.00938)	**4.09 (0.24)**	**0.774611 (0.002376)**	4.38 (1.09)	0.766945 (0.04824)	4.17 (0.73)	0.767333 (0.004712)	3.57 (0.51)	0.765691 (0.005686)	3.78 (1.25)	0.768419 (0.006721)
Dataset_4	**6.04 (0.25)**	**0.849184 (0.00472)**	6.13 (1.27)	0.838303 (0.056741)	6.15 (0.79)	0.824910 (0.00721)	6.31 (1.46)	0.840613 (0.04752)	6.28 (0.46)	0.827618 (0.02871)	6.08 (0.51)	0.832527 (0.007825)
Dataset_5	9.24 (3.89)	0.778102 (0.04363)	**10.03 (0.37)**	**0.778736 (0.008125)**	9.26 (2.88)	0.746241 (0.00673)	11.45 (1.68)	0.777406 (0.05345)	12.43 (0.939)	0.768379 (0.005384)	10.41 (0.80)	0.769342 (0.006208)
Dataset_6	5.19 (0.93)	0.643316 (0.00381)	5.62 (0.867)	0.637986 (0.008112)	**5.08 (0.35)**	**0.645446 (0.05524)**	5.26 (0.78)	0.640832 (0.004983)	4.65 (1.58)	0.642091 (0.002833)	5.16 (0.38)	0.640957 (0.008349)
Iris	2.31 (0.76)	0.606353 (0.03483)	**2.97 (0.40)**	**0.606864 (0.03234)**	2.25 (0.324)	0.569675 (0.04752)	2.14 (0.58)	0.600304 (0.004618)	2.16 (1.06)	0.566613 (0.082651)	3.05 (0.37)	0.6003725 (0.005129)
Wine	**3.16 (0.46)**	**0.582197 (0.00427)**	3.65 (0.83)	0.568391 (0.007473)	4.01 (1.35)	0.525840 (0.01213)	3.52 (1.20)	0.5457383 (0.009497)	3.88 (0.67)	0.5767342 (0.009415)	3.59 (0.46)	0.576834 (0.000812)
Breast Cancer	**2.08 (0.38)**	**0.648297 (0.00734)**	2.68 (0.64)	0.609123 (0.57813)	2.15 (0.46)	0.619219 (0.00567)	2.33 (0.67)	0.628352 (0.006782)	2.57 (0.60)	0.6004642 (0.004561)	2.10 (0.53)	0.626719 (0.01094)
Yeast Sporulation	**7.08 (0.12)**	**0.676434 (0.00072)**	6.34 (0.32)	0.558619 (0.057832)	8.36 (2.81)	0.595367 (0.00721)	7.24 (1.04)	0.604513 (0.005728)	7.22 (0.68)	0.641306 (0.04813)	6.67 (0.857)	0.613567 (0.005738)

Table 6.3 Mean value of adjusted rand index found and standard deviations (in parentheses) by six contestant algorithms over 30 independent runs on nine datasets

Dataset	Algorithms					
	MODE	DEMO	PDE	NSDE	NSGA2	MOCK
Dataset_1	**0.846199 (0.031257)**	0.828437 (0.046182)	0.719584 (0.00563741)	0.819794 (0.035285)	0.802180 (0.004782)	0.810934 (0.0059348)
Dataset_2	0.847621 (0.006312)	0.9273464 (0.0008573)	0.9372649 (0.036451)	**1.000000**	0.9378123 (0.006821)	0.946547 (0.004536)
Dataset_3	0.951786 (0.004827)	**1.000000**	0.9758732 (0.05736)	0.894635 (0.005736)	0.963841 (0.0046719)	0.878732 (0.0712523)
Dataset_4	**1.000000**	0.857463 (0.065639)	0.840953 (0.076829)	0.919843 (0.0121436)	0.957818 (0.004678)	0.978761 (0.006734)
Dataset_5	0.983785 (0.076764)	**0.993173 (0.089371)**	0.876710 (0.023376)	0.982013 (0.084372)	0.947641 (0.006646)	0.9454568 (0.0012043)
Dataset_6	0.881413 (0.05983)	0.881136 (0.078348)	**0.884930 (0.007846)**	0.880265 (0.056347)	0.881395 (0.056483)	0.910294 (0.016743)
Iris	0.738626 (0.0756779)	**0.748784 (0.067457)**	0.709036 (0.025739)	0.738960 (0.001436)	0.715898 (0.005739)	0.786574 (0.075763)
Wine	**0.875849 (0.0087642)**	0.858876 (0.0035287)	0.8265764 (0.0032429)	0.845365 (0.0065761)	0.828645 (0.0074653)	0.864764 (0.0034398)
Breast Cancer	0.956456 (0.0056453)	0.912173 (0.0043247)	0.937857 (0.0087743)	**0.950965 (0.0065682)**	0.944236 (0.006521)	0.9465731 (0.006748)

Table 6.4 Unpaired *t*-test results for silhouette index

Dataset	Std. Err	*t*	95% Conf. Intvl	Two-tailed *P*	Significance
Dataset_1	0.001	7.1968	-0.0121 to -0.0068	< 0.0001	**Extremely Significant**
Dataset_2	0.002	3.8990	-0.0129 to -0.0040	< 0.0001	**Extremely Significant**
Dataset_3	0.007	34.9267	-0.2665 to -0.2373	< 0.0001	**Extremely Significant**
Dataset_4	0.001	3.0961	-0.0051 to -0.0010	0.0037	**Very Significant**
Dataset_5	0.003	3.0684	-0.0156 to -0.0032	0.0040	**Very Significant**
Dataset_6	0.002	3.0584	-0.0109 to -0.0022	0.0041	**Very Significant**
Iris	0.009	1.3744	-0.0309 to 0.0059	0.1774	Not Significant
Wine	0.003	2.3999	-0.0118 to -0.0010	0.0214	**Significant**
Breast Cancer	0.009	1.3744	-0.0309 to 0.0059	0.1774	Not Significant
Yeast Sporulation	0.003	2.3999	-0.0118 to -0.0010	0.0214	**Significant**

solution is selected from the Pareto front in each case by the Gap statistics method described earlier in Section 6.3.3, except for MOCK. Table 6.3 is identical in spirit to Table 6.2, only difference being that this table enlists the adjusted rand index values. Missing values of standard deviations in this table indicate zero standard deviations. The Yeast sporulation data is excluded from this table as no standard nominal classification is known for this dataset. The effectiveness of the clustering on this data has been established by finding the biological interpretation of the clusters by using Gene Ontology (GO) [30] terms. Please see Section 6.4.4 for details of this issue. Tables 6.3 and 6.4 show the results of unpaired t tests

Table 6.5 Unpaired *t*-test Results for adjusted rand index

Dataset	Std. Err	*t*	95% Conf. Intvl	Two-tailed *P*	Significance
Dataset_1	0.021	2.9201	-0.1050 to -0.0189	0.0059	**Very significant**
Dataset_2	0.013	5.0453	-0.0922 to -0.0394	< 0.0001	**Extremely Significant**
Dataset_3	0.002	17.965	-0.0452 to -0.0360	< 0.0001	**Extremely Significant**
Dataset_4	0.005	6.4431	-0.0419 to -0.0219	< 0.0001	**Extremely Significant**
Dataset_5	0.009	1.3744	-0.0309 to 0.0059	0.1774	Not Significant
Dataset_6	0.003	2.3999	-0.0118 to -0.0010	0.0214	**Significant**
Iris	0.009	1.3744	-0.0309 to 0.0059	0.1774	Not Significant
Wine	0.003	2.3999	-0.0118 to -0.0010	0.0214	**Significant**
Breast Cancer	0.009	1.3744	-0.0309 to 0.0059	0.1774	Not Significant

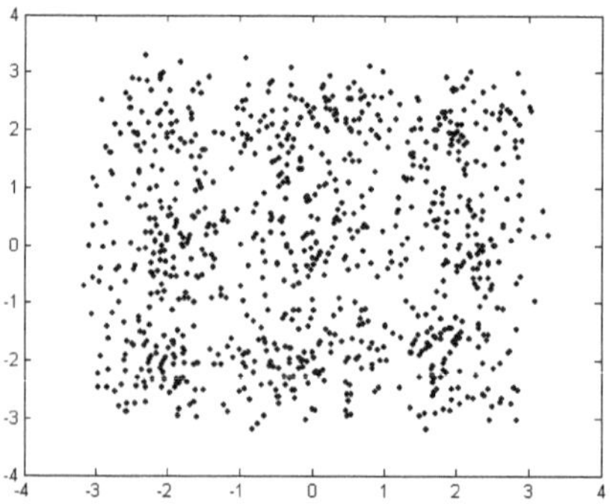

(a) The unlabelled dataset_1

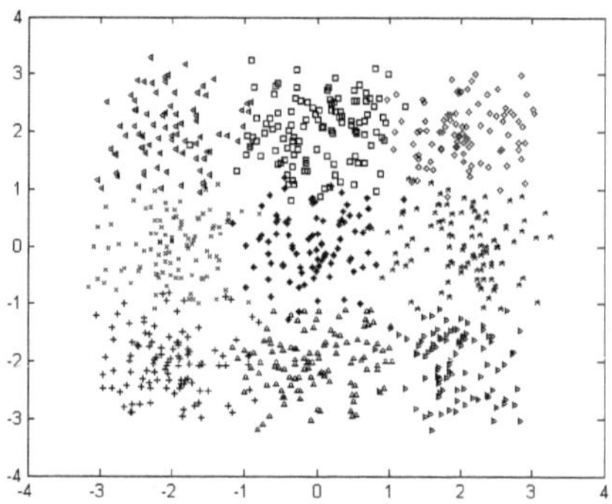

(b) Clustering with MODE

Fig. 6.1 Clustering result for artificial dataset_1

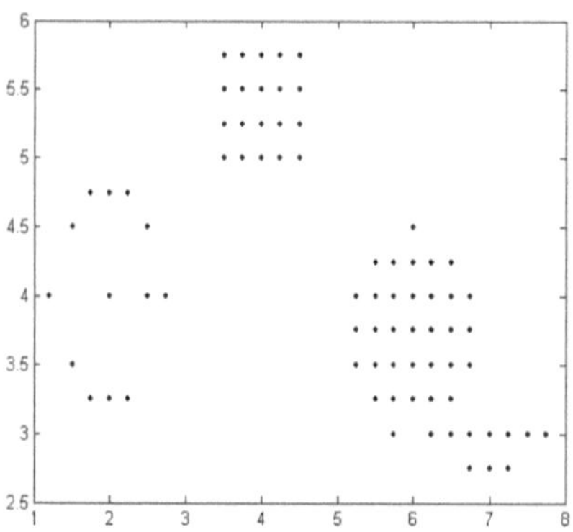

(a) The unlabelled dataset_2

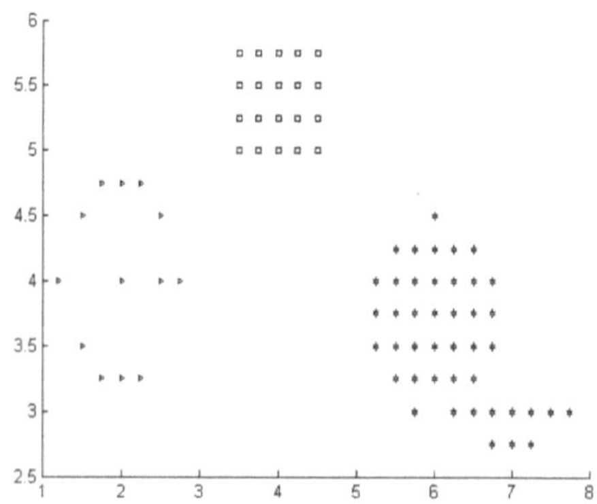

(b) Clustering with NSDE

Fig. 6.2 Clustering result for artificial dataset_2

(standard error of difference of the two means, 95% confidence interval of this difference, the t value, and the two-tailed P value) between the best and second best algorithms in terms of both average Silhouette index and adjusted rand index. For all cases in Tables 6.3 and 6.4, sample size = 30 and number of degrees of freedom = 58. The results listed in Tables 6.2 to 6.4 indicate that there is always one or more multi-objective DE variant that beats the NSGA II or MOCK in terms of mean Silhouette index and adjusted rand index in a statistically significant fashion. The six unlabelled artificial datasets and the corresponding clustered data with the best performing algorithm (which happens to be one of the four multi-objective DE variants) have been depicted in Figures 6.1 to 6.6. The corresponding Pareto optimal fronts have been provided in Figure 6.7.

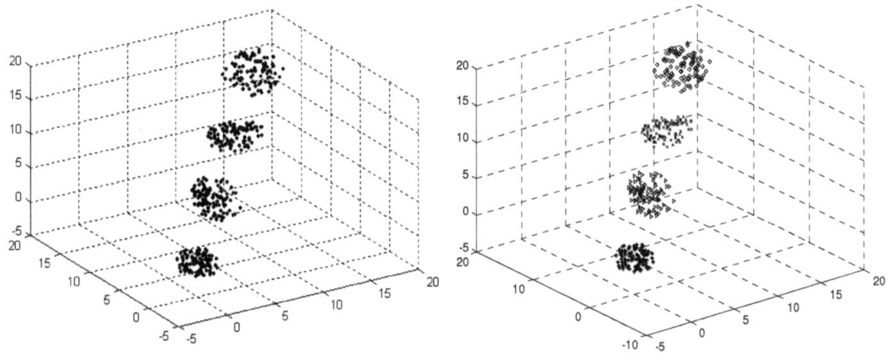

(a) The unlabelled dataset_3 (b) Clustering with DEMO

Fig. 6.3 Clustering result for artificial dataset_3

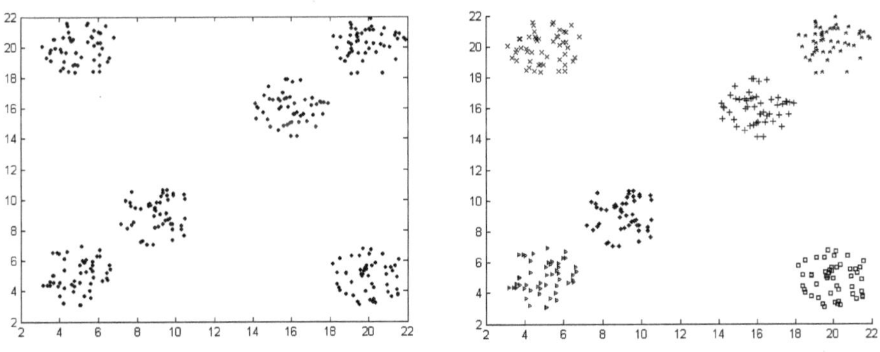

(a) The unlabelled dataset_4 (b) Clustering with MODE

Fig. 6.4 Clustering result for artificial dataset_4

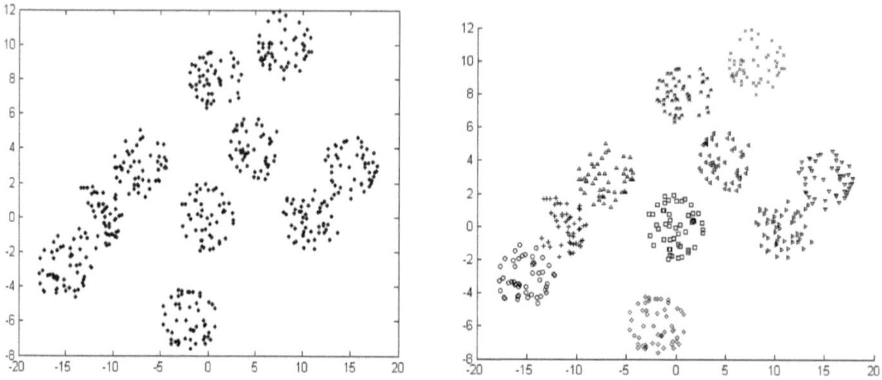

(a) The unlabelled dataset_5 (b) Clustering with DEMO

Fig. 6.5 Clustering result for artificial dataset_5

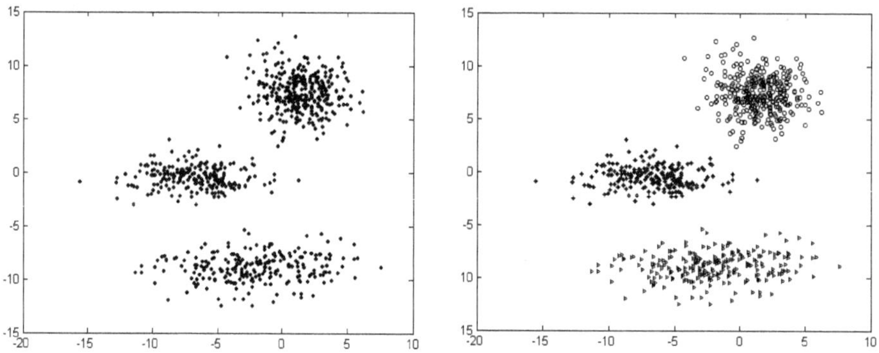

(a) The unlabelled dataset_6 (b) Clustering with PDE

Fig. 6.6 Clustering result for artificial dataset_6

6.4.4 *Significance and Validation of Microarray Data Clustering Results*

In this Section, the best clustering solution provided by different algorithms on the sporulation data of yeast has been visualized using the cluster profile plot (in parallel coordinates) [31] and the heatmap [32] plot in MATLAB 7.0.4 version. Parallel coordinates is a common way of visualizing high-dimensional geometry and analyzing multivariate data. To show a set of points in an n-dimensional space, a backdrop is drawn consisting of n parallel lines, typically vertical and equally spaced. A point in n-dimensional space is represented as a polyline with vertices on the parallel axes; the position of the vertex on the i-th axis corresponds

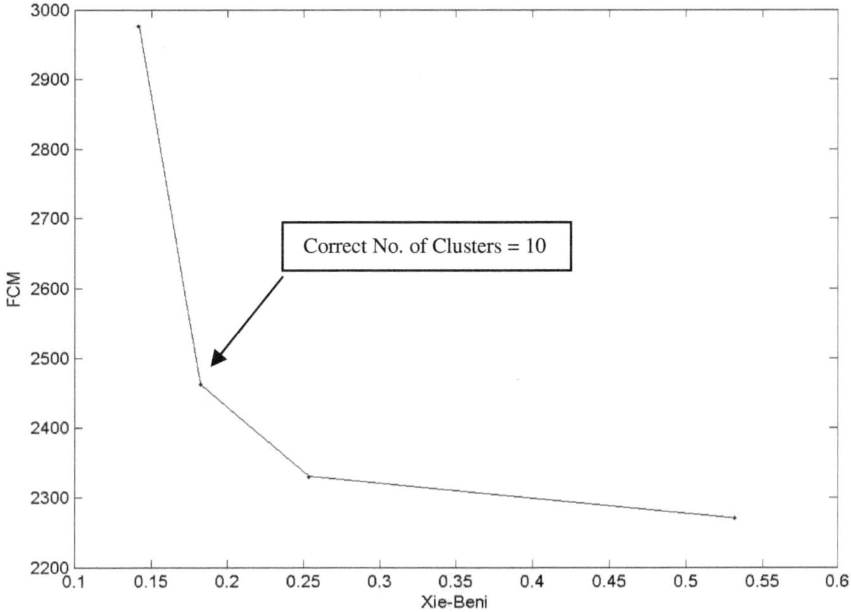

(a) Pareto optimal front (MODE) for Dataset_1

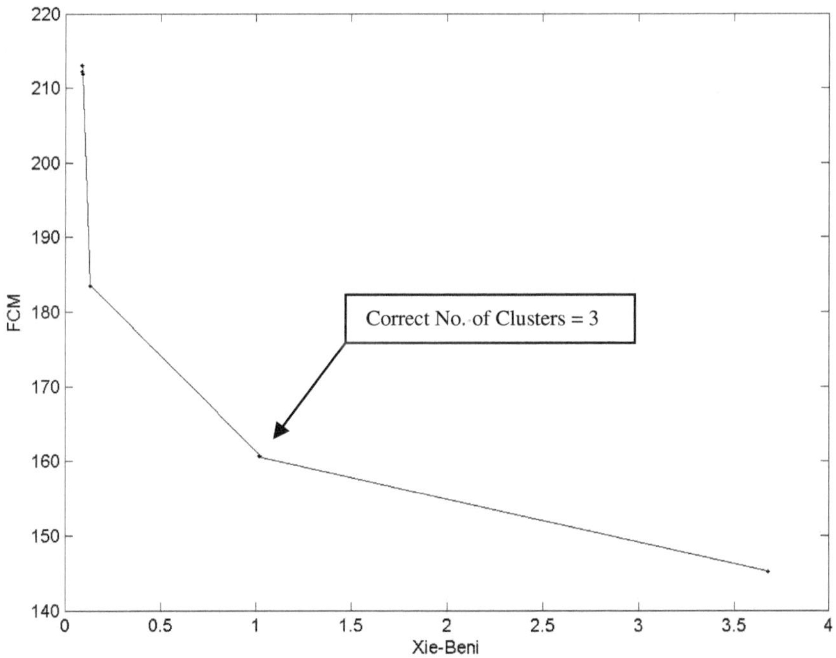

(b) Pareto optimal front (NSDE) for Dataset_2

Fig. 6.7 Best Pareto optimal fronts obtained for eight datasets under test

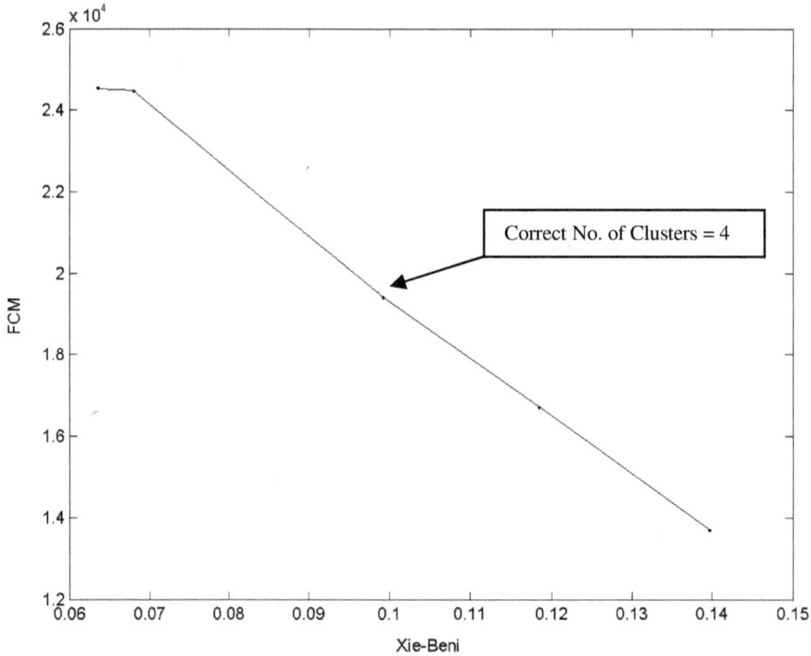

(c) Pareto optimal front (DEMO) for Dataset_3

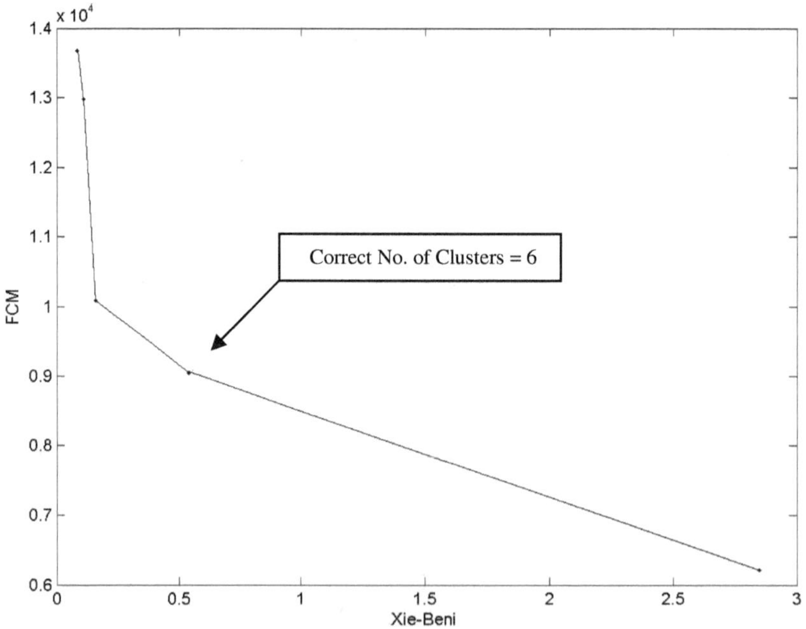

(d) Pareto optimal front (MODE) for Dataset_4

Fig. 6.7 (*continued*)

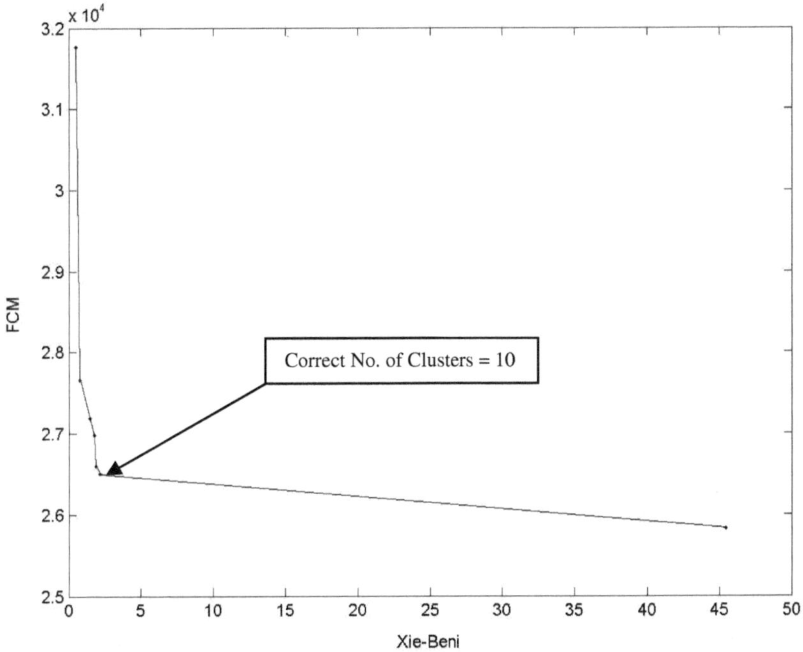

(e) Pareto optimal front (DEMO) for Dataset_5

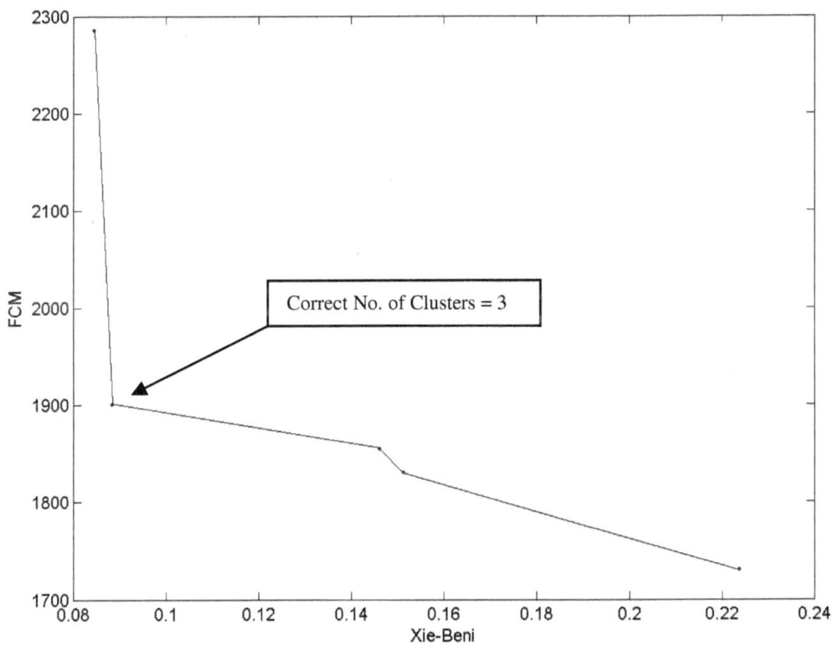

(f) Pareto optimal front (PDE) for Dataset_6

Fig. 6.7 (*continued*)

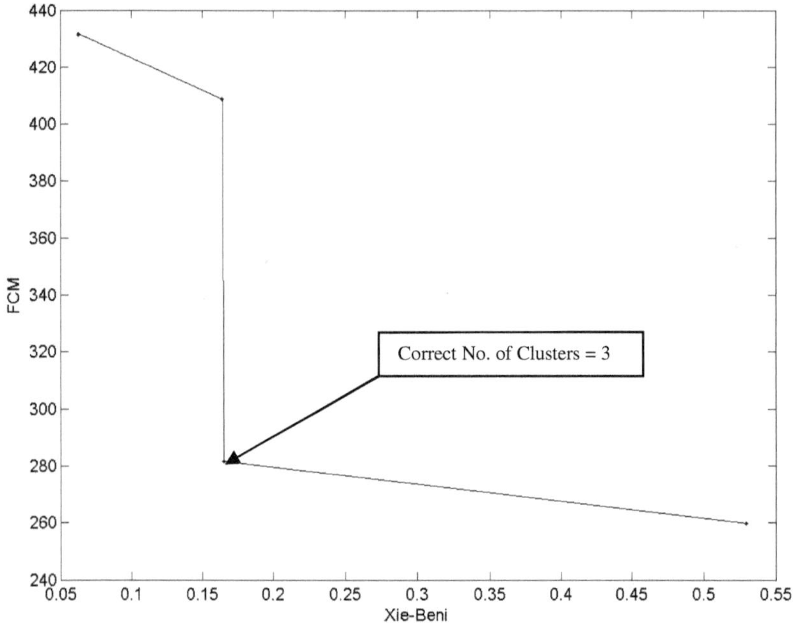

(g) Pareto optimal front (DEMO) for Iris

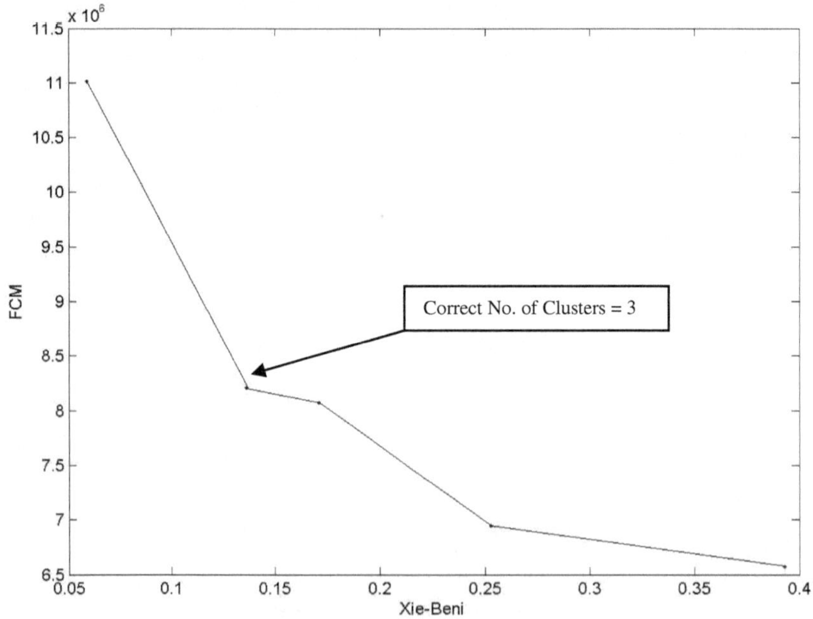

(h) Pareto optimal front (MODE) for Wine

Fig. 6.7 (*continued*)

to the i-th coordinate of the point. Cluster profile plot in parallel coordinates demonstrates how the nature of the plots of the expression values of the genes across the time points differ for each cluster. Also, the genes within a cluster are similar in nature in terms of the expression profile. Cluster profile plots of seven clusters for the best clustering result (provided by MODE) on yeast sporulation data has been shown in Figure 6.8. The blue polylines indicate the member genes within a cluster while the black polyline indicates the centroid of that gene. Cluster profile plots also demonstrate how the cluster profiles for the different groups of genes differ from each other, while the profiles within a group are reasonably similar.

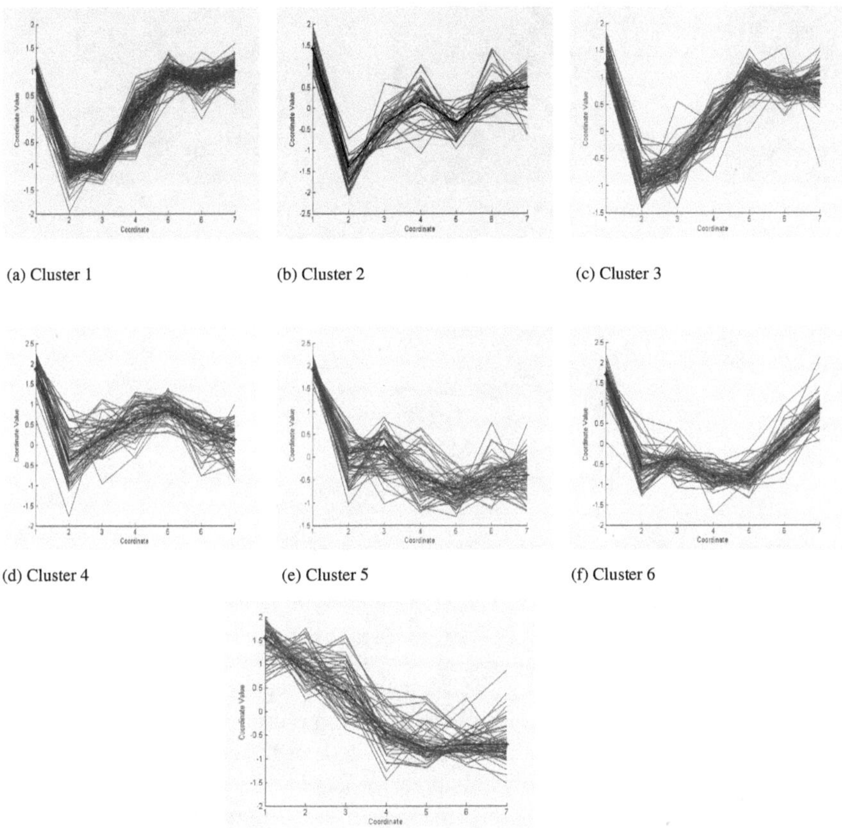

(a) Cluster 1 (b) Cluster 2 (c) Cluster 3

(d) Cluster 4 (e) Cluster 5 (f) Cluster 6

(f) Cluster 7

Fig. 6.8 Cluster profile plots for clustering solution obtained by MODE-based clustering algorithm for yeast sporulation data

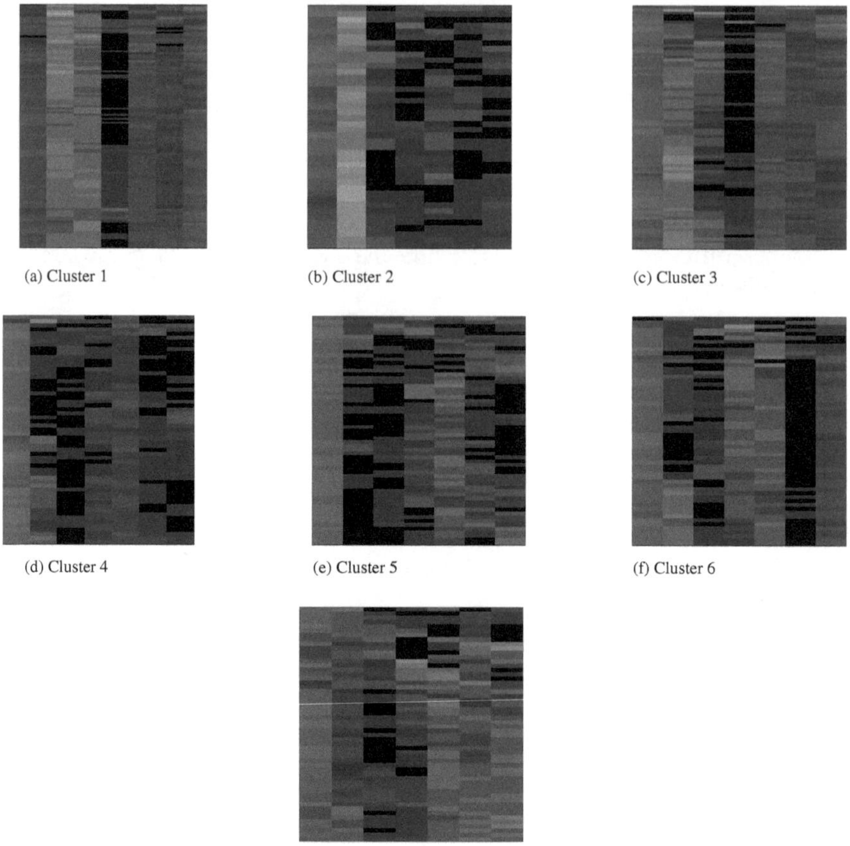

(a) Cluster 1 (b) Cluster 2 (c) Cluster 3

(d) Cluster 4 (e) Cluster 5 (f) Cluster 6

(f) Cluster 7

Fig. 6.9 Heatmaps (Eisen plots) for clustering solution obtained by MODE-based clustering algorithm for yeast sporulation data

In Heatmap (aka Eisen plot [32]), the expression value of a gene at a specific time point is represented by coloring the corresponding cell of the data matrix with a color similar to the original color of its spot on the microarray. The shades of red color represent higher expression level, the shades of green color represent low expression level and the colors towards black represent absence of differential expression values. In our representation, the genes are ordered before plotting so that the genes that belong to the same cluster are placed one after another. Figure 6.9 shows the Heatmap of the seven clusters generated by one run of the MODE algorithm for yeast sporulation data. It is evident from the Figure that the expression profiles of the genes of a cluster are similar to each other and they produce similar color patterns.

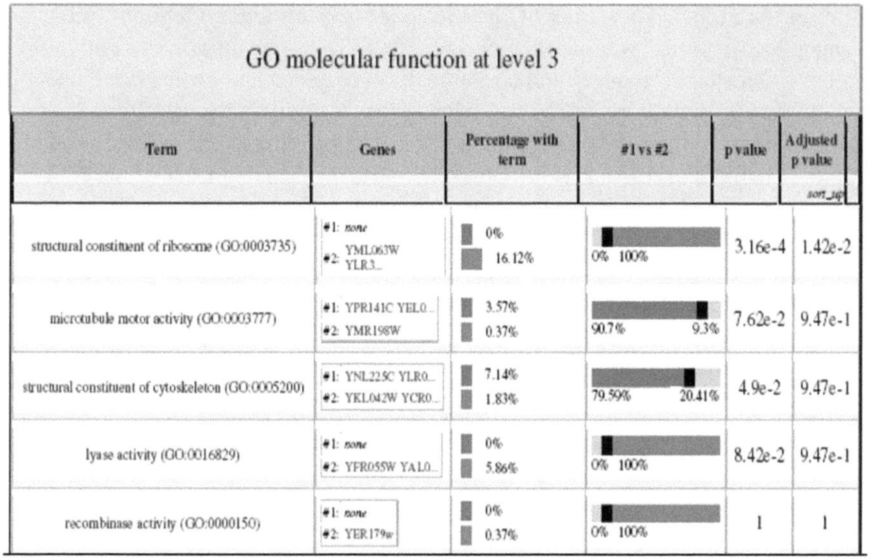

GO molecular function at level 3

Term	Genes	Percentage with term	#1 vs #2	p value	Adjusted p value
structural constituent of ribosome (GO:0003735)	#1: none #2: YML063W YLR3...	0% 16.12%	0% 100%	3.16e-4	1.42e-2
microtubule motor activity (GO:0003777)	#1: YPR141C YEL0... #2: YMR198W	3.57% 0.37%	90.7% 9.3%	7.62e-2	9.47e-1
structural constituent of cytoskeleton (GO:0005200)	#1: YNL225C YLR0... #2: YKL042W YCR0...	7.14% 1.83%	79.59% 20.41%	4.9e-2	9.47e-1
lyase activity (GO:0016829)	#1: none #2: YFR055W YAL0...	0% 5.86%	0% 100%	8.42e-2	9.47e-1
recombinase activity (GO:0000150)	#1: none #2: YER179w	0% 0.37%	0% 100%	1	1

(a)

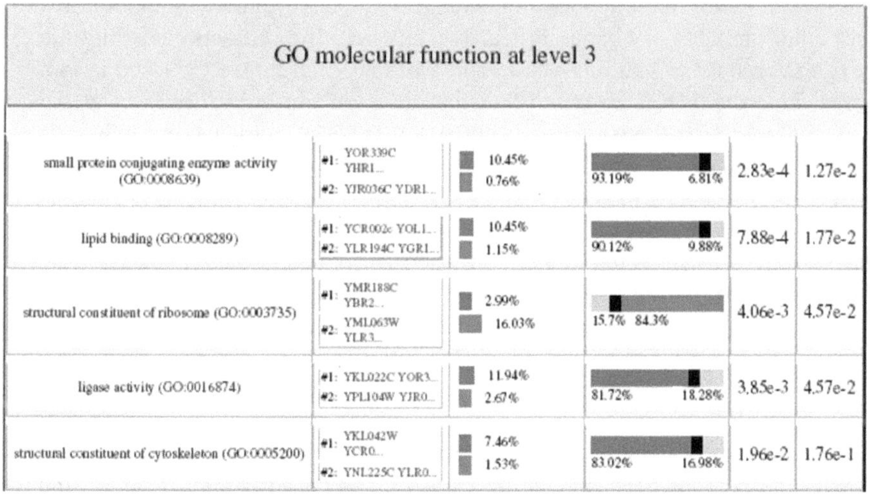

GO molecular function at level 3

small protein conjugating enzyme activity (GO:0008639)	#1: YOR339C YHR1... #2: YJR036C YDR1...	10.45% 0.76%	93.19% 6.81%	2.83e-4	1.27e-2
lipid binding (GO:0008289)	#1: YCR002c YOL1... #2: YLR194C YGR1...	10.45% 1.15%	90.12% 9.88%	7.88e-4	1.77e-2
structural constituent of ribosome (GO:0003735)	#1: YMR188C YBR2... #2: YML063W YLR3...	2.99% 16.03%	15.7% 84.3%	4.06e-3	4.57e-2
ligase activity (GO:0016874)	#1: YKL022C YOR3... #2: YPL104W YJR0...	11.94% 2.67%	81.72% 18.28%	3.85e-3	4.57e-2
structural constituent of cytoskeleton (GO:0005200)	#1: YKL042W YCR0... #2: YNL225C YLR0...	7.46% 1.53%	83.02% 16.98%	1.96e-2	1.76e-1

(b)

Fig. 6.10 Part of FatiGO result for (a) cluster 6 and (b) cluster 2 of the best multi-objective clustering algorithm on yeast sporulation dataset

Studying the functional annotations of a gene within a cluster provides a meaningful way to asses the biological significance of a cluster. Genes within the same cluster are expected to exhibit similar expressions as they should have similar functionality or contribute to the same biological processes. Here we attempt to determine the biological meanings of the clusters by using Gene Ontology (GO) terms using the popular web-based tool FatiGO [33] (www.fatigo.org) FatiGO extracts the GO terms for a query and a reference set of genes and further computes various statistics for the query set. In our experiments, a query is the set of genes of a cluster and union of the genes from the other clusters is taken as the reference set. The GO level is fixed at three. It is not possible to evaluate each cluster of the final solutions provided by all the algorithms here. So, two interesting clusters from the clustering results obtained on Yeast Sporulation data set by the best performing algorithm (MODE in this case) is examined. Figure 6.10 illustrate a part of the FatiGO results of clusters 2 and 6 of multi-objective clustering on the sporulation data. It can be observed that the percentage of genes in the query cluster is considerably different from that of the reference cluster in almost all the functionalities. This implies that the correct genes are selected to remain in the same cluster.

6.5 Conclusions

This Chapter compared the performances of four state-of-the-art multi-objective variants of DE with two other prominent multi-objective clustering algorithms. The test-suite included six hand-crafted and four real-life datasets including the gene expression data of budding yeast. The artificial datasets were chosen in two and three dimensions for the ease of visualization of clustering results and the number of clusters for them ranged from 3 to 9. The DE-variants and NSGA-II used the same objective functions based on the Xie-beni index and the FCM index. Tables 6.2 to 6.4 indicate that one or more multi-objective DE variants were always able to produce better final clustering solutions as compared to MOCK or NSGA II in terms of both adjusted Rand index and Silwhen all the algorithms were let run for an equal number of generations. Not only did they find out the correct partitions in the data but also in all cases they were able to determine an optimal number of classes with minimum standard deviations. Visualization of the yeast sporulation data clustering results with parallel coordinates and heatmap plots indicate that the MODE yielded compact and well separated clusters. Biological interpretations to the clustering solution have been given with the help of gene annotation using a web-based Gene Ontology tool (FatiGO). Experimental results indicate that DE holds immense promise as a candidate optimization technique for multi-objective clustering. Future research may extend the multi-objective DE-based clustering schemes to handle discrete chromosome representation schemes that no longer depend on cluster centroids and thus are not biased in any sense towards spherical clusters. As a scope of further research, the technique of multi-objective optimization with other cluster validity indices needs to be studied. Moreover, new ways of comparing the performance of multi-objective solutions have to be defined.

References

1. Deb, K., Pratap, A., Agarwal, S., Meyarivan, T.: A fast and elitist multiobjective genetic algorithm: NSGA-II. IEEE Transactions on Evolutionary Computation 6(2) (2002)
2. Knowles, J.D., Corne, D.W.: Approxmating the nondominated front using the pareto archived evolution strategy. Evolutionary Computation 8(2), 149–172 (2000)
3. Sawaragi, Y., Nakayama, H., Tanino, T.: Theory of multiobjective optimization. Mathematics in Science and Engineering, vol. 176. Academic Press Inc., Orlando (1985)
4. Deb, K.: Multi-Objective Optimization using Evolutionary Algorithms. John Wiley & Sons, Chichester (2001)
5. Coello Coello, C.A., Lamont, G.B., Van Veldhuizen, D.A.: Evolutionary Algorithms for Solving Multi-Objective Problems. Springer, Heidelberg (2007)
6. Suresh, K., Ghosh, S., Kundu, D., Das, S.: Clustering with multi-objective differential evolution – a comparative study. In: The International conference on Advanced Computing Technologies (ICACT 2008), Hyderabad, India, December 26-27 (to appear, 2008)
7. Abbass, H.A., Sarker, R.: The pareto differential evolution algorithm. International Journal on Artificial Intelligence Tools 11(4), 531–552 (2002)
8. Xue, F., Sanderson, A.C., Graves, R.J.: Pareto-based multi-objective differential evolution. In: Proceedings of the 2003 Congress on Evolutionary Computation (CEC 2003), Canberra, Australia, vol. 2, pp. 862–869. IEEE Press, Los Alamitos (2003)
9. Robic, T., Filipic, B.: DEMO: Differential Evolution for Multiobjective Optimization. In: Coello Coello, C.A., Hernández Aguirre, A., Zitzler, E. (eds.) EMO 2005. LNCS, vol. 3410, pp. 520–533. Springer, Heidelberg (2005)
10. Iorio, A.W., Li, X.: Solving rotated multi-objective optimization problems using differential evolution. In: Webb, G.I., Yu, X. (eds.) AI 2004. LNCS (LNAI), vol. 3339, pp. 861–872. Springer, Heidelberg (2004)
11. Xie, X., Beni, G.: Validity measure for fuzzy clustering. IEEE Trans. Pattern Anal. Machine Learning 3, 841–846 (1991)
12. Bezdek, J.C.: Cluster validity with fuzzy sets. Journal of Cybernetics (3), 58–72 (1974)
13. Tibshirani, R., Walther, G., Hastie, T.: Estimating the number of clusters in a dataset via the Gap statistic. J. Royal Statist. Soc.: SeriesB (Statistical Methodology) 63(2), 411–423 (2001)
14. Handl, J., Knowles, J.: An evolutionary approach to multiobjective clustering. IEEE Transactions on Evolutionary Computation 11(1), 56–76 (2007)
15. Bandyopadhyay, S., Maulik, U., Mukhopadhyay, A.: Multiobjective genetic clustering for pixel classification in remote sensing imagery. IEEE Transactions Geoscience and Remote Sensing (2006)
16. Chu, S., et al.: The transcriptional program of sporulation in budding yeast. Science 282, 699–705 (1998)
17. Deb, K., Agrawal, S., Pratab, A., Meyarivan, T.: A Fast Elitist Non-Dominated Sorting Genetic Algorithm for Multi-Objective Optimization: NSGA-II. In: Deb, K., Rudolph, G., Lutton, E., Merelo, J.J., Schoenauer, M., Schwefel, H.-P., Yao, X. (eds.) PPSN 2000. LNCS, vol. 1917, pp. 849–858. Springer, Heidelberg (2000)
18. Zitzler, E.: Evolutionary Algorithms for Multiobjective Optimization: Methods and Applications, Ph.D. thesis, Shaker Verlag, Aachen, Germany (1999)

19. Zitzler, E., Thiele, L.: Multiobjective Evolutionary Algorithms: A Comparative Case Study and the Strength Pareto Approach. IEEE Transactions on Evolutionary Computation 3(4), 257–271 (1999)
20. Thomsen, R.: Multimodal optimization using crowding-based differential evolution. In: Congress on Evolutionary Computation (CEC 2004), Portland, Oregon, USA, IEEE Service Center, vol. 1, pp. 1382–1389 (2004)
21. Abbass, H.A.: The self-adaptive pareto differential evolution algorithm. In: Congress on Evolutionary Computation (CEC 2002), Piscataway, New Jersey, IEEE Service Center, vol. 1, pp. 831–836 (2002)
22. Knowles, J.D., Corne, D.W.: Approximating the nondominated front using the pareto archived evolution strategy. Evolutionary Computation 8(2), 149–172 (2000)
23. Branke, J., Deb, K., Dierolf, H., Osswald, M.: Finding knees in multi-objective optimization. In: Proc. 8th Int. Conf. Parallel Problem Solving From Nature, pp. 722–731 (2004)
24. Deb, K.: Multi-objective evolutionary algorithms: Introducing bias among Pareto-optimal solutions. In: Advances in Evolutionary Computing: Theory and Applications, pp. 263–292. Springer, London (2003)
25. Mattson, C.A., Mullur, A.A., Messac, A.: Smart Pareto filter: Obtaining a minimal representation of multiobjective design space. Eng. Optim. 36(6), 721–740 (2004)
26. Hubert, L., Arabie, P.: Comparing partitions. Journal of Classification, 193–218 (1985)
27. Rand, W.M.: Objective criteria for the evaluation of clustering methods. Journal of the American Statistical Association 66, 846–850 (1971)
28. Rousseeuw, P.J.: Silhouettes: A graphical aid to the interpretation and validation of cluster analysis. J. Comput. Appl. Math. 20(1), 53–65 (1987)
29. Theodoridis, S., Koutroumbas, K.: Pattern Recognition, 2nd edn. Elsevier Academic Press, Amsterdam (2003)
30. The Gene Ontology Consortium: Gene ontology: tool for the unification of biology. Nat. Genet. 25, 25–29 (2000)
31. Keim, D.A., Kriegel, H.-P.: Visualization techniques for mining large databases: a comparison. IEEE Transactions on Knowledge and Data Engineering 8(6), 923–938 (1996)
32. Eisen, M.B., Spellman, P.T., Brown, P.O., Botstein, D.: Cluster analysis and display of genome-wide expression patterns. Proc. Natl. Acad. Sci. USA 95, 14863–14868 (1998)
33. Al-Shahrour, F., Díaz-Uriarte, R., Dopazo, J.: FatiGO: a web tool for finding significant associations of Gene Ontology terms with groups of genes. Bioinformatics 20, 578–580 (2004)

Chapter 7
Conclusions and Future Research

Abstract. This Chapter provides a critical review of the works presented in this Volume and also narrates the scope of potential applications of the different meta-heuristic-based automatic clustering schemes in data mining, high level image processing and bioinformatics. The chapter ends with a discussion on the possible evolution of the proposed methods for handling clusters of non-spherical and shell type shapes, co-clustering and the problem of integrating together a feature selection module and a clustering module under the framework of Differential Evolution (DE).

7.1 Cluster Analysis Using Metaheuristics: A Roadmap of This Volume

Many partitional clustering algorithms based on the Genetic Algorithms (GA), have been proposed in recent past to tackle the problem of finding the optimal partition of a dataset. Very few studies have, however, considered alternative stochastic search-heuristics other than GAs or Simulated Annealing (SA). A promising metaheuristic algorithm for numerical optimization, which is hardly known in the field of pattern recognition, is Differential Evolution (DE). We investigated the performance of DE on the problem of data clustering. The proposed clustering framework based on DE not only could partition the data optimally into naturally occurring clusters, but also could find the correct number of such clusters at the same time. The performance of GAs, and a few other powerful search heuristics like Particle Swarm Optimization (PSO) or Ant Colony Optimization (ACO) algorithms for a representative point evolution approach to clustering is compared with different variants of DE. The empirical results illustrate that DE is clearly and consistently superior compared to GAs, PSO as well as traditional clustering algorithms like k-means, fuzzy c-means, and agglomerative hierarchical clustering algorithms both with respect to precision as well as robustness (reproducibility) of the results. Only for simple data sets, the GA and PSO can obtain the same quality of results. Apart from superior performance, DE is easy to implement and requires hardly any parameter tuning compared to the substantial tunings essential for GAs and PSOs. The most important conclusion we reach at this point is that, DE rather than GAs should receive primary attention in devising evolutionary partitional clustering algorithms, which are offline and demand high degrees of final accuracy.

S. Das, A. Abraham, and A. Konar: Metaheuristic Clustering, SCI 178, pp. 239–247.
springerlink.com

We summarize the major contributions as follows:

1) In order to gain an insight into the search mechanism of DE, Chapter 3 proposed a simple mathematical model of the underlying evolutionary dynamics of a one-dimensional DE-population. Analysis of the model leads us to one important finding that is the similarity of the fundamental differential equation governing the expected velocity of a search-agent in DE with that of the classical gradient descent search with momentum. This suggests that DE uses a stochastic estimate of the gradient of the objective function (which was assumed to be continuous in our analysis in order to keep the mathematics less rigorous) in order to locate the optima of the function. It is due to the gradient descent type search strategy, that DE converges much faster than algorithms like GA or PSO over uni-modal benchmarks. Application of Lyapunov's stability theorems to our model reveals that the near-equilibrium behavior of a DE population is inherently stable and free from any kind of oscillatory behaviors seen in other optimization algorithms like PSO or the Bacterial Foraging Algorithm (BFA).

2) We proposed a novel framework under the directives of the DE algorithm for the crisp clustering of large real life datasets. An important feature of the proposed technique is that it is able to detect the optimal number of clusters automatically (that is, the number of clusters does not have to be known in advance) even for very high dimensional datasets where tracking of the number of clusters may be well nigh impossible. The new scheme can go on with any distance measure (although we used the most common Euclidean distance metric) and any objective function.

3) Further, the DE-based clustering scheme was extended to the fuzzy clustering problem in a kernel-induced feature space for handling overlapping and linearly non-separable data. To improve the convergence speed and accuracy of DE in kernel induced clustering problems, we proposed a new family of variants of the DE/target-to-best/1 strategy that incorporates an explorative and an exploitative mutation operator with a view to balancing their effects. The new DE variant was shown to perform statistically significantly better than several other evolutionary as well as non-evolutionary kernel clustering algorithms.

4) Multi-objective optimization approach is relatively newer in the domain of clustering. There have been two prominent journal papers [1, 2] in this direction earlier, describing two methods of multi-objective clustering, both based on different multi-objective variants of GA. We compared the four most recently developed multi-objective variants of DE under a common automatic clustering scheme that simultaneously optimizes two conflicting fuzzy objective functions. One or more of these DE -variants was found to yield statistically superior results than both the GA-based multi-objective clustering algorithms over several simple artificial datasets as well as complicated real-world datasets including the gene expression data of budding yeast.

5) Since the region based segmentation of digital images can be achieved by clustering the image pixels in their intensity space, the thesis investigates an interesting

application of the DE-based hard and fuzzy clustering schemes to the automatic segmentation of gray-scale images. In Chapter 5, it also attempted to evaluate the segmentation results objectively based on the comparison with manually segmented images via different statistical scoring functions. Land cover study from the segmentation of remote sensing satellite images was another important application presented in Chapter 5.

A downside to the clustering framework is that the parameter vectors in DE use a representation based on cluster centers owing to the real-coded nature of DE. The centroid-based encoding often biases the solution towards its underlying spherical cluster model and therefore towards the objective of over all deviation. Thus the scheme may identify poor clusters in case of certain hand-crafted datasets like Smile, Spiral, or Long as described in [2]. However, experimental results indicated that the scheme provides very good results with several real life datasets including gray scale images. Another disadvantage of the approaches proposed in Chapters 4 and 5 is the dependence of the fitness function for DE upon the choice of a suitable clustering validity index. An inefficient validity index may result into many false clusters (due to the over-fitting of data) even when the actual number of clusters in the given dataset may be very much tractable. Some ways to circumvent the two above-mentioned difficulties have been discussed in the following Sections.

7.2 Potential Application Areas for Clustering Schemes

The DE-based clustering methods may find very important applications in the following areas:

1) ***Image Color Quantization:*** One of the various forms in which image compression may be realized is by reducing the number of colors used to encode an image. With the DE-based automatic clustering techniques we can find k best colors for representing an entire image, quantizing all the colors in the image to one of the k code colors. The experimental results of the thesis seems to suggest that DE should achieve better results on automatic color quantization on images as compared to the GA or PSO based quantization schemes proposed recently [3, 4].

2) ***Microarray Data Clustering:*** The expression levels of thousands of genes can be measured at the same time using the modern microarray technology [5, 6]. DNA microarrays usually consist of thin glass or nylon substrates containing specific DNA gene samples spotted in an array by a robotic printing device. Proper selection, analysis, and interpretation of the microarray data can lead us to the answers of many important problems in experimental biology. Currently clustering has become a widely used process in microarray engineering for understanding the functional relationship between groups of genes. Clustering was used, for example, to understand the functional differences in cultured primary epatocytes relative to the intact liver [7]. In another study, clustering techniques

were used on gene expression data for tumor and normal colon tissue probed by oligonucleotide arrays [8]. We believe that the DE-based clustering techniques will be very helpful for automatically detecting clusters of co-expressed genes with similar biological functions from microarray datasets. A first study towards this direction has already been reported in Chapter 6 on yeast sporulation data.

3) ***Anomaly-based Network Intrusion Detection:*** Rapid expansion of computer network throughout the world has made security a crucial issue in a computing environment. Recently data mining methods have gained importance in addressing network security issues, including network intrusion detection [9, 10] - a challenging task in network security. Intrusion detection systems aim to identify attacks with a high detection rate and a low false alarm rate. Classification-based data mining models for intrusion detection are often ineffective in dealing with dynamic changes in intrusion patterns and characteristics. Consequently, unsupervised learning methods have been given a closer look for network intrusion detection. An important advantage of using clustering or unsupervised learning to detect network attacks is the ability to find new attacks not seen before. This implies that attack types with unknown pattern signatures can be detected. Clustering results can also assist the network security expert with labeling network traffic records as normal or intrusive. A few recent approaches towards anomaly-based intrusion detection with traditional as well as GA-based clustering techniques can be found in [11 - 14]. Experimental results indicates that the DE-based automatic clustering methods may prove very effective in this field and they are expected to yield more accurate results consuming less amount of computational time as compared to GAs.

4) ***Text Mining and Information Retrieval:*** Clustering of text documents plays a vital role in efficient document organization, summarization, topic extraction and information retrieval. Although initially used for improving the precision or recall in an information retrieval system [15, 16], more recently, clustering has been proposed for use in browsing a collection of documents [17] or in organizing the results returned by a search engine in response to a user's query [18]. Document clustering has also been used to automatically generate hierarchical clusters of documents [19]. (The automatic generation of a taxonomy of Web documents like that provided by Yahoo! (www.yahoo.com) is often cited as a goal). In recent years, it has been recognized that the partitional clustering technique is well suited for clustering a large document dataset due to their relatively low computational requirements [20, 21]. Recently Genetic Algorithm (GA) [22, 23], Self-Organizing Maps (SOM) [24], and finally Particle Swarm Optimization (PSO) [25] have been used for document clustering. The comparative performance of DE on clustering problems in this thesis indicates that DE and its variants can be very efficient for off-line clustering of large datasets of documents, compared to GAs and PSO, both in respect to precision as well as robustness of the results. A preliminary work in this direction has been reported in [26].

7.3 Future Research Directions

The works reported in this Volume may be extended in a number of ways, some of which are discussed below.

1) An interesting problem for future research is to associate the automatic clustering scheme with the choice of the most relevant features compared to the process used. Often we have a great number of features (especially for a high-dimensional dataset like the foliage images used in Chapter 4), which are not all relevant for a given operation. Thus, integrating an automatic feature-subset selection scheme with the DE-based clustering techniques will reduce the runtime complexity of the clustering algorithm (due to dimensionality reduction) to a great extent. The combined algorithm is expected to automatically project the data to a low dimensional feature subspace, determine the number of clusters and find out the appropriate cluster centers with the most relevant features at a faster pace. An interesting work reporting simultaneous feature selection and classifier design with Genetic Programming (GP) can be found in [27]. But similar works are yet to be done with evolutionary clustering algorithms.

2) Centroid based encoding of chromosomes or DE-vectors may often bias the solutions to its underlying spherical cluster models. In order to avoid that, it is intersting to study some other kind of representation scheme for the parameter vectors. For example, we may use a scheme where in each such chromosome, a gene encodes one cluster. Thus if the r-th chromosome encodes K groups $\{G_1^r, \ldots\ldots, G_i^r, \ldots\ldots, G_K^r\}$ it has K genes. Each such group or cluster G_i^r of the r-th chromosome contains n_i^r number of data points such that $\sum_{i=1}^{K} n_i^r = n$ where n is the total number of data points in the given dataset. Each gene encoding a cluster actually consists of the integral labels of those data points, which belong to that cluster. For example, the i-th gene G_i^r of the r-th

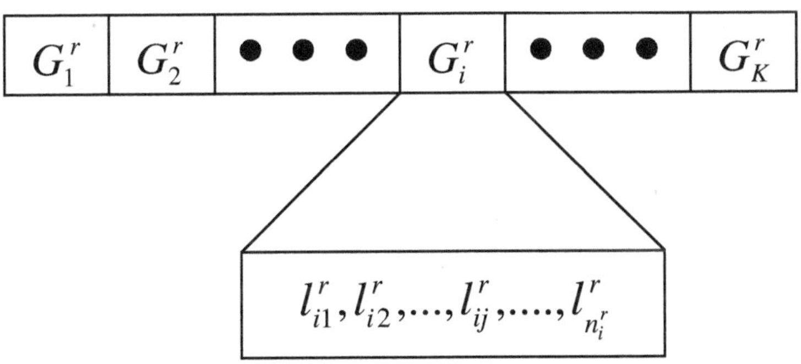

Fig. 7.1 An alternative chromosome representation scheme

chromosome contains n_i^r integers $\{l_{i1}^r, l_{i2}^r, ..., l_{ij}^r, ..., l_{n_i^r}^r\}$ such that $l_{ij}^r \in \{1, ..., n\}$ and l_{ij}^r physically means that the j-th pattern from the data set **P** = {**P**$_1$, **P**$_2$... **P**$_n$} under test, belongs to the i-th cluster encoded by the r-th chromosome. This representation scheme is illustrated in Figure 7.1.

We may use binary DE or some discrete variant of DE (please see [28] for example) to evolve such chromosomes.

3) Distance measures other than Euclidean type should be investigated in conjunction with the DE-based clustering schemes. Symmetry is considered a pre-attentive feature which enhances recognition and reconstruction of shapes and objects [29]. Almost every interesting area around us consists of some generalized form of symmetry. As symmetry is so common in the natural world, it can be assumed that some kind of symmetry exists in the clusters also. Based on this, Su and Chou [30] have proposed a symmetry-based clustering technique. Authors assigned points to a particular cluster if they present a symmetrical structure with respect to the cluster centre. A new type of non-metric distance, based on point symmetry, is proposed, which is used in a k-means based clustering algorithm, referred to as Symmetry-based k-means (SBKM) algorithm. Bandyopadhyay and Saha [31] offered certain improvements of this point-symmetric distance measure and used it to cluster overlapping and arbitrary shaped clusters with variable length GA (VGA). It is interesting to study the performance of the DE-based clustering schemes using the symmetry-based distance metrics. In mathematics, another interesting metric is Bregman divergence or Bregman distance [32], which does neither satisfy the triangle inequality nor the symmetry criteria. There are two ways in which Bregman divergences are important. Firstly, they generalize squared Euclidean distance to a class of distances that all share similar properties. Secondly, they bear a strong connection to exponential families of distributions. Banerjee *et al.* [33] have shown that there is a bijection between exponential families and Bregman divergences. It would be interesting to study the performance of DE with Bregman distances.

4) Biclustering, co-clustering, or two-mode clustering [34] is a data mining technique, which allows simultaneous clustering of the rows and columns of a matrix. Given a set of *m* rows in *n* columns (i.e., an $m \times n$ matrix), the biclustering algorithm generates biclusters - a subset of rows which exhibit similar behavior across a subset of columns, or vice versa. Biclustering has emerged as an important tool for analyzing large gene expression matrices in bioinformatics. Recently some works have been undertaken in [35 – 37] to apply both single and multi-objective evolutionary algorithms to the biclustering problem of microarray datasets. Based on the experimental results, we feel that DE can serve as a promising candidate optimization algorithm for applications in biclustering and should be investigated in this direction in future.

5) Subspace clustering [38] may be defined as the problem of automatically identifying clusters present in the subspaces of a high-dimensional data space that allows better clustering of the data points than the original space. Often in high-dimensional data, many dimensions are irrelevant and can mask existing clusters in noisy data. Feature selection removes irrelevant and redundant dimensions by analyzing the entire dataset. Subspace clustering algorithms localize the search for relevant dimensions allowing them to find clusters that exist in multiple, possibly overlapping subspaces. A preliminary work attempting to apply the evolutionary algorithms for subspace clustering of very high-dimensional data has been reported in [39]. It is interesting to extend the DE-based clustering scheme outlined in Chapter 4 towards subspace clustering.

References

[1] Handl, J., Knowles, J.: An evolutionary approach to multiobjective clustering. IEEE Transactions on Evolutionary Computation 11(1), 56–76 (2007)

[2] Bandyopadhyay, S., Maulik, U., Mukhopadhyay, A.: Multiobjective genetic clustering for pixel classification in remote sensing imagery. IEEE Transactions Geoscience and Remote Sensing (2006)

[3] Omran, M., Engelbrecht, A.P., Salman, A.: A color image quantization algorithm based on particle swarm optimization. Informatica Journal 29, 261–269 (2005)

[4] Freisleben, B., Schrader, A.: An evolutionary approach to color image quantization. In: IEEE International Conference on Evolutionary Computation, pp. 459–464 (1997)

[5] Quackenbush, J.: Computational analysis of microarray data. National Review of Genetics 2, 418–427 (2001)

[6] Special Issue on Bioinformatics, IEEE Computer 35 (July 2002)

[7] Baker, T.K., Carfagna, M.A., Gao, H., Dow, E.R., Li, O., Searfoss, G.H., Ryan, T.P.: Temporal gene expression analysis of monolayer cultured rat hepatocytes. Chem. Res. Toxicol. 14(9) (2001)

[8] Alon, U., Barkai, N., Notterman, D.A., Gish, K., Ybarra, S., Mack, D., Levine, A.J.: Broad patterns of gene expression revealed by clustering analysis of tumor and normal colon tissues probed by oligonucleotide arrays. Proc. Natl. Acad. Sci. USA 96, 6745–6750 (1999) (Cell Biology)

[9] Mykerjee, B., et al.: Network Intrusion Detection. IEEE Network 8(3), 26–41 (1994)

[10] Denning, D.: An intrusion detection model. IEEE Transactions on Software Engineering (1987)

[11] Shafi, K., Abbass, H.A.: Biologically-inspired complex adaptive systems approaches to network intrusion detection. Inf. Secur. Tech. Rep. 12(4), 209–217 (2007)

[12] Fries, T.P.: A fuzzy-genetic approach to network intrusion detection. In: Keijzer, M. (ed.) Proceedings of the 2008 GECCO Conference Companion on Genetic and Evolutionary Computation, GECCO 2008, pp. 2141–2146. ACM, New York (2008)

[13] Mark, C., Spafford, G.: Applying genetic programming to intrusion detection. In: Proceedings of 1995 AAAI Fall Symposium on Genetic Programming, Cambridge, Massachusetts, October 30, pp. 1–8 (2003), http://citeseer.nj.nec.com/crosbie95applying.html

[14] Owais, S., Snášel, V., Krömer, P., Abraham, A.: Survey: using genetic algorithm approach in intrusion detection systems techniques. In: 7th International Conference on Computer Information Systems and Industrial Management Applications (CISIM 2008), pp. 296–303. IEEE Computer Society press, USA (2008)
[15] van Rijsbergen, C.J.: Information Retrieval, 2nd edn. Buttersworth, London (1989)
[16] Kowalski, G.: Information Retrieval Systems – Theory and Implementation. Kluwer Academic Publishers, Dordrecht (1997)
[17] Cutting, D.R., Karger, D.R., Pedersen, J.O., Tukey, J.W.: Scatter/gather: a cluster-based approach to browsing large document collections. In: SIGIR 1992, pp. 318–329 (1992)
[18] Zamir, O., Etzioni, O., Madani, O., Karp, R.M.: Fast and intuitive clustering of web documents. In: KDD 1997, pp. 287–290 (1997)
[19] Koller, D., Sahami, M.: Hierarchically classifying documents using very few words. In: Proceedings of the 14th International Conference on Machine Learning (ML), Nashville, Tennessee, pp. 170–178 (July 1997)
[20] Steinbach, M., Karypis, G., Kumar, V.: A comparison of document clustering techniques. In: Text Mining Workshop, KDD (2000)
[21] Zhao, Y., Karypis, G.: Empirical and Theoretical Comparisons of Selected Criterion Functions for Document Clustering. Machine Learning 55(3), 311–331 (2004)
[22] Santimetvirul, C., Willett, P.: Non-hierarchic document clustering using a genetic algorithm. Information Research 1(1) (1995)
[23] Raghavan, V.V., Birchand, K.: A clustering strategy based on a formalism of the reproductive process in a natural system. In: Proceedings of the Second International Conference on Information Storage and Retrieval, pp. 10–22 (1979)
[24] Merkl, D.: Text mining with self-organizing maps. In: Handbook of data mining and knowledge, pp. 903–910. Oxford University Press, Inc., New York (2002)
[25] Xiaohui, C., Potok, T.E., Palathingal, P.: Document Clustering using Particle Swarm Optimization. In: IEEE Swar Intelligence Symposium, The Westin Pasadena, Pasadena, California (2005)
[26] Abraham, A., Das, S., Konar, A.: Document clustering using differential evolution. In: IEEE Congress in Evolutionary Computation, World Congress in Computational Intelligence, pp. 6248–6255. IEEE press, Los Alamitos (2006)
[27] Muni, D.P., Pal, N.R., Das, J.: Genetic programming for simultaneous feature selection and classifier design. IEEE Transactions on Systems, Man and Cybernetics-B 36(1) (2004)
[28] Pan, Q., Tasgetiren, M.F., Liang, Y.: A discrete differential evolution algorithm for the permutation flowshop scheduling problem. In: Proceedings of the 9th Annual Conference on Genetic and Evolutionary Computation, GECCO 2007, pp. 126–133. ACM, New York (2007)
[29] Attneave, F.: Symmetry information and memory for pattern. Am. J. Psychology 68, 209–222 (1995)
[30] Su, M.-C., Chou, C.-H.: A modified version of the k-means algorithm with a distance based on cluster symmetry. IEEE Transactions Pattern Analysis and Machine Intelligence 23(6), 674–680 (2001)
[31] Bandyopadhyay, S., Saha, S.: Point Symmetry Based Clustering Technique for Automatic Evolution of Clusters. IEEE Transactions on Knowledge and Data Engineering 20(11), 1–17 (2008)

[32] Bregman, L.M.: The relaxation method of finding the common points of convex sets and its application to the solution of problems in convex programming. USSR Computational Mathematics and Mathematical Physics 7, 200–217 (1967)
[33] Banerjee, A., Merugu, S., Dhillon, I.S., Ghosh, J.: Clustering with Bregman divergences. Journal of Machine Learning Research 6, 1705–1749 (2005)
[34] Van Mechelen, I., Bock, H.H., De Boeck, P.: Two-mode clustering methods: a structured overview. Statistical Methods in Medical Research 13(5), 363–394 (2004)
[35] Madeira, S.C., Oliveira, A.L.: Biclustering algorithms for biological data analysis: a survey. IEEE Transactions on Computational Biology and Bioinformatics 1(1), 24–45 (2004)
[36] Mitra, S., Banka, H.: Multi-objective evolutionary biclustering of gene expression data. Pattern Recognition 39, 2464–2477 (2006)
[37] Divina, F., Aguilar-Ruiz, J.S.: Biclustering of expression data with evolutionary computation. IEEE Trans. Knowl. Data Eng. 18(5), 590–602 (2006)
[38] Parsons, L., Haque, E., Liu, H.: Subspace clustering for high dimensional data: a review. SIGKDD Explor. Newsl. 6, 90–105 (2004)
[39] Sarafis, I.A., Trinder, P.W., Zalzala, A.M.S.: Towards effective subspace clustering with an evolutionary algorithm. In: Proceedings of the IEEE Congress on Evolutionary Computation (CEC 2003). IEEE Press, Los Alamitos (2003)

Index

FSC
www.fsc.org
MIX
Papier | Fördert
gute Waldnutzung
FSC® C083411

Kolibri 360 GmbH
Ferdinand-Jühlke-Straße 7
99095 Erfurt, Deutschland
produktsicherheit@kolibri360.de